T0192283

An Image Processing Tour of College Mathematics

Mathematical and Computational Imaging Sciences Series

Series Editors:
Chandrajit Bajaj, Guillermo Sapiro

About the Series

This series aims to capture new developments and summarize what is known over the whole spectrum of mathematical and computational imaging sciences. It seeks to encourage the integration of mathematical, statistical, and computational methods in image acquisition and processing by publishing a broad range of textbooks, reference works, and handbooks. The titles included in the series are meant to appeal to students, researchers, and professionals in the mathematical, statistical, and computational sciences and application areas as well as interdisciplinary researchers involved in the field. The inclusion of concrete examples and applications and programming code and examples is highly encouraged.

Rough Fuzzy Image Analysis
Foundations and Methodologies
Sankar K. Pal, James F. Peters

Theoretical Foundations of Digital Imaging Using MATLAB®
Leonid P. Yaroslavsky

Statistical and Computational Methods in Brain Image Analysis
Michael Ruzhansky, Makhmud Moo K. Chung

Image Processing for Cinema
Marcelo Bertalmío

Variational Methods in Image Processing
Luminita A. Vese, Carole Le Guyader

Geometric Modeling and Mesh Generation from Scanned Images
Yongjie Jessica Zhang

An Image Processing Tour of College Mathematics
Yevgeniy V. Galperin

For more information about this series please visit: https://www.routledge.com/Chapman-HallCRC-Mathematical-and-Computational-Imaging-Sciences-Series/book-series/CHCRCMATC

An Image Processing Tour of College Mathematics

Yevgeniy V. Galperin

CRC Press is an imprint of the
Taylor & Francis Group, an **informa** business
A CHAPMAN & HALL BOOK

First edition published 2021
by CRC Press
6000 Broken Sound Parkway NW, Suite 300, Boca Raton, FL 33487-2742

and by CRC Press
2 Park Square, Milton Park, Abingdon, Oxon, OX14 4RN

CRC Press is an imprint of Taylor & Francis Group, LLC

Library of Congress Cataloging-in-Publication Data

Names: Galperin, Yevgeniy V., author.
Title: An image processing tour of college mathematics / Yevgeniy V. Galperin, East Stroudsburg University of Pennsylvania, USA.
Description: First edition. | Boca Raton : Chapman & Hall, CRC Press, 2020.
| Series: Chapman & Hall/CRC mathematical and computational imaging sciences series | Includes bibliographical references and index.
Identifiers: LCCN 2020038966 (print) | LCCN 2020038967 (ebook) | ISBN 780367002022 (hardback) | ISBN 9780429400612 (ebook)
Subjects: LCSH: Mathematics--Data processing. | Image processing.
Classification: LCC QA76.95 .G35 2020 (print) | LCC QA76.95 (ebook) | DDC
6.4/2--dc23
LC record available at https://lccn.loc.gov/2020038966
LC ebook record available at https://lccn.loc.gov/2020038967

ISBN: 9780367002022 (hbk)
ISBN: 9780429400612 (ebk)

Contents

Preface

This book is intended for students who major in mathematics, computer science, or mathematics education and is based on the course I have taught at East Stroudsburg University of Pennsylvania since 2008. My aims for designing the course and for writing the book fall into five general areas.

A review of college math fundamentals

The first aim is to provide a salient line for a review of math fundamentals. In my teaching experience, I have come to believe that many third- or fourth-year students can derive a great deal of benefit from a review of the most important topics and techniques of college mathematics - from precalculus to linear algebra and mathematical statistics. But just plain reviewing for its own sake would be excruciatingly boring (and it would also be difficult to justify awarding academic credits for it). To make such a review more exciting and productive, I find it vital to provide a unifying theme that would help move the material forward and hold it together. And that is exactly what this book seeks to provide – the context of digital image processing in which a review of the key topics of college mathematics feels timely and relevant.

The topics covered include a variety of elementary functions (reviewed and explained in the context of pixel-based operations on images, basic concepts of descriptive statistics (reviewed and explained in the context of image histogram manipulations), transformations of random variables in the context of image equalization, definitions and concepts behind first- and second-order derivatives (reviewed and interpreted in the context of edge detection), as well as many of the concepts and techniques that make up traditional linear algebra courses (used extensively throughout the text and, especially, in the context of developing discrete wavelet transforms).

In addition to reviewing familiar topics, the text seeks to reinforce those that are typically glossed over or quickly forgotten due to insufficient opportunities for practical application provided within traditional courses. Such topics include piecewise-linear functions, the definition of the derivative, transformations of random variables, matrices of linear transformations, and many others.

An introduction to digital image processing

The second aim of the text is to introduce students who major in mathematics to the rich and beautiful field of digital image processing. Although not a comprehensive introduction by any means, the book does provide in-depth material on such topics as intensity transformations, spacial and frequency filtering, image enhancement and image restoration, geometric transformations, wavelet-based image compression, edge detection, noise reduction, and several others.

While there are numerous excellent comprehensive texts available on the topic of digital image processing, what this book attempts to do is to fill the gap between the mathematics curriculum and image processing courses. For example, very few linear algebra textbooks provide detailed instructions on the implementation of the projective transformation and the construction of its matrix, whereas very few texts with the focus on image processing spend any time on the fundamentals of linear transformations. This books seeks to do both and as such might be of interest to students interested both in math and in computer science.

Throughout this book, consistent effort has been made to help the reader move away from viewing college mathematics as a bunch of formulas and cookbook recipes (that need to be followed to solve textbook problems) and move towards recognizing it as a toolbox to be used as needed and as appropriate in order to solve practical problems. And digital image processing provides a diverse pool of problems where this approach can be practiced and where results and solutions are easy to visualize.

A bridge towards graduate school

The third aim proceeds from the author's own experience during the first years of graduate school. It is to strengthen and enrich the students' mathematical training beyond what is traditionally included in the college curriculum and to better prepare them for the future challenges of graduate school math or engineering courses.

For example, all too often, virtually missing from the college math experience are any references to Fourier analysis and to the operation of convolution. This text provides an elementary introduction to Fourier series and discrete Fourier transforms; an introduction that is both accessible, yet sufficient for understanding the subsequent material related to image restoration and image compression and rigorous enough to help the readers deepen their understanding of abstract mathematics. The convolution operation (which students often find very difficult to understand and master) is introduced naturally in the context of naive filtering for denoising and is subsequently used consistently throughout the rest of the text reflecting its fundamental importance in most areas of pure and applied mathematics.

The benefit derived from learning about convolution, Fourier analysis, LTI systems, transformations of random variables, and other topics covered in this book extends way beyond digital image processing as they play a prominent role in numerous areas of pure and applied mathematics, science, technology, and even social sciences. No matter what the reader is planning to do after graduating from college – whether it is working in the industry or going to graduate school – the additional experience gained in this chapter will likely prove to be of lasting practical value.

MATLAB® programming

The fourth aim of the book is to help students of mathematics ease their way into mastering the basics of scientific computer programming. One often hears that the best way to understand a topic is to explain it to a critically-minded audience. No audience is as tough as a computer, which refuses to understand you at the slightest technical inaccuracy on your part, and it is precisely from this standpoint that we approach scientific computer programming. MATLAB as a programming language is an integral component of the text. Each chapter contains a substantial number of MATLAB programming exercises, and we view the independent implementation of the mathematical techniques in software as indispensable for a fuller understanding of the concepts and techniques of college mathematics. We also view it as a valuable confidence-building experience for those students who plan to continue their studies in the fields of applied mathematics, computer science, or engineering.

Career planning

Which brings us to *the last but not least important aim* of this text, and it has to do with students' post-college careers. College graduates with degrees in mathematics are often at a disadvantage vis-a-vis their peers with degrees in other disciplines for two reasons: most of what they have learned in college is 300 to 2,500 years old, and they have little to show for it in terms of tangible and visual projects.

This book and the course it is based on seek to address both problems. Some of the methods described in this book (particularly, wavelet-based image compression) were developed less than 30 years ago by Ingrid Daubechies and her colleagues. And by completing the MATLAB exercises that follow each section, the students will accumulate an impressive software package of their own design, which can undoubtedly be helpful in making the next step in their careers.

In conclusion, I would like to express my gratitude and appreciation to my wife, Kazuko Kihara, and our children, Sayuri, Seiji, and Kenji, for their patience and understanding while I was writing and editing this text. Most importantly, thanks are due to the students of my MATH 445 (Mathematics in Modern Technology) classes for their enthusiasm and support, which inspired me to write this book.

Yevgeniy V. Galperin,
East Stroudsburg University of PA.

Chapter 1

Introduction to Basics of Digital Images

In this chapter, the reader will master the basics of working with grayscale and color images in MATLAB. Also discussed are the fundamental statistical characteristics of digital images. Although quite brief, this chapter provides the necessary technical preliminaries for all the practical exercises in the subsequent chapters.

1.1 Grayscale Digital Images

For our purposes, a grayscale digital image is just an $M \times N$ matrix

$$A = \begin{bmatrix} a_{1,1} & a_{1,2} & \cdots & a_{1,N-1} & a_{1,N} \\ a_{2,1} & a_{2,2} & \cdots & a_{2,N-1} & a_{2,N} \\ \vdots & \vdots & \ddots & \vdots & \vdots \\ a_{M-1,1} & a_{M-1,2} & \cdots & a_{M-1,N-1} & a_{M-1,N} \\ a_{M,1} & a_{M,2} & \cdots & a_{M,N-1} & a_{M,N} \end{bmatrix}$$

of discrete pixel values. The position in the upper left corner of the image corresponds to the matrix indices $(1, 1)$, and the position in the lower right corner of the image corresponds to the matrix indices (M, N). The maximal light intensity is represented by the pixel value of 255, and the absence of light is represented by the pixel value 0, with the shades of gray thus represented by the values between 0 and 255. For example, the matrix

$$\begin{bmatrix} 223 & 111 & 159 & 47 \\ 127 & 255 & 63 & 143 \\ 191 & 79 & 207 & 15 \\ 95 & 175 & 31 & 239 \end{bmatrix}$$

represents the image (magnified by the factor of 100) in Figure 1.1 below.

FIGURE 1.1: An illustration of grayscale pixel values.

As the emphasis of this text is on college mathematics, we will not discuss such broad and fascinating topics as visual perception, image formation, and many others. There is a huge body of literature covering all those areas, and we cannot possibly do any of them justice within the scope of this text. We also refer the reader to the fundamental work [11] for a detailed introduction into the topics of image sensing and acquisition, image sampling and quantization, representation of digital images, as well as examples of fields of science and technology where methods of digital image processing are used most extensively.

Section 1.1 Exercises.

1. Enter relevant terms like "image processing" and "grayscale images" into a search engine and compile a list of applications of image processing in the following fields:

    - Art and design
    - Astronomy
    - Medicine
    - Meteorology
    - Photography and motion pictures
    - Transportation and autonomous vehicles

2. Find a few black-and-white photos in your family photo album or in old newspapers. Select the ones that could benefit from image enhancement. Scan them and have them available for exercises in subsequent chapters.

1.2 Working with Images in MATLAB®

Throughout this and subsequent chapters, the reader will be asked to complete exercises that require the use of MATLAB[1]. Although this text is by no means intended to serve as an introduction to MATLAB, we will, nevertheless, provide a few basic examples of MATLAB usage, particularly when it comes to basic image manipulations.

The very first thing we would like to be able to do when working with digital images is to import them into our computing/programming environment. For that purpose, we will mostly be using the MATLAB function *imread*,

[1] MATLAB® is a registered trademark of The MathWorks, Inc. For product information, please contact:

The MathWorks, Inc.

3 Apple Hill Drive

Natick, MA 01760-2098 USA

Tel: 508 647 7000

Fax: 508-647-7001

E-mail: info@mathworks.com

Web: www.mathworks.com

which reads the image from a graphics file of any of a large number of common formats and returns a matrix of pixel values. For example, the result of executing the command

```
>> A=imread('trees.tif');
```

is the matrix A containing the pixel values (on the scale of 0 to 255) of the grayscale image *trees.tif.* The purpose of putting the semicolon at the end of the command is to prevent MATLAB from displaying all the values of the matrix A.

If we need to know the size of the image 'trees.tif', that is, the number of rows and columns of the matrix A, we can calculate it by using the MATLAB function

```
>> [M,N]=size(A),
```

which returns the values $M = 258$ and $N = 350$, indicating that the image has dimensions 258×350. Entering the MATLAB function

```
>> class(A)
```

into the command line returns the value "uint8", which tells us that the pixel values are represented by 8-bit unsigned integers, providing for $2^8 = 256$ possible shades of gray.

Naturally, we would like to use MATLAB not just to acquire images but also to display them. For that purpose, we usually use the function *imshow* which, in its basic form, displays the image represented by the specified matrix of type "uint8" or "double" with pixel values to the scale of 0 to 255 or 0 to 1 respectively. For example, Figure 1.2 below illustrates the result of executing the MATLAB command

```
>> imshow(A),
```

where A is the image matrix in the "uint8" type obtained by executing the command *A=imread('trees.tif').*

Working with matrices of type "uint8" is not always practical. In subsequent chapters, we will often need to convert the data type of the matrix that represents the image from "uint8" to other numerical types, real or even complex, in order to perform various image manipulations. For example, if we want MATLAB to perform floating-point arithmetic on pixels values, we must convert the matrix type from "uint8" to "double" by means of the command

```
>> A=double(A).
```

However, in order to properly display an image whose matrix is of the data type "double", we need to modify our usage of the *imshow* function by supplying the range of the pixel values. For example, we can use

```
>> imshow(A,[0,255])
```

FIGURE 1.2: An image displayed by the MATLAB command "imshow".

if we want the pixel values contained in A to be displayed on the scale of 0 to 255, which is not the default for images matrices of the type "double". If, on the contrary, we would like MATLAB to automatically calculate the smallest and the largest pixel values contained in A and stretch contrast as much as possible by making them black and white respectively, we can use the syntax

```
>> imshow(A,[]).
```

If we just use the function *imshow* in its basic form, all the pixels whose values exceed 1.0 will be interpreted as white and displayed accordingly.

Like other MATLAB functions, *imshow* has many parameters and options, which can be explored by accessing the Mathworks website www.mathworks.com. Alternatively, the quickest way of getting information on anything related to MATLAB is to use the *help* command. For example, typing

```
>> help imshow
```

in the MATLAB Command Window will produce the output containing a detailed general description, examples, a list of parameters, and also a list of related MATLAB functions and useful links.

To complete this section we must mention another frequently-used MATLAB function called *subplot*, which enables us to fit several images or other graphics into the same figure. Executing the command

```
>> subplot(m,n,p)
```

divides the current figure into an $m \times n$ grid and prepares to display an object in the position of that grid specified by p. We illustrate its usage by the following MATLAB code

```
A=imread('coins.png');
B=imread('trees.tif');
subplot(1,2,1);
imshow(A);
subplot(1,2,2);
imshow(B);
```

whose output is shown in Figure 1.3.

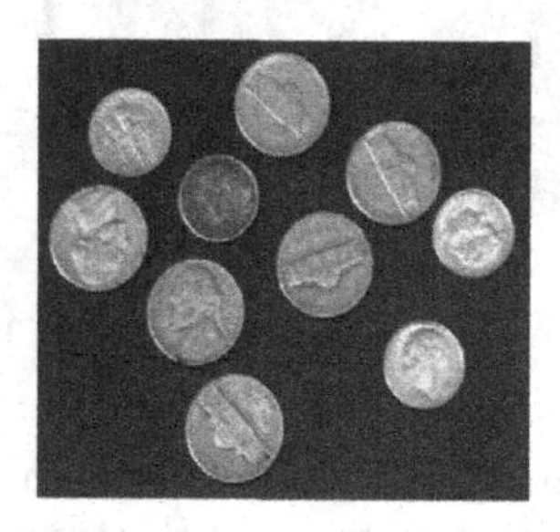

FIGURE 1.3: An illustration of the MATLAB command "subplot".

We also encourage the reader to undertake an independent exploration of the extensive features of such MATLAB functions as *image*, *imread*, *imwrite*, *imagesc*, and other image-processing functions.

Section 1.2 MATLAB Exercises.

1. Use the MATLAB function *imread* to import the photos you prepared in Exercise 2 of Section 1.1 into the MATLAB computing environment.

2. Determine the sizes of the images you imported in Exercise 1 in terms of the numbers of columns and rows and the number of pixels.

3. Display the images you imported in the previous exercise by means of the MATLAB functions *imshow* and *imagesc*. Comment on the difference in the ways the image appears on the computer screen. Try to "dress up" the images by supplying titles using the MATLAB function *title* with some of its parameters, such as *FontSize*.

4. Display six images at a time by arranging them in a 3×2 grid using the MATLAB functions *subplot* and *imshow*.

5. Construct a matrix representing the following geometric objects:

(a) A dark square against a light background;

(b) A light triangle against a dark background;

(c) A gray diamond inscribed into a light square against a dark background;

(d) Any other patterns you find interesting.

Use the MATLAB function *imshow* to display the images represented by the matrices you constructed.

6. Try to lighten or darken the images you imported or constructed in the previous exercises by adding or subtracting a fixed number to all the pixels. Be careful not to exceed the maximum allowed value of 255 and not to create any negative pixel values. In order to do this problem, you may need to explore the MATLAB constructs *for* and *if-else*. Comment on the effectiveness of this naive method of image enhancement.

1.3 Images and Statistical Description of Quantitative Data

In order to better understand the visual quality of an image, it is often useful to explore its statistical characteristics, particularly the frequency distribution of the pixel values. For example, we may wish to determine whether there is a predominance of lighter or darker pixels, and, if we want to, we may try to alter the distribution of pixel values in order to lighten or darken the image. Or, if we discover that most of the pixel values fall into the middle range of the gray scale, we may wish to change that distribution in order to increase the image contrast. In order to develop techniques that would help us achieve this type of image enhancement, we need to review a number of concepts from general statistics.

1.3.1 Image Histograms

When we collect data with the purpose of deepening our understanding of a given situation or a problem, we hope that analyzing the data will assist us in making informed decisions on a future course of action. A matrix of a digital image may also be considered a type of statistical data, and we can analyze it with the help of statistical methods. Where do we begin this data analysis?

Whereas the raw data contain all the available information, they fail to communicate any insights or to tell any stories. Presenting a list (or a matrix) of numerical values to an audience (or even to a researcher) rarely has any effect beyond eliciting yawns of boredom. In order for the data to tell a story, to exhibit patterns, and to help us draw conclusions and make decisions, it is necessary to summarize the data and to present the summary in a compelling graphical form.

A common way to summarize quantitative data is by constructing frequency distributions. One first calculates the range of the data set, which is the difference between the largest and the smallest values and then decides on the number of bins, that is, the number of subintervals to divide the data into. The subintervals are always adjacent to each other and are usually selected to be of equal width. One then counts the frequencies, that is, the numbers of data values that fall within each subinterval. Dividing frequencies by the size of the data set (the total number of data points) gives us *relative frequencies*. These bins, together with the frequencies, constitute a frequency distribution of the data.

For example, the frequency distribution of the data set

$$\{0, 2, 4, 4, 4, 4, 4, 6, 7, 10\}$$

using 4 bins is shown in Table 1.1.

TABLE 1.1: An Example of a Frequency Distribution

Bin	**Count**	**Relative Frequency**
0 to 2.5	2	0.2
2.5 to 5	5	0.5
5.0 to 7.5	2	0.2
7.5 to 10	1	0.1
Total:	**10**	**1.00**

Frequency distributions are helpful in gaining insight into the data, but they lack visual appeal. They also do not readily provide information on symmetry, the number of modes, the existence of outliers, and other important features of the distribution. To provide all of that, one illustrates frequency distributions by means of histograms - rectangular bars built on top of the subintervals, with the heights of the bars proportional to the number of data values that fall within each bin. Figure 1.4 shows a histogram for the frequency distribution in Table 1.1.

An image histogram in its simplest form is a graphical representation of the number of times each pixel value (that is, each gray level) occurs in the image. By default, an image histogram has 256 bins. One can also specify a smaller number of bins if a more general description of the distribution of pixel values is desired. We can gain a great deal of insight into the image by studying its histogram, as the reader will see in the subsequent chapters. Oftentimes the process of image enhancement, such as lightening, darkening, contrast improvement, or equalization begins with a careful study of the image histogram.

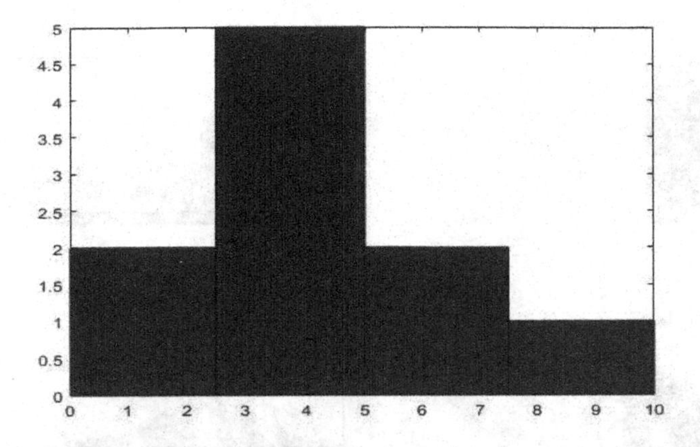

FIGURE 1.4: A histogram for Table 1.1.

In order to display an image histogram, we can use the MATLAB function *imhist*. For example, typing

```
>> imhist(A);
```

into the command window produces a histogram consisting of as many as 256 bars (one for each possible gray level), whereas typing in

```
>> imhist(A,16);
```

results in the histogram of the image A using 16 bins. Figure 1.5 shows several images alongside their histograms. The reader is invited to comment on the different shapes of the histograms that correspond to images of differing levels of darkness and contrast.

The MATLAB code used to generate Figure 1.5 is given below for the convenience of the reader.

```
A=imread('WaterFall.jpg');
A=rgb2gray(A);
subplot(3,2,1);
imshow(A,[0,255]);
subplot(3,2,2);
imhist(A);
A=imread('DWG.jpg');
A=rgb2gray(A);
subplot(3,2,3);
imshow(A,[0,255]);
subplot(3,2,4);
```

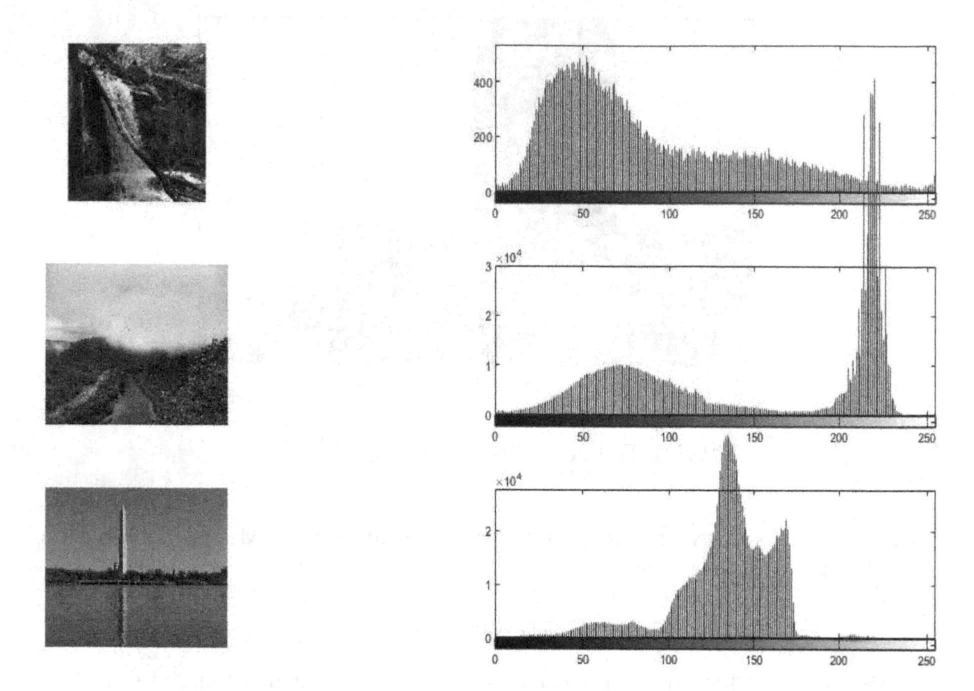

FIGURE 1.5: A selection of images with their histograms.

```
imhist(A);
A=imread('WashingtonMonument.jpg');
A=rgb2gray(A);
subplot(3,2,5);
imshow(A,[0,255]);
subplot(3,2,6);
imhist(A);
```

The photos used in this example were taken by the author and are not part of the standard MATLAB package. Note that the MATLAB function *rgb2gray* was used to convert color images to their grayscale versions. The reader is encouraged to compile a selection of images of their own and explore the relationship between the features of the images and the shapes of the image histograms.

1.3.2 Measures of Center and Spread

In some image processing applications, we may find it useful to know the average pixel value in the image we are working with. In general, whenever we are asked to determine the average of any data set, the first thing that comes to mind is to calculate the mean – that is, to add up the data and to divide the sum by their number. Unfortunately, however, the mean is not always very useful. For example, consider a somewhat unrealistic example of an opulent $10,000,000 mansion in the middle of a modest community of 49 other houses, each worth about $200,000. The mean property value in the community is

$$\bar{y} = \frac{\$10,000,000 + \$200,000 + \cdots + \$200,000}{50} = \$396,000 \text{ per house},$$

which, unfortunately, is representative of neither the opulent mansion nor of the rest of the properties. It just would not be reasonable to claim that the average house in the community is worth $396,000.

Similarly, if we are looking for a job and are trying to get a feel for the average pay at the company we are thinking about applying to, it would not do us much good to find out the mean salary. Say, for the sake of a simple example, the chief executive officer is making $1,000,000 a year and the other 19 employees are paid $40,000 a year. Then the mean company salary of

$$\bar{y} = \frac{\$1,000,000 + \$40,000 + \cdots + \$40,000}{20} = \$88,000 \text{ a year}$$

has little relation to what we are going to be paid if we get the job.

The obvious conclusion is that if the data are skewed or contain outliers, the mean is not a very useful measure of center of the data set. Instead, in such circumstances, the other measure of average – the *median* – is usually used. In order to calculate the median one first arranges the data in the ascending order and then picks the middle entry (or the average of the two middle entries) of the list.

For example, in order to calculate the median of the even-size data set

$$\{5, 1, 4, 8, 2, 6, 11, 72, 4, 7, 1, 9\},$$

we first rearrange it in the ascending order as

$$\{1, 1, 2, 4, 4, \quad 5, 6, \quad 7, 8, 9, 11, 72\},$$

and then observe that the average of the two middle entries of the list is

$$m = \frac{5+6}{2} = 5.5,$$

which is precisely the median of the given data set. In order to calculate the median of the odd-size data set

$$\{5, 1, 4, 8, 2, 6, 17, 72, 4, 7, 10\},$$

we first rewrite it as

$$\{1, 2, 4, 4, 5, \quad 6, \quad 7, 8, 10, 17, 72\},$$

and then observe that the middle entry is 6, which is precisely the median of these data.

We note that the median property value in the real estate example we considered earlier is $200,000, and it is quite reasonable to state that the average house in the community is worth that amount, just as it is reasonable to claim that the average employee earns $40,000, which is precisely the median salary in our job application example.

Continuing the latter example, if we were serious about researching the salary structure for an attractive company, we would be interested in the spread of the salaries almost as much as in the median salary. But it would not be the entire range (defined as the difference between the largest and the smallest values) that interests us most. After all, we are not applying for the lowest-salary job and we probably do not have the ambition to get hired as the CEO of the company right off the bat. Rather, we would be eyeing the **mainstream** jobs.

We could define the mainstream of a data set as its middle half – the part of the data that falls between the thresholds for the lower quarter and the upper quarter. These thresholds are called quartiles. The first quartile Q_1 is the median of the lower half of the data set and the third quartile Q_3 is the median of the upper half. The second quartile Q_2 is the median of the entire data set.

For example, to find the quartiles of the familiar data set

$$\{5, 1, 4, 8, 2, 6, 17, 72, 4, 7, 1, 10\},$$

we first rearrange it in the ascending order as

$$\{1, 1, 2, 4, 4, 5, \quad 6, 7, 8, 10, 17, 72\},$$

and then observe that $Q_1 = 3$ and $Q_3 = 9$ as the medians of the lower and upper halves respectively.

The practical relevance of the quartiles in our job application example is that we would be expecting to be hired at a salary level close to the first quartile Q_1 and to be eventually promoted to the level Q_3, making the *interquartile range*

$$IQR = Q_3 - Q_1$$

a meaningful measure of spread of the distribution of salaries.

Quartiles provide thresholds for breaking data sets up into quarters. But this is not the only possible useful way to subdivide data. More generally, the *kth percentile*, denoted by P_k, is a value with the property that $k\%$ of the data are less than or equal to it. We leave it as an exercise to determine various percentiles of the data sets we have considered.

In connection with the quartiles and the interquartile range, we mention one more concept that might prove useful in image analysis - that of an *outlier* - a value that is vastly different from most of the rest of the data. If this sounds vague and imprecise, that is because it is, and there exist several technical definitions of the term "outlier". One of them, specifically the Tukey fences definition, defines an *outlier* to be any value outside the range

$$[Q_1 - 1.5 \times IQR, \quad Q_3 + 1.5 \times IQR]$$

and a *strong outlier* to be any value outside the range

$$[Q_1 - 3 \times IQR, \quad Q_3 + 3 \times IQR].$$

For example, in the data set

$$\{5, 1, 4, 8, 2, 6, 17, 72, 4, 7, 1, 10\}$$

that we considered earlier, the value 72 is a strong outlier because it is larger than the upper Tukey fence

$$Q_3 + 3 \times IQR = 9 + 3 \times 6 = 27,$$

whereas the value 17 is not an outlier because it falls within the Tukey fences

$$[Q_1 - 1.5 \times IQR, \quad Q_3 + 1.5 \times IQR] \quad = \quad [-8, 18]$$

for the given data set.

Similar considerations apply when we analyze the sets of pixel values of digital images. The mean pixel value is usually of very little importance, whereas the median is often of genuine interest. Whereas the overall range of pixel values is, of course, of some importance, it is the interquartile range that largely determines the level of contrast in an image.

In order to calculate the median and the quartiles of a digital image, one first reshapes the 2-D matrix representing the image into a one-dimensional list of values, rearranges that list in the ascending order, and then identifies the middle entry as well as the middles of the lower and upper halves for the values of the median, Q_1, and Q_3 respectively.

For example, reshaping the matrix

$$\begin{bmatrix} 222 & 115 & 159 & 47 \\ 127 & 251 & 63 & 143 \\ 191 & 79 & 207 & 25 \\ 95 & 175 & 31 & 249 \end{bmatrix}$$

yields the list

$$\{222, 115, 159, 47, 127, 251, 63, 143, 191, 79, 207, 25, 95, 175, 31, 249\},$$

which, after rearranging, becomes

$$\{25, 31, 47, 63, \quad 79, 95, 115, 127, \quad 143, 159, 175, 191, \quad 207, 222, 249, 251\},$$

and we can look up the values

$$Q_1 = 69, \quad Q_2 = 135, \text{ and } Q_3 = 199$$

from this ordered list. There are no outliers because all the values fall within the Tukey fences

$$[Q_1 - 1.5 \times IQR, \quad Q_3 + 1.5 \times IQR] \quad = \quad [-33, 295]$$

for the given data set.

Identifying outliers and strong outliers in digital images is important in such applications as feature extraction and image restoration. For example, in Chapter 6 we will discuss how detecting what seems to be a point source of light can help reverse a motion blur.

Obviously, digital images are way too large to even consider executing these types of calculations by hand. Rather, we use MATLAB functions *reshape*, *prctile*, and *quantile*, which is discussed in detail in Section 2.5 in the context of improving the contrast of digital images. As a brief example, the MATLAB code

```
A=imread('cameraman.tif');
[M,N]=size(A);
A=reshape(A,[1,M*N]);
Q1=prctile(A,25),
Q3=prctile(A,75)
```

calculates the quartiles of the image *cameraman.tif* to be $Q_1 = 68$ and $Q_3 = 166$ respectively, making for the interquartile range of $IQR = 98$. Thus, the middle half of all the pixel values in the given image falls within the range of 98 between $Q_1 = 68$ and $Q_3 = 166$.

The exercises that follow are intended to provide a review of some of the concepts covered in this section, particularly the ones related to summarizing and presenting quantitative data. The reader will also be asked to use MATLAB to calculate statistical measures of center and spread of data sets and to plot histograms of digital images.

Section 1.3 Exercises.

1. Is it possible for almost all the students in a class to have grades above the average for that class? If your answer is yes, give an example. If your answer is no, explain why.

2. Is it possible for almost all the property values in a certain community to be below the average for that community? If your answer is yes, give an example. If your answer is no, explain why.

3. For the data representing the lengths of hospital stays in a large hospital, what would you expect to be larger: the mean or the median? Explain your answer.

4. Do you believe the range of a population can be estimated by calculating the range of a sample taken from that population? How about the interquartile range? Explain your answer.

5. For the following data sets, calculate by hand the median, quartiles, and the interquartile range. Determine whether the given data sets contain outliers.

   (a) There are thirty students taking a general statistics course and the numerical grades for the first exam are

   83, 71, 49, 96, 65, 78, 88, 82, 57, 66,
   74, 91, 85, 76, 83, 90, 69, 71, 81, 73,
   80, 79, 62, 88, 99, 46, 77, 84, 71, 64

   (b) The maximal daily wind speeds in kilometers per hour recorded atop the summit of a certain mountain over twenty days are

   184, 98, 101, 126, 150, 166, 82, 136, 124, 118,
   133, 83, 86, 101, 105, 97, 88, 131, 128, 106

   (c) The ages of the first forty-five presidents of the United States upon inauguration in the ascending order are

   42, 43, 46, 46, 47, 47, 48, 49, 49, 50, 51, 51,
   51, 51, 51, 52, 52, 54, 54, 54, 54, 54, 55, 55,
   55, 55, 56, 56, 56, 57, 57, 57, 57, 58, 60, 61,
   61, 61, 62, 64, 64, 65, 68, 69, 70

   Check your answers with the help of appropriate MATLAB functions.

6. For the data sets in the previous problem, construct frequency distributions and sketch histograms by hand using

   (a) Four bins;
   (b) Eight bins.

7. For the images described by the following matrices, calculate by hand the median, the quartiles, and the interquartile range of the pixel values. Determine whether there are any outliers.

   (a) $A = \begin{bmatrix} 161 & 158 & 158 & 160 & 159 \\ 157 & 159 & 155 & 156 & 159 \\ 160 & 161 & 160 & 160 & 158 \\ 158 & 156 & 159 & 161 & 159 \\ 161 & 160 & 160 & 161 & 162 \end{bmatrix}$

(b) $B = \begin{bmatrix} 128 & 128 & 125 & 125 & 123 & 120 \\ 126 & 127 & 125 & 125 & 121 & 116 \\ 123 & 122 & 122 & 123 & 121 & 121 \\ 119 & 120 & 120 & 120 & 120 & 119 \\ 112 & 117 & 112 & 112 & 116 & 116 \\ 109 & 112 & 110 & 109 & 109 & 109 \end{bmatrix}$

Check your answers using appropriate MATLAB functions.

MATLAB Exercises.

1. For each image you imported in Exercise 1 and each image you created in Exercise 5 of the previous section, use the *imhist* function to display three histograms side-by-side using 16, 32, and 256 bins. Comment on the distribution of pixel values and its relation to such features of the image as contrast and darkness.

2. For each image you imported in Exercise 1 and each image you created in Exercise 5 of the previous section, use the appropriate MATLAB function to calculate the range of pixel values, the median, the quartiles, and the IQR. Comment on the relation of these numerical characteristics of pixel distribution and their relation to such features of the image as contrast and darkness.

1.4 Color Images and Color Spaces

Although we will be concentrating almost exclusively on grayscale images throughout this text, some of the exercises will make references to color images as well. Therefore, this chapter will be incomplete without at least a brief mention of ways to describe, store, process, and display color images. For all our practical purposes, a color image is a stack of three matrices which contain information on the three color channels. The exact composition of those color channels differs from model to model.

Even though numerous methods and formats have been developed, we will focus on just two commonly used color models - the *RGB* model and the *YCbCr* model. In the RGB color model, the three primary colors - Red, Green, and Blue - are added together, which, at least in theory, enables us to reproduce as many as $256^3 = 16,777,216$ synthetic colors. RGB is probably the oldest color model; it predates the invention and development of color photography and is based on our understanding of human perception of colors. When we use the MATLAB command *imread*, the result (with very few exceptions) is a three-layered matrix whose layers contain the red, green, and blue channels corresponding to the RGB colorspace.

Figure 1.6 below displays the 5-color flag of the Republic of Seychelles against the backdrop of the light-blue sky, and Figure 1.7 shows separately the red, green, and blue channels, which, when added together produce the colors of the flag and the sky. We share the MATLAB code used to produce Figure 1.7 in grayscale and in color to help the reader recreate these figures.

```
A=imread('FlagOfSeychelles.jpg');
imshow(A);
%
% Showing the channels in grayscale
%
Red=A(:,:,1); Green=A(:,:,2); Blue=A(:,:,3);
figure;
subplot(1,3,1);
imshow(Red);title('(a) The red channel');
subplot(1,3,2);
imshow(Green); title('(b) The green channel');
subplot(1,3,3);
imshow(Blue); title('(c) The blue channel');
%
% Showing the channels in color
%
figure;
Red=A;Green=A;Blue=A;
Red(:,:,2:3)=zeros(M,N,2);
Green(:,:,1)=zeros(M,N,1); Green(:,:,3)=zeros(M,N,1);
Blue(:,:,1:2)=zeros(M,N,2);
subplot(1,3,1);
imshow(Red);title('(a) The red channel');
subplot(1,3,2);
imshow(Green);title('(b) The green channel');
subplot(1,3,3);
imshow(Blue);title('(c) The blue channel');
```

The code shown above is certainly not the most efficient, but we sacrificed efficiency for the sake of transparency in the hope that the reader will be encouraged to use it and to modify it. We also encourage the reader to experiment with separating their own pictures into the red, green, and blue channels and to explore the sometimes surprising ways familiar colors are comprised of the three primary colors.

FIGURE 1.6: The flag of Republic of Seychelles.

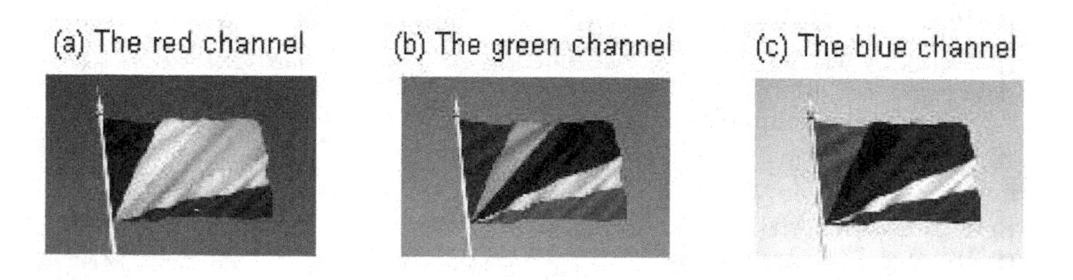

FIGURE 1.7: The Red, Green and Blue channels of the flag of Republic of Seychelles.

The RGB color model is most commonly used for displaying images, but it is not very efficient for storage and transmission due to its inherent redundancy. The reason is that we perceive the brightness and color separately and are much more sensitive to small distortions in brightness than to those in color. In most cases, we can readily sacrifice the latter for the sake of increasing the speed of transmission.

For that reason, several color models that separate light intensity and color into different channels have been developed. One of the most commonly used models of this kind is **YCbCr**. The luminance channel Y contains the information on the light intensity and is stored with high resolution, whereas the two chrominance channels - Cb and Cr - represent the difference from pure Blue and pure Red and can be compressed at a very high compression ratio.

In order to convert an image from the RGB color space to the YCbCr color space, we use the MATLAB function *rgb2ycbcr*. Figure 1.8 shows the three YCbCr channels in the flag of the Republic of Seychelles displayed side-by-side. The MATLAB code used to generate Figure 1.8 is shown below to help the reader reconstruct it on their own.

```
A=imread('FlagOfSeychelles.jpg'); B=rgb2ycbcr(A);
Y=B(:,:,1); Cb=B(:,:,2); Cr=B(:,:,3);
subplot(1,3,1);
imshow(Y); title('(a) The Y channel');
subplot(1,3,2);
imshow(Cb); title('(b) The Cb channel');
subplot(1,3,3);
imshow(Cr); title('(c) The Cr channel');
```

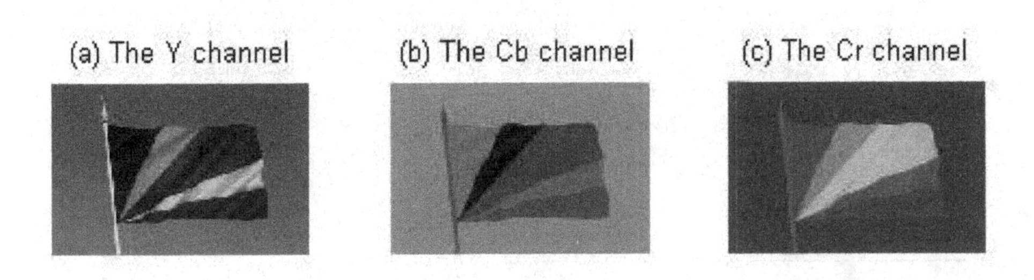

FIGURE 1.8: The Y, Cb, and Cr channels of the flag of Republic of Seychelles.

In the exercises at the end of this chapter, the reader will be asked to experiment with a selection of images of their choice by expanding them into the separate color channels using both the RGB and the YCbCr color models.

Section 1.4 MATLAB Exercises.

1. Import several color images into the MATLAB computing environment using the function *imread* and display them using the functions *imshow* and *imagesc*. Comment on the difference in the image appearance on the computer screen when using the two display functions. Try to "dress up" the images by supplying titles using the MATLAB function *title* with some of its parameters, such as *FontSize*.

2. For each of the images you imported perform the following operations:
   (a) Display the read, green, and blue channels side-by-side.
   (b) Display the Y, Cb, and Cr channels side-by-side.
   (c) Comment on ways the appearance of individual channels is related to the overall appearance of the image.
3. Compile the results of the previous MATLAB exercises into a PowerPoint presentation.

Chapter 2

A Library of Elementary Functions

In this chapter, the reader will discover that a number of image-enhancing transformations can be performed using elementary means, such as linear, piecewise-linear, power, exponential, and logarithmic functions.

2.1 Introduction

Imagine that you are going through your old family album. Back in the day when there was no digital photography and when exposed 35mm film had to be sent to the lab for processing and printing, there was no way to predict the quality of the images in advance. As a result, some of the photos turned out overexposed (which makes them appear too light) and some turned out underexposed (which makes them appear too dark). Some of the images lack contrast, and some turned out blurry. But they still represent precious memories and insights into family history. So, what can we do to "improve" them - to lighten them up if they are underexposed, to darken them if they are overexposed, or to enhance contrast? Does the mathematical toolbox assembled over the first couple of years of college provide any instruments to achieve that goal?

In this chapter, we will see how even the simplest elementary functions studied in algebra and pre-calculus courses can help us achieve remarkable results in our endeavor of restoring old photos.

2.2 Power Functions and Gamma-Correction

Suppose that you have found a photo (shown in Figure 2.1) of your grandfather cross-country skiing at Mount Van Hoevenberg Olympic Trail System near Lake Placid, NY, when he was taken there by his parents during a family vacation. Unfortunately, the picture seems badly under-exposed. Is there anything you can do to help reconstruct the view your grandfather was enjoying during his trip to the Adirondack Mountains High Peaks Area?

FIGURE 2.1: An underexposed photo of a winter forest.

In the same album, you found a photo (shown in Figure 2.2) that your grandmother took while hiking the Living Room Trail in the mountains overlooking Salt Lake City, UT, when she was visiting her friends at Utah State University. This time, the image seems overexposed. What can you do to restore the beautiful view of the Wasatch Mountain Range that overlooks Salt Lake City?

In order to work out a possible strategy, we take a look at the histograms of the two images. The histogram of the winter scene in the woods is shown

FIGURE 2.2: An overexposed photo of a mountain view.

in Figure 2.3. We can see that most of the pixels have grayscale values in the lower range between 0 and 120 or even lower. What is the best way we go about "stretching" the interval that contains the majority of pixel values?

On the other hand, the histogram of the mountain view (Figure 2.4) shows that most pixels in that image have values in the range of 120 to 255. How can we achieve a "stretching" of that interval?

Can elementary functions be of any help to us? Figure 2.5 displays a graph of what is probably the most often encountered function in elementary mathematics, $f(x) = x^2$, together with a graph of its inverse, $g(x) = \sqrt{x}$. Also displayed are graphs of their close relatives, with their domains restricted to $[0, 1]$. On the graphs, we can see that $f(x) = x^2$ effectively "shrinks" the intervals of small values of x while "stretching" the intervals of large values of x. For example, $f(x) = x^2$ maps the interval $[0, 1/2]$ onto the shorter interval $[0, 1/4]$ while mapping the interval $[1/2, 1]$ onto the longer interval $[1/4, 1]$. Conversely, $g(x) = \sqrt{x}$ effectively "stretches" the intervals of small values of x while "shrinking" the intervals of large values of x. For example, $g(x) = \sqrt{x}$ maps the interval $[0, 1/4]$ onto the longer interval $[0, 1/2]$ while mapping the interval $[1/4, 1]$ onto the shorter interval $[1/2, 1]$.

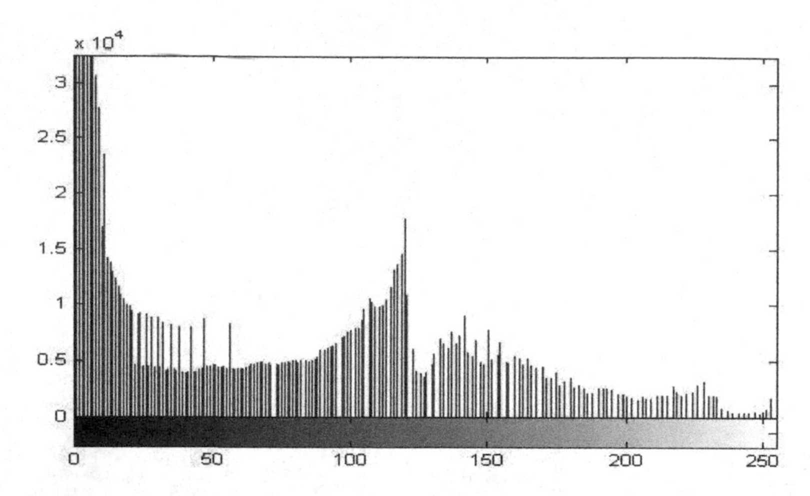

FIGURE 2.3: Histogram of the underexposed photo in Figure 2.1.

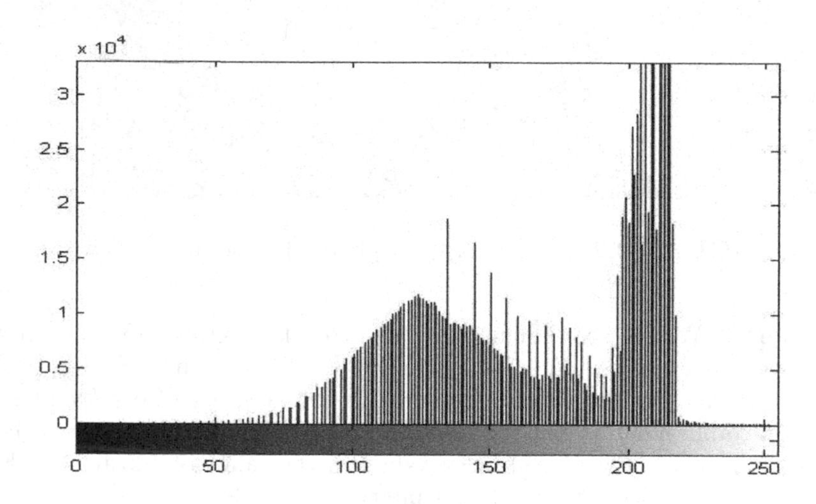

FIGURE 2.4: Histogram of the overexposed photo in Figure 2.2.

But didn't we wish we could "stretch" the interval of the lower pixel values for the underexposed image of the winter forest? Well, we have found the very instrument we need for the job in our elementary math toolbox. We first divide the pixel values by 255 in order to normalize them by putting them into the range of 0 to 1. Next, we raise the (normalized) pixel values to a positive power $0 < \gamma < 1$ (for example, $\gamma = 1/2$). Finally, we multiply the transformed pixel values by 255 and plot the resulting image.

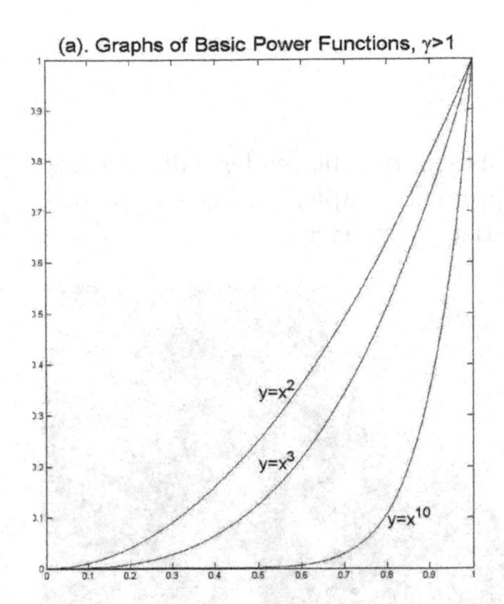

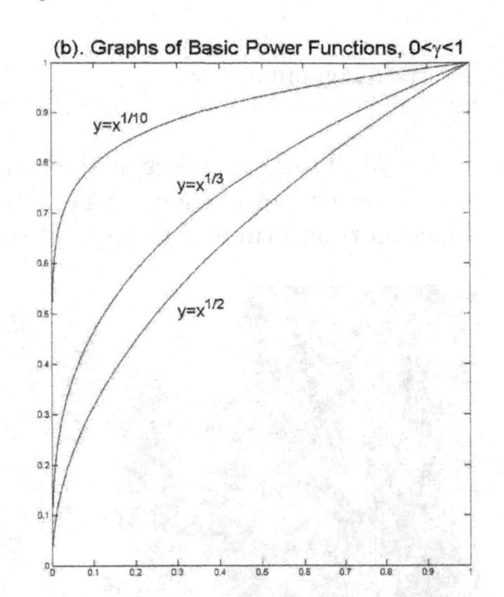

FIGURE 2.5: The graphs of $y = x^\gamma$ for several different values of γ.

Formally, let A be the original image (normalized by dividing every pixel by 255), and let B be the normalized enhanced version of the image. Then every pixel in B can be calculated using the formula

$$B(m,n) = A(m,n)^\gamma. \tag{2.1}$$

If it is preferable to have the pixel values occupy the standard range from 0 to 255 after completing the exponential-function transform, all the pixels in B need to be multiplied by the factor of 255. This last step is optional, because, as we have already mentioned in Section 1.2, for images of the class "double", the *imshow* command interprets the pixel values on the scale from 0 to 1 by default. The end result of all these transformations is displayed in Figure 2.6, and a possible code for generating it is given below.

```
A=imread('WinterForest.jpg');
gamma=.5;
[M,N]=size(A);
for m=1:M
   for n=1:N
      B(m,n)=(A(m,n)/255)∧ gamma;
   end
end
B=255*B; imshow(B,[0,255]);
```

The double *for* loop was shown just to illustrate the use of this structure as the same result could be achieved with just a one-line command

```
B=A.∧gamma;
```

In the MATLAB exercises at the end of this section, the reader will be asked to create a more efficient MATLAB program for implementing the power-function transform and to apply it to a variety of images.

FIGURE 2.6: An enhanced version of the photo in Figure 2.1.

We can employ a similar strategy to "stretch" the interval of higher pixel values of the overexposed image. Just like in the previous example, we first divide the pixel values by 255 to normalize them, but this time we raise the (normalized) pixel values to a positive power $\gamma > 1$ (for example, $\gamma = 2$). Finally, if we choose to, we can multiply the transformed pixel values by 255. The end result of all those transformations is displayed in Figure 2.7.

We hope that the reader will agree that the quality and visual appeal of the resulting images represent a significant improvement over the original photos. Most importantly, we invite the reader to appreciate the fact that even the simplest mathematical tools (like basic power functions) can be of great practical use and can help achieve significant practical results.

FIGURE 2.7: An enhanced photo of the mountain view.

There is nothing special about the values $\gamma = 2$ and $\gamma = 1/2$ in the examples above. Any value $\gamma > 1$ can be used to darken the image, and any value $0 < \gamma < 1$ can be used to lighten it. In the exercises below, the reader is encouraged to experiment with their own favorite images (like the ones imported in MATLAB Exercises following Section 1.2) and with various values of γ.

This method can also be applied to color images by first converting the color image to the YCbCr color space followed by a power-function transformation of the Y channel.

Section 2.2 Exercises.

The exercises in this section provide a review of basic properties and applications of power functions. While the main point of this section is the application of the power-function transformation to digital images, we hope that the reader is motivated to explore other practical uses of this class of functions.

1. Sketch the graphs of a few simple power functions such as

   (a) $y = x, \quad y = x^{1.5}, \quad y = x^2, \quad y = x^3, \quad \text{and} \quad y = x^4,$

(b) $y = x, \quad y = x^{1/2}, \quad y = x^{1/3}, \quad \text{and} \quad y = x^{1/4}$

on the same set of axes. Try to experiment with different domains, such as $[0, 1]$, $[0, 3]$, and $[-1, 1]$. Comment on the relative shapes of the graphs. Is the domain $[-1, 1]$) appropriate for all of the functions?

2. Suppose that the function is described by the formula $y = cx^{\gamma}$. Determine the values of the parameters c and γ if the graph of the function passes through the points

   (a) (1,3) and (2,12),

   (b) (1,5) and (4,10).

3. The surface area S of a sphere is a function of its radius. It is described by the formula $S(r) = 4\pi r^2$. The volume V of a sphere is also a function of its radius. It is described by the formula $V(r) = \frac{4}{3}\pi r^3$.

   (a) Sketch the graphs of the surface area and the volume as functions of the radius. Comment on the concavity of the graphs.

   (b) What will happen to the surface area and the volume if you double the radius? If you triple the radius? If you quadruple the radius?

   (c) Find the formula for the *ratio* of the volume of the sphere to the surface area of the sphere and sketch a graph of this ratio as a function of the radius.

   (d) Comment on the physical properties of small and large objects (like particles of dust, raindrops, and rocks) that tend to result from the different ratios of their volumes and surface areas. In particular, comment on the way those objects tend to fall through the air. Also comment on the danger of spontaneous combustion and explosion in grain storage facilities.

4. The surface area S of a cylinder is a function of its radius r and its height h. It is described by the formula $S = 2\pi r^2 + 2\pi r h$. The volume of a cylinder is also a function of its radius and its height. It is described by the formula $V = \pi r^2 h$.

   (a) Sketch the graphs of the surface area and the volume as functions of the radius (with the height held constant). Comment on the concavity of the graphs.

   (b) Sketch the graphs of the surface area and the volume as functions of the height (with the radius held constant). Comment on the shape of the graphs.

   (c) What will happen to the surface area and the volume if you double the radius? If you triple the radius? If you double the height? If you triple the height?

(d) Find the *ratio* of the volume of the cylinder to its surface area.

(e) Comment on the physical properties of small and large objects (like small twigs and logs of firewood) and explain how to best light a campfire.

5. The tidal force, T, of a celestial object on the Earth's oceans is proportional to the mass of that object and is inversely proportional to the cube of the distance to that object.

   (a) Write a formula that describes this relationship.

   (b) Look up the masses of the Sun and the Moon, as well as their distances from the Earth, and compare the tidal forces exerted on the Earth's oceans by the Sun and the Moon.

   (c) Look up the masses of the planets of the Solar System and their smallest distances from the Earth. Do you think any of them have an appreciable effect on the ocean tides (as compared to the Sun and the Moon)?

MATLAB Exercises.

1. Find a few photos that are either too dark or too light. If necessary, use the MATLAB function *rgb2gray* to convert a color image to a grayscale image.

2. Write a MATLAB function that would lighten (or darken) the specified image using the specified value of γ. The function should accept both the matrix of the image and the value of γ as inputs. Ideally, it should work with any grayscale image and with any color image. In the latter case, the image should be converted into the YCbCr color space, and the transformation should be applied to the Y channel.

3. Write a MATLAB program that will do the following:

   (a) Import an image of your choice into the MATLAB computing environment.

   (b) Plot the original image and its histogram

   (c) Call the MATLAB function you created in Exercise 2 to lighten (or darken) the original image. Experiment with different values of γ and try to find the optimal one.

   (d) Display the enhanced version of the image together with its histogram.

4. Test your program using several different images (grayscale and color).

5. Compile the results of the previous MATLAB exercises into a PowerPoint presentation.

2.3 Exponential Functions and Image Transformations

Power functions are certainly not the only tools suitable for the task of lightening or darkening a grayscale image. As we saw in the previous section, darkening could be accomplished by raising the (normalized) pixel values to a positive power $\gamma > 1$, that is, by applying a function that has the property of being concave up. Another well-known class of functions that are concave up is the class of exponential functions. In Figure 2.8(a) we can see a sample of graphs of exponential functions with different positive bases, and we can observe the effect the base has on the shape of the graph.

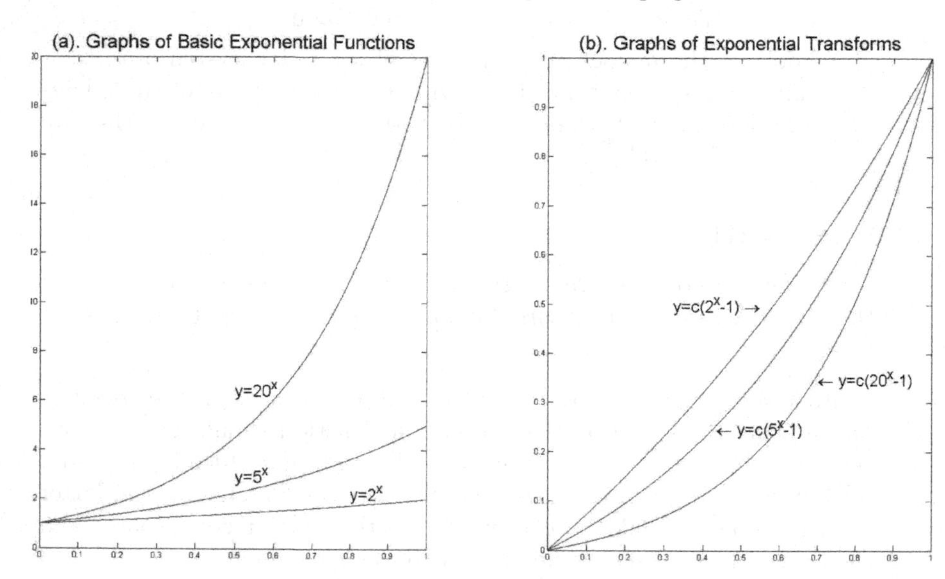

FIGURE 2.8: Graphs of several exponential functions and the corresponding transforms.

What immediately gets our attention, however, is that unlike the graphs of the power functions, the graphs of exponential functions do not pass through the origin, which is going to be a problem if we apply an exponential function to the pixel values. Fortunately, this deficiency can be easily remedied by shifting the graph one unit down.

We would also like our function to map the interval $[0, 1]$ into the interval $[0, 1]$. Putting all our wishes together, we would like to have a function described by the general formula

$$f(x) = c(b^x - 1) \tag{2.2}$$

that satisfies the conditions

$$f(0) = 0 \text{ and } f(1) = 1. \tag{2.3}$$

The second condition in (2.3) implies that the constant c has to equal

$$c = \frac{1}{b-1}, \tag{2.4}$$

as the reader can verify on their own. Figure 2.8(b) shows several graphs of exponential functions of this class for different values of the parameter b.

How does one apply the transformation (2.2) to digital images? Let us denote the overexposed image that we would like to enhance by A, and its enhanced version by B. First, all the pixel values in A must be divided by 255 in order to bring them into the interval $[0, 1]$. We can then make use of the function (2.2) to calculate every pixel value $B(n, n)$ in B from the corresponding pixel $A(m, n)$ by means of the formula

$$B(m, n) = c\left(b^{A(m,n)} - 1\right). \tag{2.5}$$

Finally, if it is preferable to have the pixel values occupy the standard range from 0 to 255 after completing the exponential-function transform, all the pixels in B need to be multiplied by the factor of 255. This last step is optional, because, as the reader remembers, for images of the class "double", the *imshow* command interprets the pixel values on the scale from 0 to 1.

If we choose to, we can avoid dealing with normalization altogether and use instead the function of the form

$$f(x) = k[(1+\alpha)^x - 1] \tag{2.6}$$

with the constant k determined by the condition

$$f(255) = 255,$$

which yields

$$k = \frac{255}{(1+\alpha)^{255} - 1}. \tag{2.7}$$

The two methods are equivalent, and we leave it to the reader to establish the relationships between the constants c and k and between the parameters b and α. In order to apply the transformation (2.6) to digital images, we calculate every pixel $B(n, n)$ in B from the corresponding pixel $A(m, n)$ by using the formula

$$B(m, n) = k[(1+\alpha)^{A(m,n)} - 1]. \tag{2.8}$$

We are now ready to make practical use of the new method of darkening or lightening our images. Figure 2.9 shows a badly overexposed photo of Mt. Washington in the White Mountains of New Hampshire and the effect of using the exponential transform of the form (2.5) with the base $b = 1000$ on the image and on the image histogram. We can see that the correction performed by the exponential transform is similar to that performed by a power-function transform with the parameter $\gamma > 1$ discussed in the previous

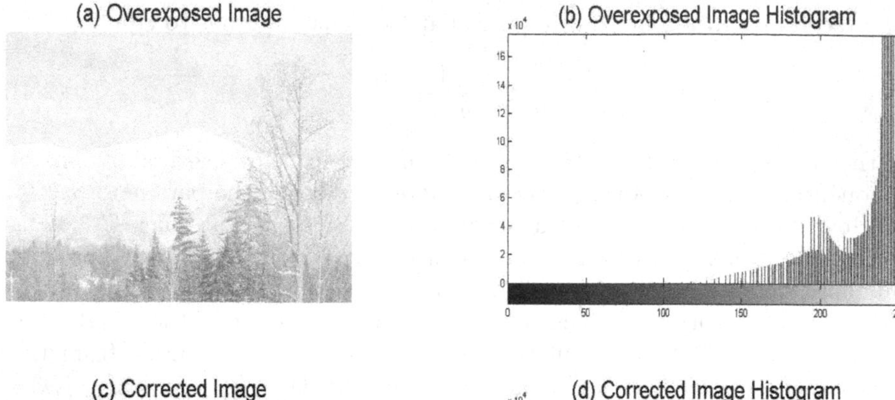

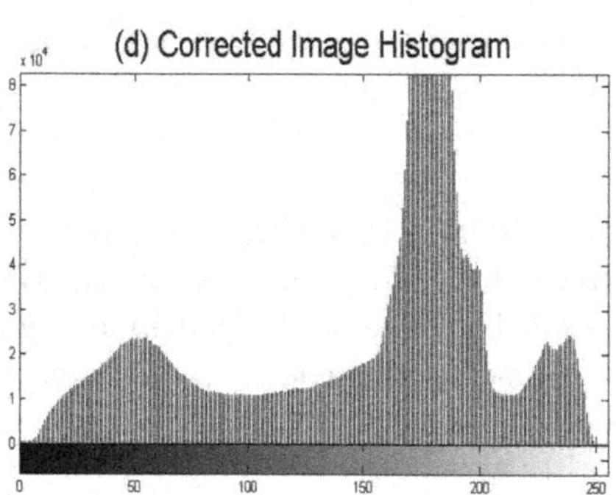

FIGURE 2.9: An overexposed image corrected by the exponential transform.

section. However, the exponential transform tends to be more effective in correcting overexposed images that have most of the pixel values crowded in a small interval near the maximal intensity.

Section 2.3 Exercises.

The exercises in this section are intended to encourage the reader to review some of the basic properties and common applications of exponential functions. The reader is also asked to verify all the formulas related to the exponential-function transform. While the main point of this section is the application of the exponential-function transform to digital images, we hope that the reader is motivated to explore other practical applications of this class of functions.

1. Sketch the graphs of a few simple exponential functions, such as

   (a) $y = 1.1^x$, $y = 1.5^x$, $y = 2^x$, $y = e^x$, $y = 3^x$, and $y = 4^x$,

   (b) $y = (1/2)^x$, $y = (1/3)^x$, and $y = (1/4)^x$

   on the same set of axes. Try to experiment with different domains, such as $[-1, 1]$, $[-2, 2]$, and $[-3, 3]$. Comment on the relative shapes of the graphs. Which ones of the functions represent growth and which ones represent decay?

2. Suppose that the function is described by the formula $y = cb^x$. Determine the values of the parameters c and b if the graph of the function passes through the points
   (a) (0,3) and (2,12),
   (b) (1,6) and (2,18),
   (c) (1,15) and (3,500),
   (d) (2,200) and (5,25).

3. Suppose that the population P of a certain metropolitan area is 12 million people at the beginning of the year 2020 and that it is increasing by 2% every year.
   (a) Write a formula that gives the population as a function of time (in years since the beginning of 2020). Use this formula to predict the population in 2025, 2030, and 2040.
   (b) Sketch the graph of the population as a function of time and comment on the shape of the graph and on the rate at which the population is increasing.
   (c) Try to determine how many years it will take for the population to double and to triple.

4. Suppose that you decided to invest $1,000 at 3% interest rate. Write a formula for the value, A of your account as a function of time if the interest is compounded
   (a) annually,
   (b) quarterly,
   (c) monthly,
   (d) daily,
   (e) continuously.

   Determine the effective annual percentage yield and calculate the value of your account in ten years under all those scenarios.

5. Repeat the previous exercise for the interest rates of 2%, 4%, and 6%. Compile your results in a PowerPoint presentation and comment on the effect of the interest rate and the mode of compounding on the account balance after the term of ten years.

6. According to the United States Census Bureau, between 1990 and 2020, the U.S. population increased from 250 million to 330 million.

(a) Using this information, try to write a formula that gives the U.S. population as a function of time (in years since the beginning of 2000). Use your formula to predict the U.S. population in 2030.

(b) Determine the average annual growth rate of the U.S. population during the time periods between 1990 and 2020, between 1990 and 2000, between 2000 and 2010, and between 2010 and 2020.

(c) Sketch a graph of the U.S. population as a function of time. Comment on the shape of the graph and on the rate at which the U.S. population is increasing.

(d) Try to make a prediction when the U.S. population will reach 500 million.

7. Suppose that it takes five hours for a certain bacterial culture to double in size.

   (a) Write a formula that gives the size of the bacterial culture, P as a function of time, t.

   (b) Sketch the graph of the size of the bacterial culture as a function of time. Comment on the shape of the graph and on the rate at which the size of the culture is growing.

   (c) Try to determine how many hours it will take for the bacterial culture to triple in size.

8. The value of a used car of a certain model tends to decrease approximately 20% per year.

   (a) Given that a new car of this model is sold for $\$30,000$, write a formula for the value V of the car as a function of time t (in years). Use this formula to predict the value of a 5-year-old car of this model.

   (b) Sketch the graph of the value of the car as a function of time. Comment on the shape of the graph and on the rate at which the value of the car is decreasing.

   (c) Try to determine how many years it will take a car of this model to lose half of its value.

9. Chlorine is often used to maintain water quality in swimming pools. According to regulations, chlorine concentration is supposed to be between 1.5 and 2.5 parts per million. It has been estimated that during the summer, the dissipation rate of chlorine is 1 percent per hour.

   (a) Write a formula for the value C of chlorine concentration t hours after being replenished to the maximal concentration of 2.5 ppm.

   (b) How often would you recommend replenishing chlorine to make sure its concentration does not drop below 1.5 ppm?

10. Verify the formulas (2.4) and (2.7).

11. Derive the formulas that establish a relationship between the parameters b and α in the expressions (2.2) and (2.6) respectively.

12. Derive the formulas that establish a relationship between the constants c and k in the formulas (2.4) and (2.7) respectively.

MATLAB Exercises.

1. Find a few pictures that seem to be overexposed with most of the pixels appearing extremely bright. If necessary, use the MATLAB *rgb2gray* function to convert a color image to a grayscale image.

2. Write a MATLAB function that would darken the specified image using the exponential transform with a specified value of the base b. Your function should accept the image and the base b as inputs. Ideally, your function should work with any grayscale image and with any color image. In the latter case, the image must be converted into the YCbCr color space, and the transformation must be applied to the Y channel. Feel free to use either (2.5) or (2.8) to implement the exponential transform.

3. Write a MATLAB program that will do the following:

    (a) Import an overexposed image of your choice into the MATLAB computing environment.

    (b) Plot the original image and its histogram.

    (c) Call the MATLAB function you created in Exercise 2 to darken the original image. Experiment with different bases and try to find the optimal value for the base.

    (d) Display the enhanced image together with its histogram.

4. Test your program using several different images (grayscale and color).

5. Compile the results of the previous MATLAB exercises into a PowerPoint presentation.

2.4 Logarithmic Functions and Image Transformations

In the previous section, we saw that in order to darken an overexposed image, we had a choice of mathematical instruments and were not limited to just the power-function transforms with the parameter $\gamma > 1$. We could alternatively use the exponential-function transform, and the latter had certain advantages, particularly for images where most of the pixel values concentrated in a narrow interval of near-maximal intensity. Similarly, when it comes to lightening an image, there is no reason to feel limited to just the power-function transform with the parameter $0 < \gamma < 1$.

As we saw in Section 2.2, what made such power functions suitable for the purpose of lightening an image is that their graphs are concave down. Can we think of any other elementary functions with the same property? The logarithmic function quickly comes to mind. In Figure 2.10(a), we can compare graphs of a sample of logarithmic functions with different positive bases and can observe the effect the base has on the shape of the graph.

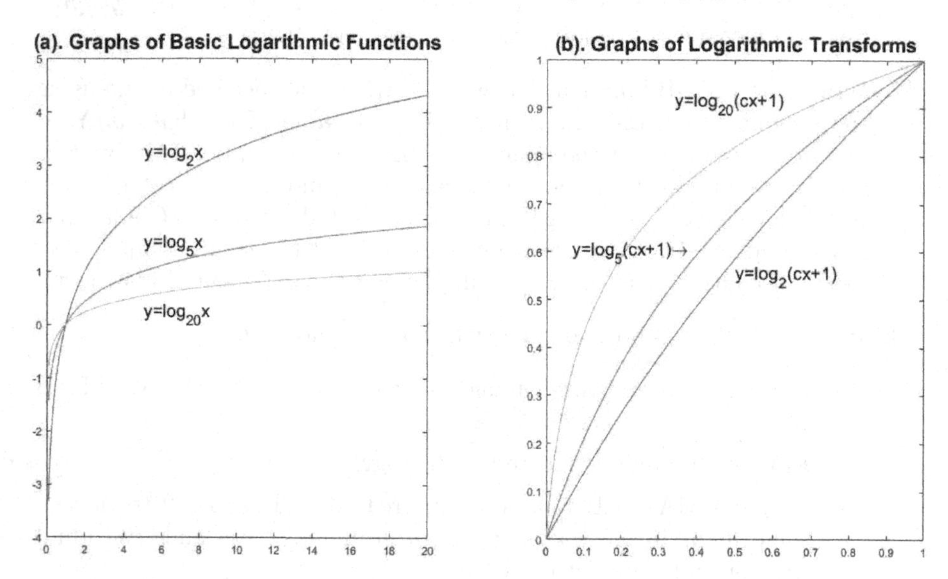

FIGURE 2.10: Graphs of logarithmic functions and transforms.

We immediately recognize a technical issue of the kind we have already dealt with in the previous section in the context of exponential functions. Unlike the graphs of the power functions, the graphs of the logarithmic functions do not pass through the origin. Nor do they pass through the point $(1, 1)$. The former deficiency can be easily remedied by shifting the graph one unit to the left, which is accomplished by adding 1 to the expression the logarithm is applied to. In order to ensure that our function maps the interval $[0, 1]$ onto the interval $[0, 1]$, we need to impose further conditions.

Following the method we used in the previous section, we could try to construct a logarithmic function described by the formula

$$f(x) = c \log_b(x + 1) \tag{2.9}$$

and satisfying the conditions

$$f(0) = 0 \text{ and } f(1) = 1. \tag{2.10}$$

Unfortunately, this approach **is not going to work**. Indeed, the second condition in (2.10) implies that the constant c has to equal

$$c = \frac{1}{\log_b 2}.$$

Substituting this expression into (2.9) yields

$$f(x) = \log_2(x + 1), \tag{2.11}$$

which is independent of b, thus depriving us of the ability to vary the parameter that, we hope, controls the degree to which the image is lightened when the transform is applied.

So, what are we to do? One possible idea would be to recall that logarithmic and exponential functions are inverses of each other. We can then construct and use the inverse of the function (2.2) from the previous section as a candidate for the logarithmic transform. We can easily determine that the inverse of

$$f(x) = c(b^x - 1)$$

is

$$f^{-1}(x) = \log_b(x/c + 1). \tag{2.12}$$

Therefore, it would seem reasonable to construct a suitable function for our logarithmic transform in the form

$$g(x) = \log_b(\sigma x + 1) \tag{2.13}$$

with the parameter σ satifsying the condition

$$\sigma = b - 1. \tag{2.14}$$

Figure 2.10(b) shows several graphs of the function g for different values of the parameter b. Similarly to the approach taken in Section 2.3, in order to apply the function of the form (2.13), all the pixel values need to be divided by 255 in order to bring them into the interval $[0, 1]$ and then, after completing the transform, they need to be multiplied by the factor of 255 (if it is desired to have the pixel values occupy the range from 0 to 255). Formally, the pixel values of the (normalized) enhanced image B are calculated from the pixel values of the (normalized) original image A using the formula

$$B(m, n) = \log_b(\sigma A(m, n) + 1). \tag{2.15}$$

If we choose to, we can avoid the normalization steps and use instead the function of the form

$$g(x) = \log_u(kx + 1) \tag{2.16}$$

with the constant k determined by the condition

$$f(255) = 255,$$

which yields

$$k = \frac{u^{255} - 1}{255}. \tag{2.17}$$

The two methods are equivalent, and we leave it to the reader to establish the relationships between the constants σ and k and between the parameters b and u.

Figure 2.11 shows a photo of the part of the night sky that contains the Big Dipper and the effect of using the logarithmic transform with the base $b = 1000$. We can see that the correction performed by the logarithmic transform is similar to that performed by a power-function transform with the parameter $0 < \gamma < 1$ discussed in the previous section. However, the logarithmic transform tends to be more effective in bringing to light the otherwise hidden features in an image where intensity levels differ vastly, but most of the pixel values are very small.

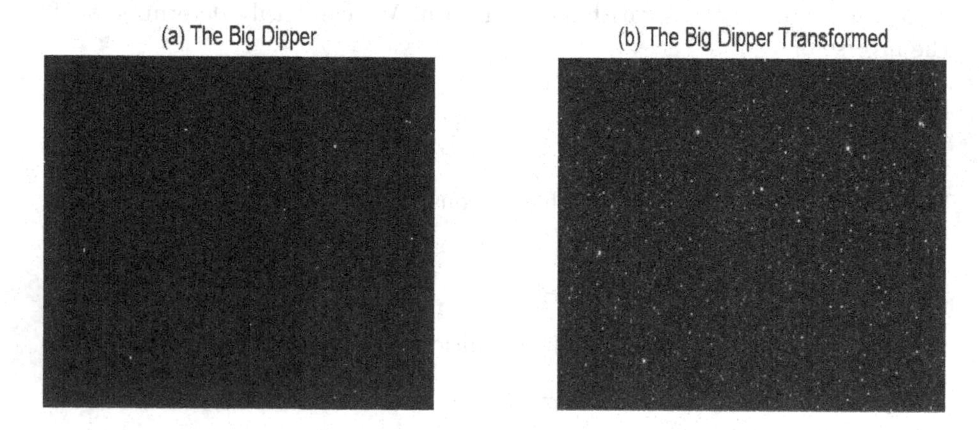

FIGURE 2.11: A view of the Big Dipper and its logarithmic transform.

Unfortunately, logarithms tend to have a somewhat dubious reputation among college students. This section has given us an opportunity to appreciate one more way in which they can be useful. In the exercises at the end of this section, the reader is encouraged to review the basic properties of logarithms and to explore a variety of practical applications of logarithmic functions. In MATLAB exercises, the reader will be asked to applying the logarithmic transform to images that contain hidden features in order to bring them to light.

Section 2.4 Exercises.

1. Verify the formulas (2.12), (2.14), and (2.17).

2. Derive a relationship between the constants σ and k in the formulas (2.14) and (2.17) respectively.

3. Sketch the graphs of a few simple logarithmic functions with different bases, such as

   (a) $y = \log_{1.1} x$, $y = \log_{1.5} x$, and $y = \log_2 x$,
   (b) $y = \ln x$, $y = \log_5 x$, and $y = \log_{10} x$

   on the same set of axes. Comment on the relative shapes of the graphs.

4. Use the definition of logarithms to evaluate the following expressions without using a calculator:

   (a) $\log_2 64$, (c) $\log_2 2^{3.12}$, (e) $\log_{10}(0.0001)$,
   (b) $\log_2(1/32)$, (d) $\log_{10} 1000$, (f) $\log_{10}(10^{4.51})$.

5. Use appropriate properties of exponents to prove the following rules of logarithms:

   (a) The Product Rule, which states that $\log_b(xy) = \log_b x + \log_b y$.
   (b) The Quotient Rule, which states that $\log_b\left(\frac{x}{y}\right) = \log_b x - \log_b y$.
   (c) The Power Rule, which states that $\log_b x^p = p \log_b x$.
   (d) The Change-of-Base Formula

   $$\log_b x = \frac{\log_a x}{\log_a b}$$

   provided that $a \neq 1$ and $b \neq 1$.

6. Use online resources to research the history of the slide rule and its uses for multiplication, division, powers, and even for trigonometry.

7. Use the basic properties of logarithms to expand the following expressions:

   (a) $\log_2\left(x^2 y^3 \sqrt{z^2 + 1}\right)$,
   (b) $\log\left(\frac{x^3\sqrt{y^2 z^5}}{\sqrt{x^2+y^2}}\right)$.

8. Solve the following equations using properties of logarithms:

   (a) $3 \cdot 2^t = 4$,

   (b) $5 \cdot 3^t = 7 \cdot 2^{t-1}$,
   (c) $\ln x - \ln(x+1) = 4$.

9. Suppose that a function is described by the formula $y = \log_b(cx+1)$. Determine the values of the parameters b and c if the graph of the functions passes through the points
   (a) (1,3) and (2,12),
   (b) (1,5) and (4,10).

10. Use logarithms to determine the doubling time of your account if the interest rate is 4% compounded
    (a) annually,
    (b) quarterly,
    (c) monthly,
    (d) daily ,
    (e) continuously.

11. Repeat the previous problem for the interest rates of 2%, 3%, and 6%.

12. Rewrite all the formulas you created for Problems 3 through 8 from Section 2.3 using base e.

13. Use logarithms to calculate precise answers for relevant parts of Problems 6 – 9 from Section 2.3.

14. The magnitude M of an earthquake on the Richter sale is defined by the formula

$$M = \log_{10}\left(\frac{I}{I_0}\right),$$

    where I is the intensity of the earthquake measured by a seismograph, and I_0 is the minimal reference intensity.
    (a) Determine the magnitude of an earthquake if its intensity I is 5,000 greater than the reference intensity I_0.
    (b) How much more intense was the 2011 Tohoku earthquake in Japan, which had magnitude 9.0 on the Richter scale, than the 1994 Northridge earthquake in Los Angeles County, which had magnitude 6.7 on the Richter scale?

15. Sound level N is measured in *decibels* using the formula

$$N = 10\log_{10}\frac{I}{I_0},$$

    where I is the intensity of the sound and I_0 is the threshold of hearing.

For example, if the intensity of the sound is 100 times greater than the threshold of hearing, then $I/I_0 = 100$, and the sound level is

$$N = 10 \log_{10} 100 = 20 \text{ dB}.$$

(a) If one truck speeding down the highway produces the sound level of 80 dB, then what is the sound level of 10 such trucks speeding down the highway in close proximity to each other?

(b) How much louder are 5 police whistles sounded at the same time as compared to one police whistle when measured in decibels?

Section 2.4 MATLAB Exercises.

1. Find a picture in which most of the pixel values are very small, but some of the pixels appear extremely bright. If necessary, use the MATLAB *rgb2gray* function to convert a color image to a grayscale image.

2. Write a MATLAB function that would bring out the hidden features of the specified image using the logarithmic transform with the specified value of the base b. Your function should accept the image and the base as inputs. Ideally, your function should work with any grayscale image and with any color image. In the latter case, the image must be converted into the YCbCr color space, and the transformation must be applied to the Y channel.

3. Write a MATLAB program that will do the following:
   (a) Import an image into the MATLAB computing environment.
   (b) Display the original image and its histogram.
   (c) Call the MATLAB function you created in Problem 2 to bring out the hidden features in the original image. Experiment with the parameters of your logarithmic transform and try to determine the optimal values of those parameters.
   (d) Display the enhanced image together with its histogram.

4. Test your program using several different images (grayscale and color).

5. Compile the results of the previous MATLAB exercises into a PowerPoint presentation.

2.5 Linear Functions and Contrast Stretching

Let us take a look at the image on the left side of Figure 2.12. What is our first impression? We don't find the image exciting. It is not that the image

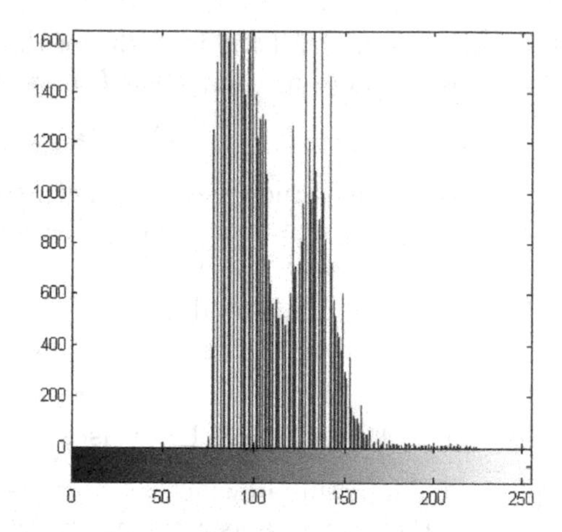

FIGURE 2.12: A low-contrast image and its histogram.

is too light or too dark. The problem is that it appears to be very gray and, we might say, it lacks contrast. A look at the histogram of the pixel values (the right side of Figure 2.12) reveals that the image only takes advantage of a small part of the available range of pixel values. Instead of being spread between 0 and 255, most of the pixel values appear to fall into the fairly narrow interval between, roughly, 75 and 160.

To render this general description of the distribution of the pixel values more precise, we calculate its five-number summary, which consists of the smallest value, the quartiles (including the median), and the largest value. As already discussed in Section 1.3, we first use the MATLAB commands

```
>> [M,N]=size(A)
```

and

```
>> V=reshape(A,[1,M*N]);
```

to rearrange all the pixel values in the image A into a one-dimensional array denoted by V and follow up with the MATLAB commands

```
>> MinA=min(V), MaxA=max(V)
```

to determine that the darkest and the brightest pixels have the values 74 and 224 respectively. We can observe that the narrowness of this range is part of the explanation for the overall lack of contrast in the image. Finally, we use the commands

```
>> Q1=prctile(V,25), Q2=prctile(V,50), Q3=prctile(V,75)
```

to calculate the quartiles

$$Q_1 = 89 \text{ and } Q_2 = 131$$

of the set of pixel values, which tells us that the entire middle half of the pixels is squished in a very narrow range. In fact, the interquartile range (IQR) of the pixel values is only

$$IQR = Q_3 - Q_1 = 131 - 89 = 42,$$

which goes further to explain why the image appears so gray and boring.

The strategy for improving the image seems apparent, and it includes two objectives – a modest one and a more ambitious one.

- At the very least, we must somehow "stretch" the current overall interval $[74, 224]$ of the pixel values so that it occupies the entire available range of 0 to 255.
- We would also like to "stretch" the narrow range between the quartiles $Q_1 = 89$ and $Q_3 = 131$ so that it occupies the middle half (that is, 63 to 191) of the possible grayscale values that are available to us.

How should we approach this task? To reach the first objective, we would like to map 74 to 0 and 224 to 255 respectively. That is, for our low-contrast image, we would need to design a monotone function f with the property

$$f(74) = 0 \text{ and } f(224) = 255. \tag{2.18}$$

To achieve the second goal, we would need to map the smallest pixel value (denoted by $MinA$) to 0, the first quartile Q_1 to 63, the third quartile Q_3 to 191, and the largest pixel value (denoted by $MaxA$) to 255. Formally, for the specific image we are working with, we must design a monotone function g with the property

$$g(74) = 0, \quad g(89) = 63, \quad g(131) = 191, \quad \text{and} \quad g(224) = 255. \tag{2.19}$$

Well, there is no reason why we should not be able to accomplish both tasks once we decide on the type of functions we should try to design. Reaching into our mathematical toolbox, the simplest tools we can find there that might serve the purpose are linear functions (for the overall contrast stretching) and piecewise-linear functions (for the more ambitious second goal).

Since we probably have not used those functions since our course in college algebra, our first topic will have to be a review of linear functions, including their graphs and equations. After that, we will be prepared to discuss piecewise-linear functions. Having reviewed all of the above, we will apply those instruments in our image-enhancing endeavors.

We recall that a *linear function* f is described by the equation

$$f(x) = mx + b, \tag{2.20}$$

where m is a constant parameter called "the slope", and b is a constant parameter called "the y-intercept". The graphs of linear functions are straight lines; when the slope m is positive, the line rises as we move from the left to the right, whereas when the slope m is negative, the line descends as we move towards the right. It is evident from the defining formula (2.20) that for every unit change in the variable x, the variable y changes by m units. Thus, the slope m can be interpreted as the *rate of change* of y with respect to x.

For the purposes of this chapter, it is important for us to be able to find equations of lines that pass through two given points. Let us recall that the slope m of the line that passes through the points (x_1, y_1) and (x_2, y_2) is

$$m = \frac{\Delta y}{\Delta x} = \frac{y_2 - y_1}{x_2 - x_1}, \tag{2.21}$$

provided that $x_1 \neq x_2$. The familiar "point-and-slope formula"

$$y - y_1 = m(x - x_1) \tag{2.22}$$

for the equation of the straight line then turns into the "two-points" formula

$$y = \frac{y_2 - y_1}{x_2 - x_1}(x - x_1) + y_1. \tag{2.23}$$

For example, in order to find an equation of the straight line passing through the points $(2, 10)$ and $(5, 31)$, we first use (2.21) to calculate the slope

$$m = \frac{31 - 10}{5 - 2} = \frac{21}{3} = 7$$

and then use the point-and-slope formula (2.22) to write

$$y - 10 = 7(x - 2),$$

which simplifies to

$$y = 7x - 4.$$

Alternatively, one could use the "two-point formula" (2.23) straight away to obtain the same result by setting up and simplifying

$$y = \frac{31 - 10}{5 - 3}(x - 2) + 10 = 7x - 4.$$

Armed with the technical skills to write equations of linear functions, we can complete the procedure of stretching the overall contrast of a given image. Specifically, for the image in Figure 2.12, we use the "two-point formula" (2.23) to construct a linear function satisfying (2.18), which gives us the transform

$$y = \frac{255 - 0}{224 - 74}(x - 74),$$

which we can apply to all the pixels of the image.

For a general image, we let a and b denote the smallest and the largest pixel values. In order to satisfy the condition (2.18), a line has to pass through the points $(x_1, y_1) = (a, 0)$ and $(x_2, y_2) = (b, 255)$. Therefore, its equation is

$$y = \frac{255 - 0}{b - a}(x - a) + 0 = \frac{255}{b - a}(x - a)$$

in accordance with the "two-point formula" (2.23).

Figure 2.13 shows the effect of this transformation on the image in Figure 2.12 and on its histogram. It certainly constitutes a dramatic improvement

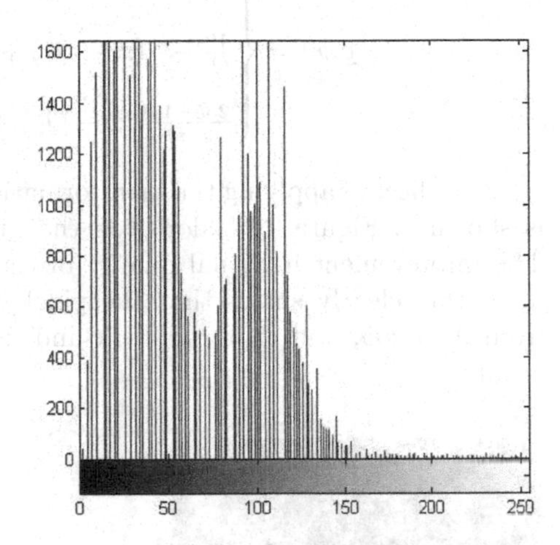

FIGURE 2.13: The image with extended overall contrast and its histogram.

over the original image. Nevertheless, we can't help thinking that it appears too dark and could benefit from an application of a power-function transform discussed in Section 2.2. Alternatively, such a correction could have been incorporated into the contrast stretching procedure; we could have used a power function instead of a linear function to map the interval $[a, b]$ onto the interval $[0, 255]$. The reader will be asked to implement both of those approaches in MATLAB Exercises at the end of this section.

Having constructed a linear function that helped us achieve the initial modest goal of stretching the pixel values to occupy the entire available range between 0 and 255, we can proceed to the more ambitious goal of shifting the quartiles of the pixel values to form regular intervals as specified in (2.19). To achieve that, we want to construct a piecewise-linear function g satisfying

$$g(a) = 0, \quad g(Q_1) = 63, \quad g(Q_3) = 191, \quad \text{and} \quad g(b) = 255,$$

which we repeated here for the convenience of the reader. This can be done in three steps:

1. We first need to construct an equation of the line that passes through the points $(a, 0)$ and $(Q_1, 63)$.

2. Next, we need an equation of the line passing through the points $(Q_1, 63)$ and $(Q_3, 191)$.

3. Finally, we have to construct an equation of the line through the points $(Q_3, 191)$ and $(b, 255)$.

Applying the "two-points-formula" (2.23) three times, we obtain the following piecewise-linear function:

$$g(x) = \begin{cases} \frac{63}{Q_1-a}(x-a), & a \leq x \leq Q_1 \\ \frac{191-63}{Q_3-Q_1}(x-Q_1)+63, & Q_1 \leq x \leq Q3 \\ \frac{255-191}{b-Q_3}(x-Q_3)+191, & Q_3 \leq x \leq b. \end{cases} \tag{2.24}$$

The effect of applying the transformation (2.24) on the image in Figure 2.12 is shown in Figure 2.14 alongside the histogram of the transformed image. The improvement in visual quality of the image is quite significant, and the histogram clearly shows that the pixel values now occupy the entire range from 0 to 255, and that the dark and light pixels are balanced much more evenly.

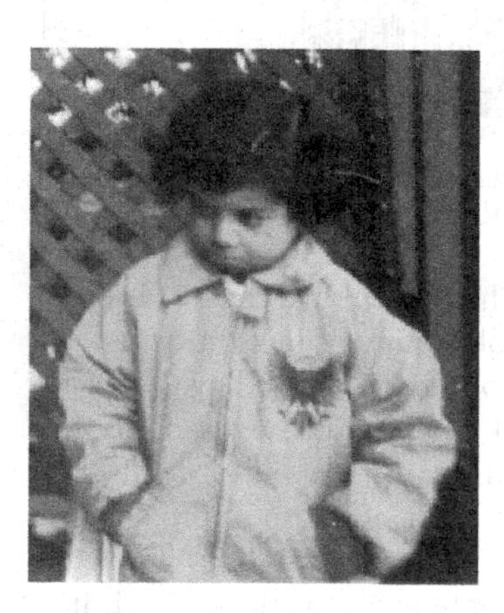

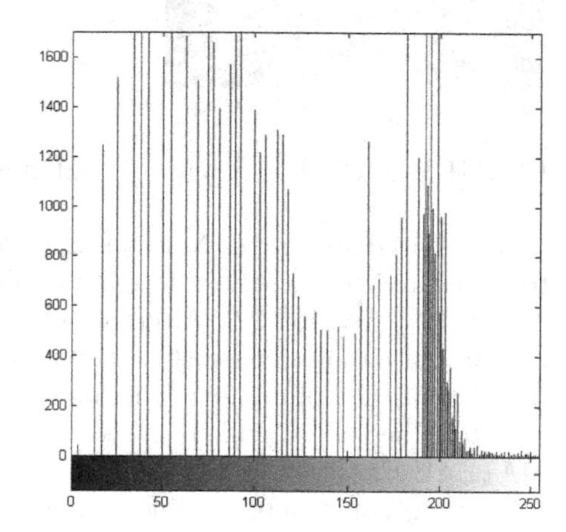

FIGURE 2.14: The image with extended contrast and equalized quartiles alongside its histogram.

The reader might be wondering whether there is any way to further improve on the methods just discussed. Indeed, there is nothing special about the quartiles. One could just as easily have used the octiles or any other set of evenly-spaced percentiles (which the reader will be asked to do in the exercises at the end of this section).

Also, whereas it is often desirable to stretch contrast, sometimes we may choose to do exactly the opposite and reduce it. Doing so may occasionally enhance the visual quality of an image by endowing it with dreamy atmosphere or romantic appeal. Indeed, suffice it to think of winter landscapes, foggy mountain scenery, or paintings by James McNeil Whistler to be convinced that sharp contrasts is not always what makes a picture beautiful.

Fortunately, contrast adjustment is just a generalization of contrast stretching discussed in earlier examples. If the pixel values of our image occupy the range from a to b, and we would like them to occupy a different range from c to d, the simplest thing we could do is construct a linear function f with the property

$$f(a) = c \text{ and } f(b) = d$$

and apply this function to all the pixel values in our image. If, in addition, we wish to lighten up or darken the image, we may combine the transformation performed by the linear function f with gamma correction discussed in Section 2.2. In the exercises at the end of this section, the reader will be encouraged to experiment with all those different approaches.

Section 2.5 Exercises.

These exercises are intended to help the reader review the basics of linear and piecewise-linear functions, including their graphs, formulas, and practical applications. The reader will also be asked to apply the methods of contrast manipulation discussed in this section to images of their choice.

1. Write an equation of the line that passes through the following pair of points:

   (a) (1,2) and (4,8),

   (b) (1,10) and (3,-4),

   (c) (2,-1) and (9,7),

   (d) (-1,5) and (7,15).

   Write your answers in the slope-intercept form. Sketch the line (both by hand and using the MATLAB function *plot*).

2. Find the equation of the line that passes through the point $(1, 2)$ and is

   (a) Parallel to the line defined by $y = 3x - 5$.

   (b) Perpendicular to the line defined by $y = 5x - 3$.

   Sketch the graphs of the equations you found (both by hand and using the MATLAB function *plot*).

3. Suppose that a phone company charges \$50 per month for a plan that includes unlimited calls, unlimited texting, and 2 gigabytes of data. Any extra data usage costs \$15 per gigabyte.

(a) Write a formula for the cost C as a function of the number D of gigabytes used.

(b) How much data usage would result in a bill of $117.50?

4. Residents of the town of Oak Grove who are connected to the municipal water supply are billed a fixed yearly surcharge plus a charge for each cubic foot of water used. A household using 1500 cubic feet was billed $98, while one using 500 cubic feet was billed $68.

   (a) Write an equation for the total cost C of a resident's water as a function of volume W of water used.

   (b) What are the charge per cubic foot and the surcharge for being connected to the municipal water supply?

   (c) What amount of water usage would lead to a bill of $120?

5. The overall parking charge is comprised of a flat fee for entering the parking lot and the hourly fee. Neither of them seems to be posted, but we observed that parking for five hours costs $21.00, and parking for eight hours costs $27.00.

   (a) Write a formula for the total cost C of parking as a function of the number of hours t of parking.

   (b) What is the flat fee and what is the hourly rate?

   (c) How many hours of parking would lead to a charge of $35.00?

6. Under a graduated income tax structure, the tax rate changes for different portions of the income after deductions. Consider a graduated tax where the first $10,000 of income is taxed at the 10% rate, and any income over $10,000 is taxed at the 20% rate. For example, if your income after deductions is $50,000, then your taxes under this plan are

$$\begin{aligned} \text{Tax} &= 10\% \text{ of } \$10,000 + 20\% \text{ of } (\$50,000 - \$10,000) \\ &= 0.1 \times \$10,000 + 0.2 \times \$40,000 \\ &= \$1000 + \$8,000 = \$9,000. \end{aligned}$$

   Write a formula for the tax T as a function of income I after deductions.

7. Next, consider a graduated tax plan where the first $10,000 of income are taxed at the 10% rate, the next $20,000 of income are taxed at the 15% rate, and any income over $30,000 is taxed at the 20% rate. Write a formula for the tax T as a function of income I after deductions.

8. A personal trainer recommended that a client walk at a moderate pace of about 3 miles per hour for 20 minutes and then jog at 6 miles per hour for the next 10 minutes. Construct a piecewise-linear function that gives the total distance D as a function of time t.

Section 2.5 MATLAB Exercises.

1. Prepare a selection of pictures that appear gray and boring due to insufficient contrast. If necessary, convert a color image to a grayscale image.
2. Write a MATLAB program that will do the following:
   (a) Plot the original picture alongside its histogram.
   (b) Calculate the quartiles of the original image by using the code given in this section and in Section 1.3.
   (c) Create a linear function that maps the smallest and largest pixel values of the original image to 0 and 255 respectively and apply it to the pixels of the image.
   (d) Create a piecewise-linear function that maps the smallest value, the quartiles Q_1, Q_2, and Q_3, and the largest value of the pixels of the original image to the equally-spaced set 0, 63, 127, 191, 255 and apply this function to the pixels of the image.
   (e) Repeat the previous exercise, but this time skip the median Q_2.
   (f) Display the enhanced versions of the image alongside their histograms. Comment on the relationship between the features of the histograms and the visual quality of the images.
   (g) Test your program using several different pictures. Ideally, it should work with any rectangular grayscale and color images.
3. Repeat the entire Problem 2 but this time experiment by combining linear (or piecewise-linear) functions with power-function transforms covered in Section 2.2.
4. Compile the results of the previous two exercises into a PowerPoint presentation.
5. Proceeding along the line of Problem 2(d), perform a similar procedure of image enhancement by contrast stretching using *octiles* instead of quartiles. Compare the image quality resulting from the two approaches.
6. This time find pictures that could be improved by **reducing** its contrast. If necessary, convert color images to grayscale images.
7. Write a MATLAB program that will do the following:
   (a) Plot the original picture alongside its histogram.
   (b) Calculate the quartiles of the original image by using the code given in this section and in Section 1.3.
   (c) Ask the user to input the desired values for the maximum, the minimum, and the quartiles.

(d) Create a linear function that maps the smallest and largest pixel values of the original image to the maximum and minimum values requested by the user and apply it to the pixels of the image.

(e) Create a piecewise-linear function that maps the five-number summary of the original image to the one requested by the user and apply this function to the pixels of the image.

(f) Repeat the previous exercise, but this time skip the median Q_2.

(g) Display the enhanced versions of the image alongside their histograms. Comment on the relationship between the features of the histograms and the visual quality of the images.

(h) Test your program using several different pictures. Ideally, it should work with any rectangular grayscale and color images.

8. Repeat the entire Problem 7 but this time experiment by combining linear (or piecewise-linear) functions with power-function transforms covered in Section 2.2.

9. Compile the results of the previous MATLAB exercises into a PowerPoint presentation.

2.6 Automation of Image Enhancement

In Sections 2.2, 2.3, and 2.4, we saw how elementary functions can help us enhance the visual appeal of overexposed or underexposed photos, to help bring out hidden features in certain types of images, and to bring life to other images by improving their level of contrast. However, selecting the values of parameters in power, exponential, and logarithmic functions involved in the process of image enhancement was left to trial and error. Can we do better than that? Any chance there might be a simple tool in our mathematical toolbox that would enable us to automate this process of parameter selection?

In this section, we make an attempt to automate the selection of the value of γ in power-function transforms used in Section 2.2. We first make the observation that in addition to lightening(darkening) the image, this image-enhancing transform also increases image contrast by stretching the intervals around the most commonly encountered pixel value, that is, near the most prominent mode of the image histogram.

To make use of this observation, we denote the most prominent mode by x_0 and choose the parameter γ in such a way that the contrast-stretching effect of the transformation

$$f(x) = x^{\gamma} \tag{2.25}$$

is maximized at the point x_0. An argument can be made that this contrast-stretching effect is related to the derivative

$$m(\gamma) = f'(x_0) = \gamma x_0^{\gamma-1}.$$

Thus, we have a standard calculus problem: find the absolute maximum of the function $m(\gamma)$ on the interval $0 < \gamma < \infty$. To solve it, we first calculate the derivative

$$m'(\gamma) = x_0^{\gamma-1} + \gamma \ln(x_0) x_0^{\gamma-1}$$

with the help of the Product Rule, and then set $m'(\gamma)$ equal to zero, which gives us

$$x_0^{\gamma-1} + \gamma \ln(x_0) x_0^{\gamma-1} = 0.$$

Solving this equation yields the critical point

$$\gamma = -\frac{1}{\ln(x_0)}, \tag{2.26}$$

which is the candidate for the optimal value of the power in the transformation (2.25) based on the criterion suggested in this section.

In the exercises below, the reader is encouraged to implement this approach for the power-function transform and to try similar approaches to automate the choice of parameters for other types of transforms. This is also an opportune moment to review the meaning and calculation of derivatives and to appreciate their usefulness in solving optimization problems.

Section 2.6 Exercises.

1. Verify all the calculations in this section leading up to the formula (2.26).

2. Explain why the contrast-stretching effect of the power-function transformation $f(x) = x^\gamma$ at a point $x = x_0$ is related to the derivative

   $$f'(x_0) = \gamma x_0^{\gamma-1},$$

   as stated in the text.

3. Denote by $C(r)$ the total cost of paying off a loan taken out at the annual interest rate of $r\%$. What are the units and practical meaning of $C'(r)$?

4. Denote by $C(a)$ the cost of building a house with the total area of x ft^2. What are the units and practical meaning of $C'(x)$?

5. Denote by $R(a)$ the company's revenue as a function of advertisement expenses. What are the units and practical meaning of $R'(a)$?

6. Denote by $E(x)$ the elevation (in meters) of the Delaware River x miles upstream from the point it flows under the Benjamin Franklin Bridge in Philadelphia, PA. What are the units and practical meaning of $E'(x)$?

7. Denote by $T(t)$ the temperature of the coffee cup left on the table for t minutes. What are the units and practical meaning of $T'(t)$?

8. For each of the following functions find the global maximum and the global minimum over the specified interval:

   (a) $f(x) = 4x - x^2 + 5$ on $\mathbb{R}$,

   (b) $f(x) = x^4 - 8x^2$ on $[-3, 1]$,

   (c) $f(x) = xe^{-2x}$ on $[0, \infty)$,

   (d) $f(x) = x^3e^{-2x}$ on $[0, \infty)$,

   (e) $f(x) = xe^{-x^2/2}$ on $[0, 2]$.

Open-Ended Exercises.

1. Try to imitate the approach of this section to automate the choice of the optimal base for the exponential transform discussed in Section 2.3.

2. Likewise, try to imitate the approach of this section to automate the choice of the optimal base for the logarithmic transform discussed in Section 2.4.

MATLAB Exercises.

1. In Section 2.2, you used a selection of photos that were either too dark or too light. In this section, you will be asked to use those same images for comparison.

2. Write a MATLAB function that would use the formula (2.26) to calculate the optimal value of γ in the power-function transform of a specified image. The function should accept the matrix of the image as an input. Ideally, it should work with any grayscale image and with any color image. In the latter case, the image should be converted into the YCbCr color space, and the function should work with the Y channel.

3. Write a MATLAB program that will do the following:

   - Import the image you have selected into the MATLAB computing environment.
   - Plot the original image and its histogram.

- Call the MATLAB function you created in the previous problem to determine the optimal value of γ and use the function you created in Exercise 2 of Section 2.2 to lighten (or darken) the original image.
- Perform several other transforms of the original image using values of γ selected by trial and error.
- Display the enhanced versions of the image together with their histograms.

4. Test your program using several different images (grayscale and color).
5. Compile the results of the previous MATLAB exercises into a PowerPoint presentation.
6. Comment on the effectiveness (or otherwise) of the method developed in this section. Can you identify a class of images for which this method seems effective and a class of images for which it does not?

Chapter 3

Probability, Random Variables, and Histogram Processing

In this chapter, the reader will review the basics of discrete and continuous random variables and their transformations. Those techniques will be applied to develop methods of image equalization and image histogram matching.

3.1 Introduction

In Chapter 2, we have developed a variety of methods of enhancing the visual quality of digital images. We have learned to darken overexposed images, to lighten underexposed ones, and to improve contrast of those images that appear gray and boring. To achieve those goals, we relied on transforms that use elementary functions studied in pre-calculus courses, such as linear, piecewise-linear, power, and logarithmic functions.

What all those methods have in common is the type of the effect they have on the image histogram. They tend to render skewed histograms more symmetric, while the histograms that are highly concentrated near a central value tend to become more uniform. And that observation might suggest a new idea: why not approach the topic of image enhancement from the point of view of histogram manipulation? Why not start the process by specifying what kind of a histogram we would like our image to have? We could then proceed to design a transformation that would produce a version of our image with the desired histogram.

In this chapter, we will see how this goal can be achieved with the help of techniques studied in college probability courses. But first, we need to brush up on the basics of random variables and their transformations.

3.2 Discrete and Continuous Random Variables

We recall that a random variable X can be loosely defined as a numerical quantity assigned to the outcome of a random experiment. For example, if we toss three different coins, we can define X to be the number of heads. Random variables are described by their probability distributions. Given a random variable, its probability distribution consists of the set of its possible values together with a measure of probabilities of those values.

In our three-coin example, the possible values of X are 0, 1, 2, and 3. To calculate the probabilities of those values, we note that the sample space $\mathcal{S}$ of the experiment consists of the equally likely outcomes

$$\mathcal{S} = \{HHH, HHT, HTH, HTT, THH, THT, TTH, TTT\},$$

and, consequently,

$$\begin{aligned} P(X=0) &= 1/8, \\ P(X=1) &= 3/8, \\ P(X=2) &= 3/8, \\ P(X=3) &= 1/8, \end{aligned}$$

which, together, comprise the probability distribution of X.

To alleviate notation, it is customary to use the so-called *probability function* $p(x)$ defined by

$$p(x) \equiv P(X = x). \tag{3.1}$$

We must also comment on the important matter of notation: the difference between the upper-case X and the lower-case x. In probability theory, the upper-case X is the shorthand for the complete verbal description of the given random variable, whereas the lower-case x stands for a specific numerical value of X.

We also recall from a course in probability theory that there are three types of random variables: discrete, continuous, and mixed. For a discrete random variable, we can arrange its values in a list (finite or infinite), whereas the values of a continuous random variable form an interval.

The random variable X from the coin-tossing example is obviously discrete, since the set of its values is $\{0, 1, 2, 3\}$. Other familiar examples of discrete random variables include the following:

- The number of successes out of a fixed-length series of attempts (a binomial random variable).

- The number of interviews the company has to conduct in order to find a qualified applicant (a geometric random variable).

- The number of interviews the company has to conduct in order to find **three** qualified applicants (a negative binomial random variable).
- The number of math majors on the four-member committee randomly selected from a mixed-major class of size forty (a hypergeometric random variable).
- The number of accidents at a certain intersection during a certain time interval given the expected number of accidents (a Poisson random variable).

Continuous random variables are often associated with measuring physical quantities. The following examples immediately come to mind:

- the waiting time until the phone rings (often modeled by an exponential random variable),
- the waiting time until the **third** phone call (often modeled by a gamma random variable),
- the fraction of the stocked produce that gets sold during a particular day (often modeled by a beta random variable),
- the number of miles driven before the first breakdown (sometimes modeled by a normal random variable),
- the amount of apples in an apple storage facility about which we do not know anything at all (usually modeled by a uniform random variable),

and many others.

So, what does all of that have to do with digital images? For any digital image A, we can define an associated random variable in the following manner: pick a pixel at random. The random variable X will be defined as the value of that randomly selected pixel. The possible values of X are the integers from 0 to 255 (corresponding to the possible grayscale values), and the probabilities of those values are provided by the image histogram.

What type of a random variable is the value X of the pixel randomly selected from a grayscale image? Is it discrete or continuous? The answer might appear obvious: it is discrete because the possible values form a list (of integers from 0 to 255). And the computer would happily accept this answer because computers work with discrete quantities. However, as humans, we find lists of that kind way too long. We have neither sufficient patience nor systematic analytic techniques to deal with discrete random variables that can take on so many different values.

Therefore, it would also seem reasonable to consider (just for matters of convenience) the pixel value to be a **continuous** random variable with its values comprising the interval $[0, 255]$. And then, we can bring all the techniques of differential and integral calculus to bear on the problem. As is often the case in applied mathematics, it is completely up to us to decide whether it is more promising and convenient to consider the given random variable discrete or continuous. And we will treat the pixel value X as continuous when we need methods of calculus to derive formulas and as discrete when implementing the image processing techniques we have developed.

Obviously, the manner in which we described the probability distribution of the random variable in the coin-tossing example will not work for continuous random variables. We need something more universally applicable. With that in mind, we recall the following definition:

Definition 3.1 *Suppose that X is a random variable (discrete or continuous). Then the* ***cumulative distribution function (CDF)*** *of X is defined by*

$$F(x) \equiv P(X \leq x). \tag{3.2}$$

We leave it to the reader to verify that the cumulative distribution function defined by (3.2) is non-negative and non-decreasing with $\lim_{x\to-\infty} F(x) = 0$ and $\lim_{x\to\infty} F(x) = 1$. But for us, the most exciting property of the cumulative distribution function is its universality:

> The cumulative distribution function is defined for both discrete and continuous random variables in precisely the same way!

It follows from the fundamental properties of probability that the probability function $p(x)$ of a *discrete* random variable X that takes on the values $\{x_k\}$ can be used to calculate its cumulative distribution function by means of the formula

$$F(x) = \sum_{x_k \leq x} p(x_k). \tag{3.3}$$

It might also be worth noting that

$$p(x_k) = F(x_k) - F(x_{k-1}).$$

To illustrate the definition of the cumulative distribution function F for a discrete random variable, we calculate it for our coin-tossing example. It is obvious that $F(x) = 0$ whenever $x < 0$. For $0 \leq x < 1$,

$$F(x) \equiv P(X \leq x) = P(X = 0) = 1/8.$$

For the case $1 \leq x < 2$, we have

$$F(x) \equiv P(X \leq x) = P(X = 0) + P(X = 1) = 1/8 + 3/8 = 1/2.$$

If $2 \leq x < 3$, then

$$F(x) \equiv P(X \leq x) \quad = \quad P(X = 0) + P(X = 1) + P(X = 2)$$

$$= p(0) + p(1)) + p(2)$$
$$= 1/8 + 3/8 + 3/8 = 7/8,$$

and, finally, if $x \geq 3$, then

$$\begin{aligned} F(x) \equiv P(X \leq x) &= P(X = 0) + P(X = 1) + P(X = 2) + P(X = 3) \\ &= p(0) + p(1) + p(2) + p(3) \\ &= 1/8 + 3/8 + 3/8 + 1/8 = 1. \end{aligned}$$

It is customary to express the cumulative distribution function in the piecewise form, such as

$$F(x) = \begin{cases} 0, & x < 0 \\ 1/8, & 0 \leq x < 1 \\ 1/2, & 1 \leq x < 2 \\ 7/8, & 2 \leq x < 3 \\ 1, & 3 \leq x. \end{cases}$$

In order to provide an example of the cumulative distribution function of a *continuous* random variable, let us suppose that we got to the bus stop at 12:00 and we know for sure that the bus is supposed to arrive any time between 12:00 and 12:30. Any arrival time within that time interval is just as likely as any other. It would seem reasonable to define the probability of the bus arrival within the interval $[a, b]$ as

$$P(a \leq X \leq b) = \frac{b - a}{30},$$

where the continuous random variable X denotes the amount of time we had to wait for the bus after getting to the bus stop at 12:00. It follows that

$$F(x) = P(X \leq x) = \begin{cases} 0, & x < 0 \\ x/30, & 0 \leq x < 30 \\ 1, & 30 \leq x. \end{cases}$$

An examination of the graphs of the cumulative distribution functions in the previous two examples suggests the general rule: whereas the cumulative distribution functions of continuous random variables are continuous, the CDFs of discrete random variables are piecewise-constant with jump discontinuities.

As we have already mentioned, the randomly selected pixel value can be treated either as a discrete or as a continuous random variable depending on the situation and our goals. We have seen that a complete description of the probability distribution of a *discrete* random variable is provided by its probability function $p(x)$. The analogue of the probability function for continuous random variables is the *probability density function*:

Definition 3.2 *Suppose X is a continuous random variable with the cumulative distribution function $F(x)$. Then its* ***probability density function (PDF)*** *is denoted by $f(x)$ and is defined by*

$$f(x) = F'(x).$$

The graph of the probability density function of a random variable X is called "the density curve" of X.

It follows from Definition 3.2 and from the Fundamental Theorem of Calculus that the probability density function f of the continuous random variable X can be used to calculate its cumulative distribution function by means of

$$F(x) = \int_{t=-\infty}^{x} f(t)dt. \tag{3.4}$$

It is also evident that either f or F can be used to calculate the probability that X falls into the interval $[a, b]$ by means of

$$P(a \leq X \leq b) = F(b) - F(a) = \int_{a}^{b} f(t)dt.$$

It is just as easy to visualize the density curves of continuous random variables as it is to visualize probability functions of their discrete counterparts. For example, if we denote the randomly chosen pixel values by X, then the image histogram provides a graphical representation of the probability function $p(x)$, whereas the smooth outline of the image histogram is precisely the density curve of X (should we choose to treat it as a continuous random variable). For the convenience of the reader, this concept is illustrated in Figure 3.1.

The MATLAB code used to generate Figure 3.1 is given below for the convenience of the reader.

```
A=imread('WaterFall.jpg');
A=rgb2gray(A);
subplot(1,2,1);
imshow(A,[0,255]);
subplot(1,2,2);
imhist(A);
hold on;
[M,N]=size(A);
[counts,bins]=imhist(A);
plot(0:255,counts,'Linewidth',2);
```

We specifically call the reader's attention to the MATLAB command

```
>> hold on;
```

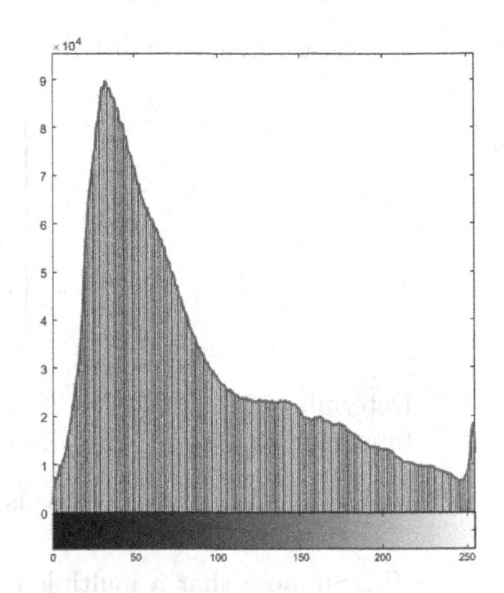

FIGURE 3.1: An image histogram with the density curve.

which holds the image histogram, so that we can add the density curve to it.

Section 3.2 Exercises.

1. Determine the cumulative distribution function F of the random variable X whose probability distribution is given by the following table:

x	1	2	3	4
$p(x)$	0.4	0.3	0.2	0.1

Sketch the graph of the cumulative distribution function either by hand or with the help of the MATLAB *plot* function.

2. Suppose that X is a random variable that takes on the integer values from 1 to N. Show that its probability function $p(x)$ and its cumulative distribution function $F(x)$ are related by the formula

$$p(x) = \begin{cases} F(1), & x = 1 \\ F(x) - F(x-1), & x = 2, \dots, N. \end{cases}$$

3. Determine the probability function $p(x)$ of the discrete random variable

X whose cumulative distribution function F is given by

$$F(x) = \begin{cases} 0, & x < 1 \\ 0.15, & 1 \leq x < 2 \\ 0.2, & 2 \leq x < 3 \\ 0.4, & 3 \leq x < 4 \\ 0.7, & 4 \leq x < 5 \\ 1, & x \geq 5. \end{cases}$$

4. Determine the probability functions and the cumulative distribution functions for the following discrete random variables:

   (a) X is the number of heads observed when four different balanced coins are tossed.

   (b) Suppose that a multiple-choice test consists of six questions, and each question has four possible answers. Define X to be the number of correct answers obtained by pure guessing.

   (c) Suppose that 30 percent of the applicants have all the necessary skills and experience that would qualify them for a certain position. Define X to be the number of job interviews that need to be conducted to find the first qualified applicant.

   (d) Repeat the previous part of the exercise, but this time, define X to be the number of job interviews conducted to find **three** qualified applicants.

   (e) Suppose that there are 15 math majors in a mixed-major class of size forty. Define X to be the number of math majors on the four-member committee randomly selected from the class.

   Sketch the graphs of the cumulative distribution functions either by hand or with the help of the MATLAB *plot* function.

5. Determine the value of the constant c and find the cumulative distribution function of the continuous random variable X defined by the following probability density function:

   (a) $f(x) = \begin{cases} c, & 0 \leq x \leq 10 \\ 0, & \text{otherwise.} \end{cases}$

   (b) $f(x) = \begin{cases} cx, & 0 \leq x \leq 10 \\ 0, & \text{otherwise.} \end{cases}$

   (c) $f(x) = \begin{cases} c(10 - x), & 0 \leq x \leq 10 \\ 0, & \text{otherwise.} \end{cases}$

(d) $f(x) = \begin{cases} c\sqrt{x}, & 0 \le x \le 1 \\ 0, & \text{otherwise.} \end{cases}$

(e) $f(x) = \begin{cases} c/x^2, & x \ge c \\ 0, & \text{otherwise.} \end{cases}$

(f) $f(x) = \begin{cases} ce^{-2x}, & x \ge 0 \\ 0, & \text{otherwise.} \end{cases}$

(g) $f(x) = \begin{cases} cxe^{-2x}, & x \ge 0 \\ 0, & \text{otherwise.} \end{cases}$

(h) $f(x) = \begin{cases} cx, & 0 \le x \le 1 \\ c, & 1 \le x \le 2 \\ 0, & \text{otherwise.} \end{cases}$

(i) $f(x) = \begin{cases} x, & 0 \le x \le 1 \\ 1 - x, & 1 \le x \le 2 \\ 0, & \text{otherwise.} \end{cases}$

Sketch the graphs of both the density functions and the cumulative distribution functions either by hand or with the help of the MATLAB *plot* function.

3.3 Transformation of Random Variables

In the previous section, we prepared a solid foundation by reviewing the mathematical instruments commonly used to describe random variables. We can now get back to the main objective of this chapter – image histogram equalization. Mathematically, this objective can be described as follows:

Given the grayscale image A, we associate with it the random variable X defined to be the value of a randomly selected pixel of A. We must design a transformation g so that the random variable Y defined by

$$Y = g(X)$$

is uniformly distributed. Which means that we want the probability function (or the probability density function) of Y to be constant over the interval $[0, 255]$. More generally, we may require that Y has any specified distribution, but we will begin with the simplest case of making the distribution of pixel values uniform.

In order to achieve this goal, we need to develop methods of working with transformations of random variables. In particular, we would like to figure out the relationship between the probability functions of X and Y (if they are discrete) or between the probability density functions of X and Y (if they are continuous).

The discrete case appears straightforward, almost trivial. Since Y takes on the value y if and only if X takes on the value $g^{-1}(y)$, it is evident that

$$p_Y(y) = p_X\big(g^{-1}(y)\big). \tag{3.5}$$

For example, suppose X is a geometric random variable with the probability function

$$p_X(x) = \frac{2}{3}\left(\frac{1}{3}\right)^{x-1}, \quad x = 1, 2, 3, \ldots,$$

and $Y = g(X) = 2X$. Then $X = g^{-1}(Y) = Y/2$, and, consequently, the probability function of Y is

$$p_Y(y) = p_X(y/2) = \frac{2}{3}\left(\frac{1}{3}\right)^{y/2-1}, \quad y = 2, 4, 6, \ldots$$

Although the formula (3.5) is indeed very easy to derive and to apply, it has an unfortunate drawback, quite typical of discrete settings. We recall that our ultimate goal is to learn to design the transformation g that would transform X into Y given the known probability functions p_X and p_Y. Unfortunately, it does not seem clear how to solve the equation (3.5) in the sense of obtaining a simple expression for g.

One way out of this difficulty is to reinterpret our discrete random variables as continuous. We therefore turn our attention to transformations of continuous random variables, where we can rely on the techniques of differential and integral calculus.

Before we attempt to derive any general formulas in the **continuous case**, we consider a simple example, which, hopefully, will provide a clear paradigm for working with arbitrary continuous random variables.

Example 3.1 *Suppose that the continuous random variable X is defined by its probability density function*

$$f_X(x) = \begin{cases} 2x, & 0 \le x \le 1 \\ 0, & \textit{otherwise,} \end{cases}$$

and the random variable Y is defined by

$$Y = g(X) = 3X.$$

Then, by definition, the cumulative distribution function of Y is

$$F_Y(y) \;=\; P(Y \le y) = P(3X \le y) = P\big(X \le \tfrac{y}{3}\big)$$

$$= \int_0^{y/3} 2x dx = x^2 \Big|_0^{y/3} = \frac{y^2}{9}$$

for all values $0 \leq y \leq 3$. *If* $y < 0$, *then* $F(y) = 0$ *and if* $y > 3$, *then* $F(y) = 1$. *Altogether,*

$$F_Y(y) = \begin{cases} 0, & y < 0 \\ y^2/9, & 0 \leq y \leq 3 \\ 1, & y > 3, \end{cases}$$

which implies that

$$f_Y(y) = F'_Y(y) = \begin{cases} 2y/9, & 0 \leq y \leq 3 \\ 0, & \textit{otherwise.} \end{cases} \tag{3.6}$$

While going over the calculations, an attentive reader might have noticed that integration was immediately followed with differentiation, which creates an opportunity for further simplification. With Example 3.1 completely understood, we can proceed to a more general situation.

Suppose that X is a continuous random variable with values in the interval $[a, b]$. Suppose that, as in the previous example, the random variable Y is defined by

$$Y = g(X),$$

where g is a strictly increasing (and, hence, invertible) differentiable function. Then, by definition, the cumulative distribution function of Y is

$$\begin{aligned} F_Y(y) &= P(Y \leq y) = P\big(g(X) \leq y\big) \\ &= P\big(X \leq g^{-1}(y)\big) \\ &= \int_a^{g^{-1}(y)} f_X(x) dx \end{aligned}$$

for all values $g(a) \leq y \leq g(b)$. Clearly, if $y < g(a)$, then $F(y) = 0$ and if $y > g(b)$, then $F(y) = 1$. The Fundamental Theorem of Calculus and the Inverse Function Theorem give us

$$\begin{aligned} f_Y(y) = F'_Y(y) &= \frac{d}{dy}\left[\int_a^{g^{-1}(y)} f_X(x) dx\right] \\ &= f_X\big(g^{-1}(y)\big) \cdot \frac{d}{dy}\left[g^{-1}(y)\right] \\ &= f_X\big(g^{-1}(y)\big) \cdot \frac{1}{g'\big(g^{-1}(y)\big)} \end{aligned}$$

for all values of y for which $g(a) \leq y \leq g(b)$ and $f_Y(y) = 0$ otherwise.

The last expression is so important for the upcoming material that we reiterate and frame it for easy reference:

Transformation of Continuous Random Variables

Suppose that X is a continuous random variable with values in the interval $[a, b]$. Suppose that the random variable Y is defined by $Y = g(X)$, where g is a strictly increasing differentiable function. Then

$$f_Y(y) = f_X\big(g^{-1}(y)\big) \cdot \frac{1}{g'\big(g^{-1}(y)\big)} \tag{3.7}$$

for all values of y for which $g(a) \leq y \leq g(b)$ (and $f_Y(y) = 0$ otherwise).

The following example is intended to illustrate this transformation formula.

Example 3.2 *Suppose that the continuous random variable X is defined by its probability density function*

$$f_X(x) = \begin{cases} 2(1-x), & 0 \leq x \leq 1 \\ 0, & \textit{otherwise,} \end{cases}$$

and the random variable Y is defined by

$$Y = g(X) = X^2.$$

For this example,

$$g^{-1}(Y) = \sqrt{Y}, \quad g'(x) = 2x, \quad g'\big(g^{-1}(y)\big) = 2\sqrt{y},$$

and, consequently, (3.7) *yields*

$$f_Y(y) = 2(1-\sqrt{y}) \cdot \frac{1}{2\sqrt{y}} = 4\big(\sqrt{y} - y\big)$$

for $0 \leq y \leq 1$.

Revisiting Example 3.1, where $f_X(x) = 2x$ for $0 \leq x \leq 1$ and $Y = g(X) = 3X$, we can also use 3.7 to obtain

$$f_Y(y) = 2 \cdot \frac{y}{3} \cdot \frac{1}{3} = \frac{2y}{9}$$

for $0 \leq y \leq 3$, which agrees completely with the result we got in (3.6).

Having developed a method to calculate probability density functions of transformations of continuous random variables, we recall that in image processing applications, it is actually the transformation g itself that we would like to design. Unlike the equation (3.5), which is not easy to solve for g, the transformation formula (3.7) can help us in this endeavor.

For example, in the image equalization problem, we want the pixel-value distribution of the transformed image to be uniform. Formally, this means

$$f_Y(y) = \frac{1}{255}, \quad \text{for} \quad 0 \leq y \leq 255,$$

assuming that the grayscale pixels take the values in the interval $[0, 255]$. Substituting this expression for $f_Y(y)$ into (3.7), we obtain

$$\frac{1}{255} = f_X\big(g^{-1}(y)\big) \cdot \frac{1}{g'\big(g^{-1}(y)\big)}$$

and, consequently,

$$g'(x) = 255 f(x),$$

which gives us the desired transformation g in the form

$$g(x) = 255 \cdot F_X(x). \tag{3.8}$$

In certain practical applications, we must solve the "opposite" problem – start with the random variable X that is uniformly distributed on the interval $[a, b]$ and turn it into the random variable Y with the specified probability density function f_Y. In that case, switching the roles of X and Y in (3.8) gives us

$$g(x) = \frac{1}{b-a} F_Y^{-1}(x), \tag{3.9}$$

which is often used to simulate samples of random populations with specified probability distributions. Finally, if we want to turn a random variable X with a known density function f_X into a random variable Y with any specified density function f_Y, we can combine (3.8) and (3.9) to obtain

$$g(x) = F_Y^{-1}\big(F_X(x)\big). \tag{3.10}$$

Unfortunately, both formulas (3.9) and (3.10) are only useful if we have convenient expressions for both F_X and F_Y^{-1}, which is not usually the case in image processing applications. Therefore, in the general setting, it is often more practical to derive a formula for g directly from (3.7). Solving it for g' and recalling that $g^{-1}(y) = x$, we obtain

$$g'(x) = \frac{f_X(x)}{f_Y(x)},$$

and integrating this last expression yields a solution for the transformation g in the form

$$g(x) = \int_{-\infty}^{x} \frac{f_X(t)}{f_Y(t)} dt, \tag{3.11}$$

which the reader is encouraged to verify.

In the next section, we will apply the formulas (3.8) and (3.11) to the problems of image equalization and histogram matching.

Section 3.3 Exercises.

1. Verify all the steps leading up to (3.7).
2. Verify the solutions (3.8) and (3.11) for the transformation function g.
3. Consider the random variable X whose probability distribution is given by the following table:

x	1	2	3	4
$p(x)$	0.4	0.3	0.2	0.1

   Find the probability distributions of the following transformations of X:

   (a) $Y = 2X + 1$, (b) $Y = X^2$.

4. Define X to be the difference of two fair dice. Determine the probability distributions of the following transformations of X:

   (a) $Y = 5X + 1$, (b) $Y = X^3$.

5. For each discrete random variable X considered in Exercise 4 of Section 3.2, determine the probability distributions of the following transformations of X:

   (a) $Y = 2X + 1$, (b) $Y = X^2$.

6. Suppose that the random variable U is uniformly distributed on the interval $[0, 1]$. Determine the probability density functions of the following transformations of U:

   (a) $Y = 2X + 1$, (c) $Y = X^2$,
   (b) $Y = 1 - 2X$, (d) $Y = \sqrt{X}$.

7. For each continuous random variable X considered in Exercise 5 of Section 3.2, determine the probability distributions of the following transformations of X in two ways: by following the method of Example 3.1 and by using the formula (3.7).

(a) $Y = 2X + 1$,
(b) $Y = 1 - 2X$,
(c) $Y = X^2$,
(d) $Y = X^3$,
(e) $Y = e^X$,
(f) $Y = \sqrt{X}$.

8. For each continuous random variable X considered in Exercise 5 of Section 3.2, find the function g so that the random variable Y defined by $Y = g(X)$ is uniformly distributed

   (a) on the interval $[0, 1]$,
   (b) on the interval $[0, 255]$.

9. Suppose that the random variable U is uniformly distributed on the interval $[0, 1]$. For each of the random variable X considered in Exercise 5 of Section 3.2, determine the function g so that $X = g(U)$.

10. Suppose U is uniformly distributed on the interval $[0, 1]$.

    (a) Show that $X = -2 \ln X$ has a chi-square distribution with 2 degrees of freedom.

    (b) Determine the transformation g so that $Y = g(U)$ has a chi-square distribution with n degrees of freedom.

11. Suppose that Z has a standard normal distribution and $X = Z^2$. Derive the probability density function of X and try to determine what type of a random variable it is.

12. Beta distribution is often used to model random percentages and proportions. Suppose that X has a beta distribution with parameters α and β, which means that its probability density function is

$$f_X(x) = cx^{\alpha-1}(1-x)^{\beta-1}, \quad 0 \le x \le 1.$$

Find the probability density function f_Y of the random variable Y given by $Y = 1 - X$ and determine what type of a random variable it is.

3.4 Image Equalization and Histogram Matching

In the previous section, we have reviewed the basic methods of working with transformations of discrete and continuous random variables. Having derived the fundamental formula (3.7) and its consequences (3.8) and (3.11), we return to the main theme of this chapter – image equalization.

Once again, we recall that given a grayscale image, we can associate with it the random variable X defined to be the value of a randomly selected pixel. As we have already mentioned in earlier sections, image equalization essentially comes down to designing a transformation g so that the random variable Y defined by

$$Y = g(X)$$

is uniformly distributed.

We also recall that the pixel value can be thought of either as a continuous or as a discrete random variable, and that the cumulative distribution function $F(x)$ is defined exactly the same way for either type. Therefore, in view of (3.3), we can recast (3.8) in the form

$$g(x) = 255 \cdot F_X(x) = 255 \sum_{t \leq x} p(t), \tag{3.12}$$

which can be easily applied to grayscale digital images. The result of the application of this transformation to the familiar image *pout.tif* is shown in Figure 3.2 (side by side with the 16-bin histograms of the original and equalized images).

The reader might find it interesting to compare the results obtained by using the transformation (3.12) to those obtained in Section 2.5 using the method of mapping quartiles. We must also mention here that not all types of images are suitable for equalization, as it often produces unnatural and undesirable effects and should, therefore, be used with great caution. In the MATLAB exercises at the end of this section, the reader will be asked to equalize a selection of images of their choice and to draw their own conclusions regarding the types of images that can benefit from equalization.

Image equalization is but a special case of the more general technique called histogram matching – transforming an image so that its histogram takes on a specified shape. Whereas equalization guarantees that all shades of gray are represented with the same frequencies, we might be more interested in emphasizing certain ranges of pixel values. For example, we might feel that darkening the image might make it more dramatic. Or we might wish to increase the image contrast beyond what equalization can provide. Or we might want to do the opposite and make the image appear graying (in order to convey the mood created by dreary weather).

To illustrate the concept of histogram matching, we consider the underwater image in Figure 3.3(a), which appears very gray and seems to be a perfect candidate for equalization. The equalization procedure based on applying the transformation (3.12) to all the pixel values certainly makes a significant improvement in the visual quality of the image (Figure 3.3(b)), yet we may still wish to increase image contrast even further. For that, we would want the image histogram to assume a V-shape with the majority of pixels becoming either very dark or very bright, with very few remaining in the middle range.

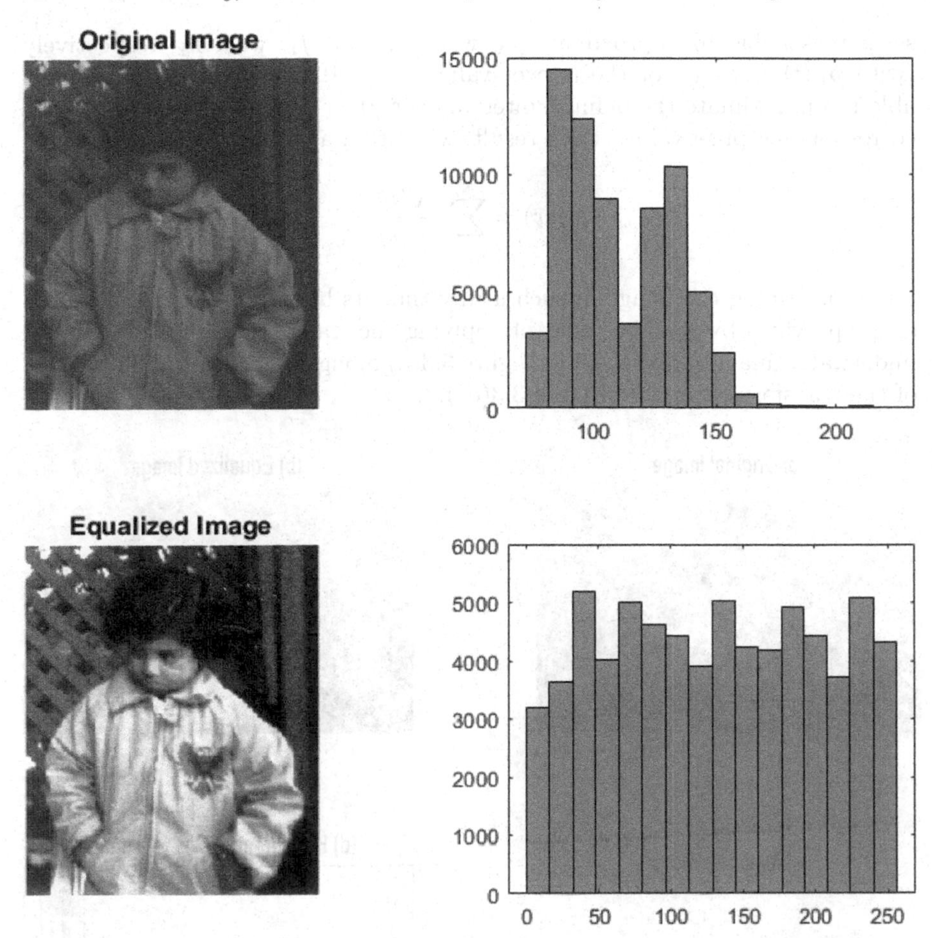

FIGURE 3.2: An example of image equalization.

Mathematically, we need to construct the transformation g so that the random variable Y defined by $Y = g(X)$ has a V-shaped probability distribution. The formula (3.11) provides the very instrument we need to accomplish that goal (albeit in the continuous setting). However, in order to apply (3.11), we must first design a reasonable V-shaped probability density function f_Y. A piecewise-linear function would probably do for our first attempt at histogram matching. We can easily construct the formula of the function of that type whose graph connects the points $(0, 7/4)$, $(128, 1/4)$, and $(255, 7/4)$. We leave it to exercise to verify that the desired formula is

$$f_Y(y) = \begin{cases} \frac{7}{4} - \frac{3/2}{128}y, & 0 \leq y < 128 \\ \frac{3/2}{128}y - \frac{5}{4}, & 128 \leq y \leq 255. \end{cases} \tag{3.13}$$

Next, we turn to the fundamental transformation formula (3.11). It would seem reasonable to approximate f_X with p_X and f_Y with p_Y respectively (with $p_Y(t) = f_Y(t)$ for the integer values of t). It would also seem reasonable to approximate the definite integral over $[0, x]$ with the summation over corresponding pixel values. As a result, we obtain a plausible expression

$$g(x) = \sum_{t=0}^{x} \frac{p_X(t)}{p_Y(t)} \tag{3.14}$$

for transforming the image in such a way that its histogram conforms to the shape provided by f_y. The result of applying the transformation (3.14) to the underwater image is displayed in Figure 3.3(c) alongside the 16-bin histogram of the transformed image (Figure 3.3(d)).

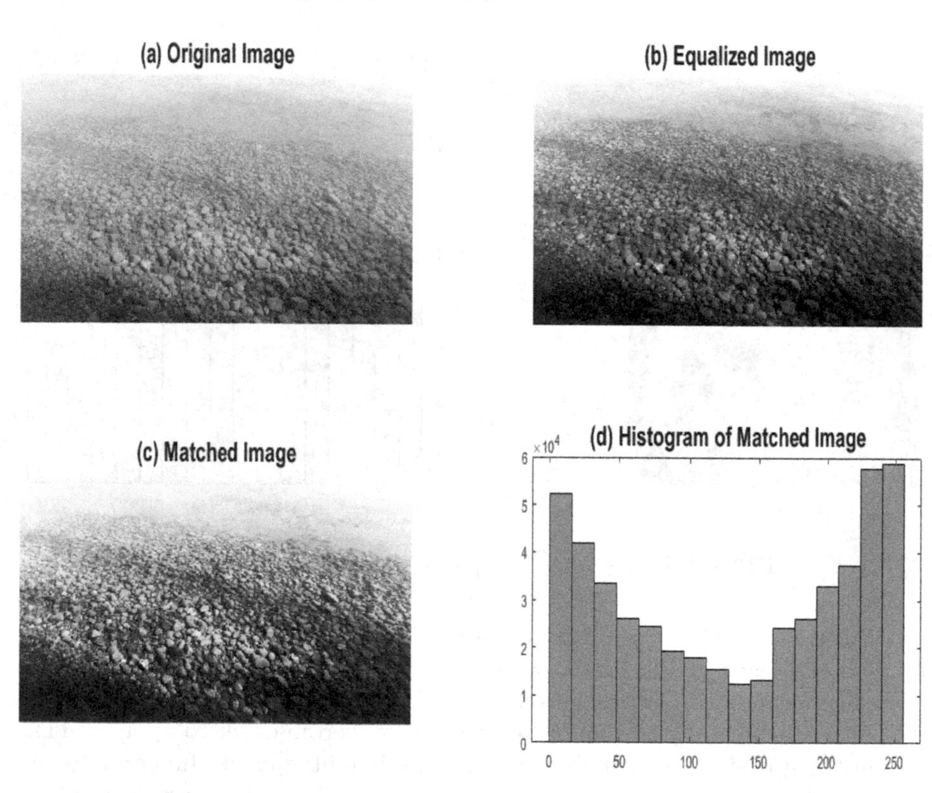

FIGURE 3.3: An example of image histogram matching.

As already mentioned, increasing contrast is not the only possible aim of histogram matching. We might also want to make the image appear somewhat grayish in order to convey the atmosphere of dreary weather. For that, we would need to design a $\wedge$-shaped distribution f_Y and substitute it into the transformation (3.14). Or, we might choose to emphasize the dark colors, in which case we will try to make the image histogram skewed right. A descending linear or exponential f_Y would then need to be designed. In the MATLAB exercises that follow, the reader will be asked to accomplish all of those goals.

Section 3.4 MATLAB Exercises.

1. Prepare a selection of pictures that appear gray and boring due to insufficient contrast. If necessary, convert a color image to a grayscale image.
2. Write a MATLAB function that would equalize the specified image by using the formula (3.12).
3. Create a selection of V-shaped and U-shaped piecewise-linear, power, and exponential density functions that can be used to increase contrast by means of a histogram-matching procedure.
4. Create a selection of $\wedge$-shaped and $\cap$-shaped piecewise-linear, power, and exponential density functions that can be used to reduce contrast by means of a histogram-matching procedure.
5. Create a selection of linear, power, and exponential density functions that can be used to emphasize darker or lighter pixel ranges by means of a histogram-matching procedure.
6. Create the corresponding MATLAB functions that would match the specified image to the probability distributions created in MATLAB Exercises 3, 4, and 5.
7. Write a MATLAB program that will do the following:
    (a) Ask the user which image is to be loaded into the MATLAB computing environment.
    (b) Plot the original picture alongside its histogram.
    (c) Ask the user what type of action is needed: equalization or histogram matching.
    (d) If the user has selected image equalization, call the function created in MATLAB Exercise 2 to equalize the image.
    (e) If the user has selected histogram matching, ask the user to choose the shape to match. Afterwards, call the appropriate function created in MATLAB Exercises 3, 4, and 5 to transform the image to give its histogram the selected shape.
    (f) Display the enhanced version of the image alongside its histogram. Comment on the relationship between the features of the histograms and the visual quality of the images.
    (g) Test your program using several different pictures. Ideally, it should work with any rectangular grayscale and color images.
8. Compile the results of the previous MATLAB exercises into a PowerPoint presentation.

Chapter 4

Matrices and Linear Transformations

In this chapter, the reader will discover that a number of image transformations, such as blending, masking, rotation, and change of perspective can be performed with the help of matrix arithmetic, linear transformations, and homogeneous coordinates.

4.1 Basic Operations on Matrices

Most of us use our smartphones and digital cameras to take pictures, and, consequently, we no longer rely on the process of turning negatives into prints. However, in spite of the evolution in photography that has seen an almost complete replacement of chemical film with digital devices, there are still areas of art photography, where negatives are produced from digital images and used in creating cyanotypes, platinum prints, and many other works of art.

We recall that grayscale images are just matrices of pixel values from 0 to 255. In order to produce the negative of an image, we have to "reverse" the brightness of every pixel. To do that, we can try to use matrix subtraction, which is defined for any two **identical-size** $M \times N$ matrices

$$A = \begin{bmatrix} a_{11} & a_{12} & \cdots & a_{1N} \\ a_{21} & a_{22} & \cdots & a_{2N} \\ \vdots & \vdots & \ddots & \vdots \\ a_{M1} & a_{M2} & \cdots & a_{MN} \end{bmatrix} \quad \text{and} \quad B = \begin{bmatrix} b_{11} & b_{12} & \cdots & b_{1N} \\ b_{21} & b_{22} & \cdots & b_{2N} \\ \vdots & \vdots & \ddots & \vdots \\ b_{M1} & b_{M2} & \cdots & b_{MN} \end{bmatrix}$$

by

$$B - A = \begin{bmatrix} b_{11} - a_{11} & b_{12} - a_{12} & \cdots & b_{1N} - a_{1N} \\ b_{21} - a_{21} & b_{22} - a_{22} & \cdots & b_{2N} - a_{2N} \\ \vdots & \vdots & \ddots & \vdots \\ b_{M1} - a_{M1} & b_{M2} - a_{M2} & \cdots & b_{MN} - a_{MN} \end{bmatrix}.$$

Applying this formula, we subtract the matrix A of pixel values from the matrix B of the same size, whose entries all equal 255, which corresponds to the maximal brightness. The result is shown in Figure 4.1, where a chessboard with the initial position on it and its negative are displayed side by side.

FIGURE 4.1: The negative of an image.

For the convenience of the reader, the code used to generate Figure 4.1 is shown below:

```
A=imread('Chess.jpg');
A=rgb2gray(A);
B=255*ones(size(A));
C=B-A;
subplot(1,2,1);
imshow(A,[]);
subplot(1,2,2);
imshow(B,[]);
```

Suppose that, for some reason, we would like to take the mirror image of a picture and turn it 90° counterclockwise – all in one step. Believe it or not, there exists a simple matrix operation that has that precise effect on the image it is applied to, and it is called "*transposition*".

Formally, the transpose of an $M \times N$ matrix

$$A = \begin{bmatrix} a_{11} & a_{12} & \cdots & a_{1N} \\ a_{21} & a_{22} & \cdots & a_{2N} \\ \vdots & \vdots & \ddots & \vdots \\ a_{M1} & a_{M2} & \cdots & a_{MN} \end{bmatrix}$$

is the $N \times M$ matrix denoted by A^T and given by

$$A^T = \begin{bmatrix} a_{11} & a_{21} & \cdots & a_{M1} \\ a_{12} & a_{22} & \cdots & a_{M2} \\ \vdots & \vdots & \ddots & \vdots \\ a_{1N} & a_{2N} & \cdots & a_{MN} \end{bmatrix}.$$

Informally, transposition flips the matrix across its diagonal or, equivalently, swaps its rows for its columns. Figure 4.2 illustrates the effect of matrix transposition on digital images.

FIGURE 4.2: An image with its transpose.

Next, suppose that we would like to superimpose one image over another. Of course, numerous sophisticated ways of blending images have been developed, but we just need a simple naive method that could be implemented by means of matrix addition, which is defined for two $M \times N$ matrices

$$A = \begin{bmatrix} a_{11} & a_{12} & \cdots & a_{1N} \\ a_{21} & a_{22} & \cdots & a_{2N} \\ \vdots & \vdots & \ddots & \vdots \\ a_{M1} & a_{M2} & \cdots & a_{MN} \end{bmatrix} \quad \text{and} \quad B = \begin{bmatrix} b_{11} & b_{12} & \cdots & b_{1N} \\ b_{21} & b_{22} & \cdots & b_{2N} \\ \vdots & \vdots & \ddots & \vdots \\ b_{M1} & b_{M2} & \cdots & b_{MN} \end{bmatrix}$$

by

$$A + B = \begin{bmatrix} a_{11} + b_{11} & a_{12} + b_{12} & \cdots & a_{1N} + b_{1N} \\ a_{21} + b_{21} & a_{22} + b_{22} & \cdots & a_{2N} + b_{2N} \\ \vdots & \vdots & \ddots & \vdots \\ a_{M1} + b_{M1} & a_{M2} + b_{M2} & \cdots & a_{MN} + b_{MN} \end{bmatrix}.$$

Applying this formula to the images of the tiger and the polar bear (the top and the middle of Figure 4.3), we obtain the image shown in the lowest part of Figure 4.3, where the tiger and the bear can be seen enjoying each other's company. Of course, in order to perform matrix addition, we first have to make sure that the two images are of the same size either by trimming the larger one or by extending the smaller one.

Below is the MATLAB code we used to generate Figure 4.3.

```
A=imread('Tiger.jpg');
A=rgb2gray(A);
B=imread('Bear.jpg');
B=rgb2gray(B);
subplot(3,1,1);
imshow(A,[]);
subplot(3,1,2);
imshow(B,[]);
[Ma,Na]=size(A); [Mb,Nb]=size(B);
M=min(Ma,Mb);
N=min(Na,Nb);
C=A(1:M,1:N)+B(1:M,1:N);
subplot(3,1,3);
imshow(C,[]);
```

Most of the interesting transformations of digital images can be performed by means of matrix multiplication. As a reminder, the product of a $K \times M$ matrix A and an $M \times N$ matrix B is the $K \times N$ matrix C given by

$$c_{kn} = \sum_{m=1}^{M} a_{km} b_{mn}$$

for all $k = 1, \ldots, K$ and all $n = 1, \ldots, N$. We remind the reader that matrix multiplication is, in general, not commutative. For example,

$$\begin{bmatrix} 1 & 2 \\ 3 & 4 \end{bmatrix} \cdot \begin{bmatrix} 0 & 1 \\ -1 & 1 \end{bmatrix} = \begin{bmatrix} 1 \cdot 0 + 2 \cdot (-1) & 1 \cdot 1 + 2 \cdot 1 \\ 3 \cdot 0 + 4 \cdot (-1) & 3 \cdot 1 + 4 \cdot 1 \end{bmatrix} = \begin{bmatrix} -2 & 2 \\ -4 & 7 \end{bmatrix},$$

whereas the product of the same two matrices multiplied in the other order is

$$\begin{bmatrix} 0 & 1 \\ -1 & 1 \end{bmatrix} \cdot \begin{bmatrix} 1 & 2 \\ 3 & 4 \end{bmatrix} = \begin{bmatrix} 0 \cdot 1 + 1 \cdot 3 & 0 \cdot 2 + 1 \cdot 4 \\ -1 \cdot 1 + 1 \cdot 3 & -1 \cdot 2 + 1 \cdot 4 \end{bmatrix} = \begin{bmatrix} 3 & 4 \\ 2 & 2 \end{bmatrix},$$

and we can see that the products taken in the two different orders disagree in every single matrix component.

Whereas square matrices can be multiplied together in either order (provided that they have the same size), this is not the case for rectangular matrices. Care must be taken that the number of columns of the "left" matrix matches the number of rows of the "right" matrix.

FIGURE 4.3: The matrix sum of two images.

Multiplication of square matrices is associative and distributive with respect to matrix addition. The additive and multiplicative identities for the $N \times N$ matrices are defined by

$$Z_N = \begin{bmatrix} 0 & 0 & \cdots & 0 \\ 0 & 0 & \cdots & 0 \\ \vdots & \vdots & \ddots & \vdots \\ 0 & 0 & \cdots & 0 \end{bmatrix} \quad \text{and} \quad I_N = \begin{bmatrix} 1 & 0 & \cdots & 0 \\ 0 & 1 & \cdots & 0 \\ \vdots & \vdots & \ddots & \vdots \\ 0 & 0 & \cdots & 1 \end{bmatrix}$$

respectively.

In this section, we will provide a few illustrations of the use of matrix multiplication in image processing, and there will be a lot more to come in later chapters. For example, multiplying an image *on the left* by the matrix

$$H = \begin{bmatrix} 1/2 & 1/2 & 0 & 0 & \cdots & 0 & 0 \\ 0 & 0 & 1/2 & 1/2 & \cdots & 0 & 0 \\ \vdots & \vdots & \vdots & \vdots & \ddots & \vdots & \vdots \\ 0 & 0 & 0 & 0 & \cdots & 1/2 & 1/2 \end{bmatrix}$$

results in the vertical compression of the image by the factor of 2, whereas multiplying it on the left by the matrix

$$G = \begin{bmatrix} -1/2 & 1/2 & 0 & 0 & \cdots & 0 & 0 \\ 0 & 0 & -1/2 & 1/2 & \cdots & 0 & 0 \\ \vdots & \vdots & \vdots & \vdots & \ddots & \vdots & \vdots \\ 0 & 0 & 0 & 0 & \cdots & -1/2 & 1/2 \end{bmatrix}$$

has the effect of extracting the horizontal edges present in the image. The original image is displayed in Figure 4.4(a), and both the vertical compression and the horizontal edges are shown together in Figure 4.4(b).

On the other hand, multiplying an image by the transpose H^T of the same matrix H *on the right* results in the horizontal compression by the same factor 2, and multiplication of the image on the right by G^T results in bringing out the vertical edges, both of which are shown in Figure 4.4(c)). The combined effect of multiplication on the left and on the right is shown in Figure 4.4(d).

Based on the definition of matrix multiplication and on the example just presented, we can draw the general conclusion that multiplication of an image on the left by another matrix affects the columns of the image, whereas multiplication on the right affects the rows. We will make extensive use of this observation in Chapter 7. While designing an appropriate matrix intended to produce a specific effect, one must always keep in mind that its size must match either the number of the rows or the number of the columns in the image, depending on the way it is going to be applied.

Sometimes, we wish to extract a specific part of our image (and mask the rest). It might be a horizontal band, a vertical band, or even a rectangle

(a) The Original Image

(b) Vertical Compression

(c) Horizontal Compression

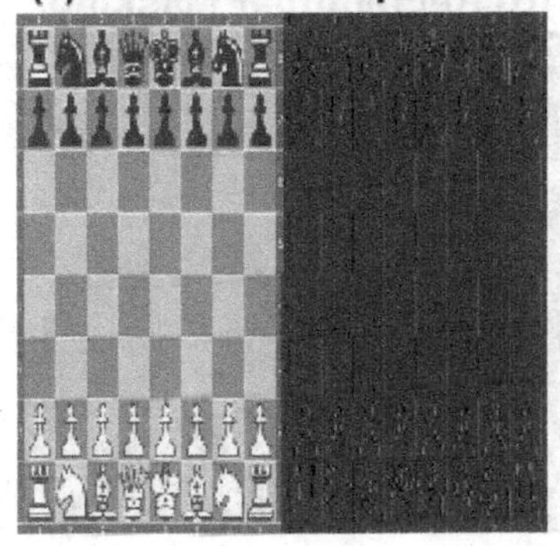

(d) The Combined Effect

FIGURE 4.4: Effects of multiplication of an image by certain matrices.

somewhere in the middle of the picture. Can matrix multiplication be used to perform this operation?

The answer, as we might expect, is yes. Suppose we would like to extract the middle horizontal half of our $M \times N$ image A (or, equivalently, mask the top and bottom quarters of the picture). Since this operation is going to affect the columns of the image (by cutting off a quarter of the column at each end), we are going to have to multiply A *on the left*. The matrix to do this job is

$$H = \left[\begin{array}{c|c|c} Z_{M/4} & Z_{M/2} & Z_{M/4} \\ \hline Z_{M/2} & I_{M/2} & Z_{M/2} \\ \hline Z_{M/4} & Z_{M/2} & Z_{M/4} \end{array}\right],$$

and it consists of the identity matrix of the size $N/2$ in the middle and zeroes all around it. It is important to note that the size of H matches the "height" of A. Figure 4.5(b) illustrates the masking effect of executing the matrix multiplication $H \cdot A$.

FIGURE 4.5: Image masking by matrix multiplication.

Alternatively, if we wanted to select the middle *vertical* part of the image, we would have multiplied A *on the right* by the matrix

$$V = \left[\begin{array}{c|c|c} Z_{N/4} & Z_{N/2} & Z_{N/4} \\ \hline Z_{N/2} & I_{N/2} & Z_{N/2} \\ \hline Z_{N/4} & Z_{N/2} & Z_{N/4} \end{array}\right],$$

which is extremely similar to H with the only difference being that its size matches the "width" of A. The reader can guess that in order to extract a rectangle in the middle of the image, one can perform both operations in succession, which can be formally described by

$$A \to H * A * V.$$

The combined effect of both multiplications is shown in Figure 4.5(d).

Naturally, from the point of view of writing computer code, there are easier ways to mask parts of the image. We discussed this particular approach in order to take advantage of the opportunity to illustrate matrix multiplication and to make a segue towards the next topic: partitioned matrices.

We are already comfortable with the idea of matrices being partitioned into rows and columns. The last example demonstrated that it can be very convenient to think of them as being partitioned into more flexible blocks, such as blocks of additive and multiplicative identities. Whenever a matrix has blocks with special characteristics, properties, or structures, it might prove worthwhile to partition the matrix into those blocks.

There are no particular rules or secrets regarding partitioned matrices. The basic principle is that the block entries can be treated just like scalar ones. Addition or subtraction of two partitioned matrices (with identical partitions) can be performed the usual way as long as the sum is defined for each pair of blocks. Likewise, partitioned matrices can be multiplied together by the usual row-times-column rule as if the block entries were scalars, provided that the following three conditions are satisfied:

- the length of the rows of blocks in the "left" matrix matches the height of the columns of blocks in the "right" matrix,
- the product is defined for each pair of blocks that needs to be multiplied in the course of carrying out the row-times-column multiplication,
- all the resulting products that need to be added together are of the same size.

In the exercises at the end of this section, the reader will be asked to practice calculations with partitioned matrices and to establish various properties related to products and inverses of partitioned matrices.

Section 4.1 Exercises.

1. Show that matrix addition is commutative and associative.
2. Show that matrix multiplication is distributive with respect to matrix addition.
3. If possible, calculate $A+B$, $A-B$, AB, and BA for the following pairs of matrices:

(a) $A = \begin{bmatrix} 3 & 2 \\ 8 & 5 \end{bmatrix}$ and $B = \begin{bmatrix} 2 & 4 \\ -4 & -6 \end{bmatrix}$,

(b) $A = \begin{bmatrix} 3 & 2 \\ 8 & -12 \end{bmatrix}$ and $B = \begin{bmatrix} 2 & 4 \\ -4 & 8 \end{bmatrix}$,

(c) $A = \begin{bmatrix} 2 & -3 \\ 1 & 0 \\ -3 & 1 \end{bmatrix}$ and $B = \begin{bmatrix} 3 & 2 \\ 8 & 5 \end{bmatrix}$,

(d) $A = \begin{bmatrix} 1 & -2 & 1 \\ 0 & 1 & 2 \end{bmatrix}$ and $B = \begin{bmatrix} 2 & 4 \\ -4 & -6 \end{bmatrix}$.

Use MATLAB to check all your answers.

4. Suppose that the matrix A is partitioned as

$$A = \left[\begin{array}{cc|cc} a_{11} & a_{12} & a_{13} & a_{14} \\ a_{21} & a_{22} & a_{23} & a_{24} \\ \hline a_{31} & a_{32} & a_{33} & a_{34} \\ a_{41} & a_{42} & a_{43} & a_{44} \end{array}\right] = \left[\begin{array}{c|c} A_{11} & A_{12} \\ \hline A_{21} & A_{22} \end{array}\right].$$

Verify that

$$A^T = \left[\begin{array}{c|c} A_{11}^T & A_{21}^T \\ \hline A_{12}^T & A_{22}^T \end{array}\right].$$

Try to generalize this example and formulate a similar rule for transposes of any partitioned matrices.

5. If possible, solve the following partitioned-matrix equations for the matrices X, Y, and Z.

(a) $\begin{bmatrix} X & 0 \\ Y & Z \end{bmatrix} \begin{bmatrix} A & 0 \\ B & C \end{bmatrix} = \begin{bmatrix} I & 0 \\ 0 & I \end{bmatrix}$,

(b) $\begin{bmatrix} X & Y \\ 0 & Z \end{bmatrix} \begin{bmatrix} A & B \\ 0 & C \end{bmatrix} = \begin{bmatrix} I & 0 \\ 0 & I \end{bmatrix}$,

(c) $\begin{bmatrix} X & Y \\ Z & 0 \end{bmatrix} \begin{bmatrix} A & B \\ C & 0 \end{bmatrix} = \begin{bmatrix} I & 0 \\ 0 & I \end{bmatrix}$,

(d) $\begin{bmatrix} 0 & X \\ Y & Z \end{bmatrix} \begin{bmatrix} 0 & A \\ B & C \end{bmatrix} = \begin{bmatrix} I & 0 \\ 0 & I \end{bmatrix}$,

(e) $\begin{bmatrix} A & B \\ 0 & I \end{bmatrix} \begin{bmatrix} X & Y & Z \\ 0 & 0 & I \end{bmatrix} = \begin{bmatrix} I & 0 & 0 \\ 0 & 0 & I \end{bmatrix}$.

6. Design a matrix H which, when multiplying an image on the left, would

   (a) select the top half of the image and mask the lower half,

   (b) select the bottom half of the image and mask the upper half,

   (c) select the middle horizontal half of the image.

7. Design a matrix V which, when multiplying an image on the right, would

   (a) select the right half of the image and mask the left half,

   (b) select the left half of the image and mask the right half,

   (c) select the middle vertical half of the image.

MATLAB Exercises.

1. Write a MATLAB function that would create the negative of the specified image (grayscale or color). With color images, your function should calculate the negative of each channel separately. Test your function with a selection of images of your choice.

2. Write three MATLAB functions that would extract the following parts of the specified image (while masking the rest of the image):

   (a) the middle vertical half,

   (b) the middle horizontal half,

   (c) the middle rectangular part.

   Test your functions with a selection of images of your choice.

3. Write similar MATLAB functions to select other regions (for example, the middle horizontal third, the upper half, the left vertical quarter, etc.) of the specified image by means of multiplication by a suitable matrix.

4. Write a MATLAB program that would do the following:

   (a) Ask the user which image is to be selected.

   (b) Ask the user to specify the type of masking that is desired.

   (c) Use the appropriate MATLAB function you created in MATLAB Exercise 2 to produce the desired masking.

   Test your program with several different images (grayscale and color).

5. Compile the results of the previous MATLAB exercises into a PowerPoint presentation.

4.2 Linear Transformations and Their Matrices

Suppose that you have taken a great picture of an amazing sunset view of what you are convinced is the most beautiful beach in the world. Unfortunately, upon reviewing your pictures, you realize that the camera must have been held at a wrong angle and, as a result, the horizon is not oriented horizontally. Is there a way to rotate the image to straighten the horizon?

The answer is yes, and the mathematical instruments we need for the job are *linear transformations*. From a course in linear algebra, we recall that a mapping T from $\mathbb{R}^n$ to $\mathbb{R}^m$ is called a **linear transformation** if

$$T(\mathbf{u}+\mathbf{v}) = T(\mathbf{u}) + T(\mathbf{v}) \tag{4.1}$$

and

$$T(\alpha\mathbf{v}) = \alpha T(\mathbf{v}) \tag{4.2}$$

for all $\mathbf{u}, \mathbf{v} \in \mathbb{R}^n$ and all scalars α.

Before we dive into the multidimensional setting, let us recall whether we have seen any linear transformations among the functions of a single variable (which can be thought of as mappings from $\mathbb{R}^1$ to $\mathbb{R}^1$) studied in pre-calculus. Let us check them one by one:

- The power function $f(x) = x^\gamma$ is certainly not a linear transformation (unless $\gamma = 1$), both because

  $$f(\alpha x) = (\alpha x)^\gamma = \alpha^\gamma x^\gamma \neq \alpha f(x)$$

  for $\alpha \neq 1$, and also because

  $$f(x_1 + x_2) = (x_1 + x_2)^\gamma \neq x_1^\gamma + x_2^\gamma = f(x_1) + f(x_2).$$

- The exponential function $f(x) = b^x$ is certainly not a linear transformation either, because

  $$f(x_1 + x_2) = b^{x_1+x_2} = b^{x_1}b^{x_2} \neq b^{x_1} + b^{x_2} = f(x_1) + f(x_2),$$

  unless $x_1 = x_2 = \log_b 2$.

- The logarithmic function $f(x) = \ln x$ is not a linear transformation because

  $$f(x_1 + x_2) = \ln(x_1 + x_2) \neq \ln x_1 + \ln x_2 = f(x_1) + f(x_2),$$

  unless $x_1 = x_2 = 2$.

- The trigonometric functions $s(x) = \sin x$ and $c(x) = \cos x$ are also not linear transformations, both because

  $$s(\alpha x) = \sin(\alpha x) \neq \alpha \sin x = \alpha s(x)$$

  (unless $\alpha = 0$) and

  $$c(\alpha x) = \cos(\alpha x) \neq \alpha \cos x = \alpha c(x)$$

  (unless $\alpha = 1$), and also because

  $$s(x_1 + x_2) = \sin(x_1 + x_2) \neq \sin x_1 + \sin x_2 = s(x_1) + s(x_2)$$

  and

  $$c(x_1 + x_2) = \cos(x_1 + x_2) \neq \cos x_1 + \cos x_2 = c(x_1) + c(x_2).$$

  In Section 6.2, the reader will have an opportunity to review the correct summation formulas for trigonometric functions.

We can see that it is not easy to find a linear transformation among the functions of a single variable. But how about linear functions? Aren't they supposed to be linear by their very definition? Let us check. For a generic linear function defined by the slope-intercept formula

$$f(x) = mx + b,$$

we have

$$f(\alpha x) = \alpha \cdot mx + b \neq \alpha mx + \alpha b = \alpha f(x)$$

and

$$f(x_1 + x_2) = m(x_1 + x_2) + b \neq mx_1 + mx_2 + 2b = f(x_1) + f(x_2)$$

unless $b = 0$. Thus,

> The only function of a single variable (among the ones we have ever encountered) that qualifies as a linear transformation is the direct proportionality
>
> $$f(x) = mx. \tag{4.3}$$

The direct counterpart of (4.3) in a multivariate setting, where we consider mappings from $\mathbb{R}^n$ to $\mathbb{R}^m$, is a matrix transformation

$$T(\mathbf{x}) = A\mathbf{x}, \tag{4.4}$$

which is linear due to the distributive properties of matrix multiplication, as the reader is encouraged to verify.

We next ask an important theoretical question: are there any other types of linear transformations, or is every linear transformation of the form (4.4) for a suitable matrix A? The amazing thing is (and it is testimony to the power of abstract mathematics) that in the course of answering this seemingly esoteric abstract question, we are going to come up with the very tool that is needed for practical applications (like rotating our photo to straighten the horizon)!

Let us assume that $T : \mathbb{R}^n \to \mathbb{R}^m$ is an arbitrary linear transformation. It is evident that we can express any given vector $\mathbf{x} \in \mathbb{R}^n$ as

$$\mathbf{x} = \begin{bmatrix} x_1 \\ x_2 \\ \vdots \\ x_n \end{bmatrix} = x_1 \begin{bmatrix} 1 \\ 0 \\ \vdots \\ 0 \end{bmatrix} + x_2 \begin{bmatrix} 0 \\ 1 \\ \vdots \\ 0 \end{bmatrix} + \cdots + x_n \begin{bmatrix} 0 \\ 0 \\ \vdots \\ 1 \end{bmatrix}.$$

Due to the linearity of T, it follows that

$$T(\mathbf{x}) = T\left(x_1 \begin{bmatrix} 1 \\ 0 \\ \vdots \\ 0 \end{bmatrix} + x_2 \begin{bmatrix} 0 \\ 1 \\ \vdots \\ 0 \end{bmatrix} + \cdots + x_n \begin{bmatrix} 0 \\ 0 \\ \vdots \\ 1 \end{bmatrix} \right) \tag{4.5}$$

$$= x_1 T\left(\begin{bmatrix}1\\0\\\vdots\\0\end{bmatrix}\right) + x_2 T\left(\begin{bmatrix}0\\1\\\vdots\\0\end{bmatrix}\right) + \cdots + x_n T\left(\begin{bmatrix}0\\0\\\vdots\\1\end{bmatrix}\right)$$

$$= x_1\mathbf{a}_1 + x_2\mathbf{a}_2 + \cdots + x_n\mathbf{a}_n = A\mathbf{x},$$

where A is the matrix whose columns $\mathbf{a}_1$, $\mathbf{a}_2$, $\ldots$, and $\mathbf{a}_n$ are the images of the basic vectors

$$\mathbf{e}_1 = \begin{bmatrix}1\\0\\\vdots\\0\end{bmatrix}, \quad \mathbf{e}_2 = \begin{bmatrix}0\\1\\\vdots\\0\end{bmatrix}, \quad \ldots, \quad \mathbf{e}_n = \begin{bmatrix}0\\0\\\vdots\\1\end{bmatrix} \tag{4.6}$$

respectively under the linear transformation T. We have, therefore, established that any linear transformation is of the form (4.4) for a suitable matrix A, called the *transformation matrix* or the *standard matrix* of the transformation T.

This result is so important that we reiterate and frame it for easy reference:

The Standard Matrix of a Linear Transformation

Any linear transformation $T : \mathbb{R}^n \to \mathbb{R}^m$ is of the form

$$T(\mathbf{x}) = A\mathbf{x}$$

for a suitable matrix A (called the *standard transformation matrix*), which is given by

$$A = \begin{bmatrix}T(\mathbf{e}_1) & T(\mathbf{e}_2) & \cdots & T(\mathbf{e}_n)\end{bmatrix}. \tag{4.7}$$

We next turn to applying the fundamental result (4.7) to image processing. In order to transform an image by rotating it, shearing it, or in any other linear fashion, we have to apply the appropriate transformation *to the pixel coordinates* as opposed to the pixel values. Since the pixel coordinates are two-dimensional, we have to design a linear transformation from $\mathbb{R}^2$ to $\mathbb{R}^2$, and all we need for that is to decide where we want the vectors

$$\mathbf{e}_1 = \begin{bmatrix}1\\0\end{bmatrix} \quad \text{and} \quad \mathbf{e}_2 = \begin{bmatrix}0\\1\end{bmatrix}$$

to be mapped to.

It is evident from Figure 4.6 that in order to perform the rotation through the angle θ in the counterclockwise direction, the point $(1, 0)$ has to be mapped

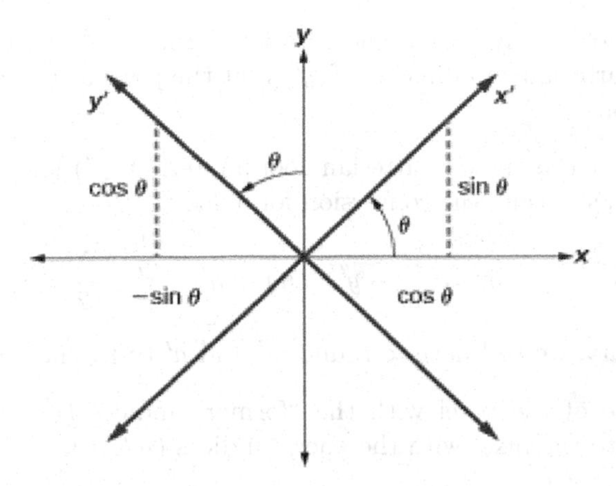

FIGURE 4.6: A rotation transformation.

to the point $(\cos\theta, \sin\theta)$, and the point $(0, 1)$ – to the point $(-\sin\theta, \cos\theta)$. The standard transformation matrix is, therefore,

$$A = \begin{bmatrix} \cos\theta & -\sin\theta \\ \sin\theta & \cos\theta \end{bmatrix}, \tag{4.8}$$

and, consequently, the "new" coordinates of the point (x, y) can be calculated using

$$\begin{bmatrix} x' \\ y' \end{bmatrix} = \begin{bmatrix} \cos\theta & -\sin\theta \\ \sin\theta & \cos\theta \end{bmatrix} \begin{bmatrix} x \\ y \end{bmatrix}. \tag{4.9}$$

Because there is a number of technical obstacles to overcome in the course of applying rotation transformations to digital images, we describe the process in full detail for the convenience of the reader.

A Step-by-Step Procedure for Implementation of Image Rotations. To be Applied to Every Pixel.

1. We recall that pixel indices range from $(1, 1)$ to (M, N), with the former indicating the upper-left corner, and the latter – the lower-right corner. As such, they do not correspond to the Cartesian coordinates of the pixels. Assuming that we intend to rotate the image around its center, we place the origin at the position with the indices $(M/2, N/2)$. Therefore, the Cartesian coordinates (x, y) for a pixel with the indices (m, n) can be calculated with the help of the conversion formulas

$$x = n - \frac{N}{2} \quad \text{and} \quad y = \frac{M}{2} - m, \tag{4.10}$$

and the reader may wish to check that these formulas work as intended.

2. We are now ready to use the rotation formula (4.9) to calculate the "new" Cartesian coordinates (x', y') that the pixel at (m, n) needs to be moved to.

3. We convert the "new" Cartesian coordinates (x', y') into the "new" indices (m', n') using the conversion formulas

$$m' = \frac{M}{2} - y' \quad \text{and} \quad n' = x' + \frac{N}{2}. \tag{4.11}$$

   If necessary, we will have to round m' and n' to the nearest integer.

4. The value of the pixel with the "former" indices (m, n) can now be assigned to the pixel with the "new" indices (m', n').

In the discussion above, we have made the simplifying assumption that both M and N are even integers. If they are odd, the only correction we have to make is to adjust the placement of the origin at the position $(\lceil N/2 \rceil, \lceil M/2 \rceil)$ and to modify the conversion formulas (4.10) accordingly. Also, depending on the image and the angle of rotation, some of the "new" indices may turn out negative. Since matrices of pixel values are only allowed to have positive indices, the size of the image may have to be increased to accommodate the "corners" of the rotated image.

For certain angles (like $\theta = 90^o$ or any other integer multiples of 90°), the four-step procedure just described is perfectly sufficient (Figure 4.7). It can also be adapted to perform other linear transformations, such as reflections and certain types of stretching, compression, and shear transformations. Unfortunately, there still remains one more annoying complication to work out.

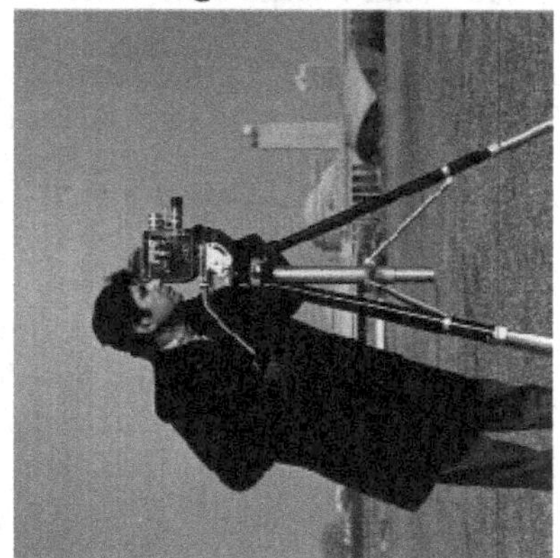

FIGURE 4.7: Rotation through the angle of 90°.

Even after we have applied the four-step process outlined above to every pixel of the image, our job is not fully completed, because the rotated image might have "holes" in it, as demonstrated in Figure 4.8(b). This annoying complication occurs due to the fact that the linear transformation (4.9) is not surjective (onto), when its domain and range consist of pairs of integers (as opposed to real numbers). As a result, it is very likely that some of the pixels in the rotated image have not been assigned any values by the transformation. Such pixels must be interpolated using the neighboring pixels that **have** been assigned new values.

(a) The Original Image

(b) Rotation w/o Interpolation

(c) The Pattern of "Holes"

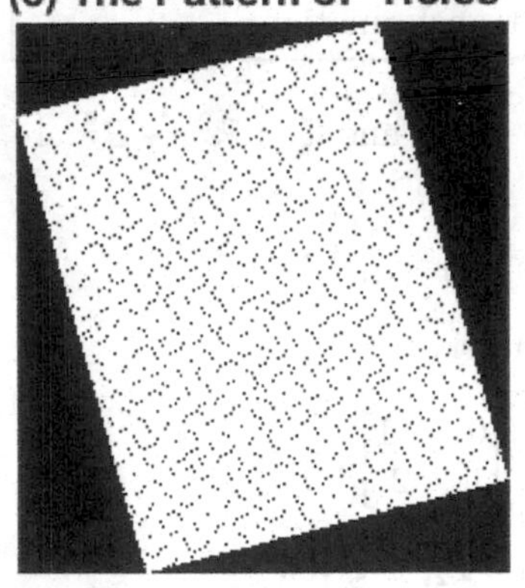

(d) Rotation with Interpolation

FIGURE 4.8: Rotation with and without interpolation.

Figure 4.8(c) shows the pattern of "holes" when rotating an image through the angle of 15°, and the result of the interpolation procedure is demonstrated in Figure 4.8(d). The reader might note a certain amount of quality loss, which explains why image processing software always asks whether the user is sure that they want to proceed with image rotation.

In this section, we focused primarily on rotation of digital images, because, in our opinion, it is both the most visually appealing and technically challenging type of linear transformations. However, the same procedure (with obvious modifications) can be followed for any other linear transformation the reader would like to experiment with.

In MATLAB exercises at the end of this section, the reader will be asked to implement image rotation and several other linear transformations, complete with extending the image and pixel interpolation where appropriate.

One of the questions we have not yet discussed apropos of linear transformations is that of transformation composition. What will happen if we apply two or more linear transformations in succession to a vector $\mathbf{x}$? Formally, suppose that

$$T_1 : \mathbb{R}^k \to \mathbb{R}^n \quad \text{and} \quad T_2 : \mathbb{R}^n \to \mathbb{R}^m$$

are linear transformations with standard matrices A_1 and A_2 respectively. By associativity of matrix multiplication, their composition $T = T_2 \circ T_1$ is given by

$$T(\mathbf{x}) = T_2\big(T_1(\mathbf{x})\big) = T_2\big(A_1\mathbf{x}\big) = A_1 A_2 \mathbf{x} = A\mathbf{x},$$

where $A = A_2 A_1$. We have thus proved the following rule for transformation composition:

Transformation Composition Rule

The standard matrix of the composition of two transformations is the product of the standard matrices of the individual transformations.

For example, if the transformations $T_1 : \mathbb{R}^2 \to \mathbb{R}^2$ and $T_2 : \mathbb{R}^2 \to \mathbb{R}^2$ are given by

$$T_1(\mathbf{x}) = \begin{bmatrix} 1 & 2 \\ 0 & 1 \end{bmatrix} \mathbf{x} \quad \text{and} \quad T_1(\mathbf{x}) = \begin{bmatrix} 0 & -1 \\ 1 & 0 \end{bmatrix} \mathbf{x},$$

then their composition $T_2 \circ T_1$ is given by the matrix

$$A_{T_2 \circ T_1} = \begin{bmatrix} 0 & -1 \\ 1 & 0 \end{bmatrix} \cdot \begin{bmatrix} 1 & 2 \\ 0 & 1 \end{bmatrix} = \begin{bmatrix} 0 & -1 \\ 1 & 2 \end{bmatrix},$$

whereas their composition in the other order, $T_1 \circ T_2$, is given by

$$A_{T_1 \circ T_2} = \begin{bmatrix} 1 & 2 \\ 0 & 1 \end{bmatrix} \cdot \begin{bmatrix} 0 & -1 \\ 1 & 0 \end{bmatrix} = \begin{bmatrix} 2 & -1 \\ 1 & 0 \end{bmatrix},$$

reminding us of the non-commutativity of matrix multiplication and also of non-commutativity of function composition.

Which brings us back to the Four-Step Procedure for image rotation. Why can't we reduce it to just two steps? After all, the first step consisted of a translation (shifting the origin of the coordinate system to the center of the picture) and a reflection (reversing the direction of the y-axis); the second step was the rotation transformation itself; and the third step was, again, a reflection combined with a translation. Why can't we just multiply all of those translation matrices and complete all three steps in one shot?

The answer, of course, is that translation is not a linear transformation, and, therefore, is not given by matrix multiplication! It is somewhat ironic that it is the simplest of all the transformations that is giving us trouble. So, unfortunately, it is not possible to combine the first three steps of the procedure into one step. Or is it? We will have to wait until the next section to find out!

Section 4.2 Exercises.

1. Suppose that $T : \mathbb{R}^n \to \mathbb{R}^m$ is a linear transformation
   (a) Show that $T(\mathbf{0}) = \mathbf{0}$.
   (b) Show that $\ker(T)$ is a subspace of $\mathbb{R}^n$.
   (c) Is the image of T a subspace of $\mathbb{R}^n$?
2. Provide a few examples of specific values of x, α, x_1, and x_2 to demonstrate that the power function $f(x) = x^\gamma$ is not a linear transformation from $\mathbb{R}$ to $\mathbb{R}$ when $\gamma \neq 1$.
3. Provide a few examples of specific values x_1, and x_2 to demonstrate that the exponential function $f(x) = b^x$ is not a linear transformation from $\mathbb{R}$ to $\mathbb{R}$.
4. Provide a few examples of specific values x_1, and x_2 to demonstrate that the logarithmic functions $f(x) = \ln x$ and $g(x) = \log x$ are not linear transformations from $\mathbb{R}$ to $\mathbb{R}$.
5. Provide a few examples of specific values of x, α, x_1, and x_2 to demonstrate that the trigonometric functions $s(x) = \sin x$ and $c(x) = \cos x$ are not linear transformations from $\mathbb{R}$ to $\mathbb{R}$.
6. Suppose that the transformation $T : \mathbb{R}^2 \to \mathbb{R}^2$ is defined by

$$T\left(\begin{bmatrix} x_1 \\ x_2 \end{bmatrix}\right) = \begin{bmatrix} x_1 + x_2 \\ x_2 - 1 \end{bmatrix}.$$

   Decide whether T is linear.
7. Verify that the mapping defined by (4.4) is, indeed, a linear transformation.

8. Verify every step in the derivation (4.6).

9. Verify the conversion formulas (4.10) and (4.11).

10. Find the standard matrices for the following transformations:

    (a) Stretch/compression by a factor of a in the direction of the x-axis.

    (b) Stretch/compression by a factor of b in the direction of the y-axis.

    (c) Horizontal shear that leaves the point $(1, 0)$ unchanged and maps the point $(0, 1)$ into $(a, 1)$.

    (d) Vertical shear that leaves the point $(0, 1)$ unchanged and maps the point $(1, 0)$ into $(1, b)$.

    (e) Reflection through the x-axis.

    (f) Reflection through the y-axis.

    (g) Reflection through the line $y = x$.

    (h) Reflection through the line $y = -x$.

    (i) Reflection through the origin.

11. Consider the linear transformation consisting of rotation through the angle of 30° counterclockwise followed by reflection through the x-axis. Construct its standard matrix in two ways:

    (a) By determining the ultimate destinations of the basic vectors $\mathbf{e}_1$ and $\mathbf{e}_2$.

    (b) By multiplying the standard matrices of the individual transformations.

12. Consider the linear transformation consisting of reflection through the x-axis followed by rotation through the angle of 30°. Construct its standard matrix in two ways:

    (a) By determining the ultimate destinations of the basic vectors $\mathbf{e}_1$ and $\mathbf{e}_2$.

    (b) By multiplying the standard matrices of the individual transformations.

13. Consider the linear transformation consisting of horizontal shear that maps the point $(0, 1)$ to $(3, 1)$ followed by a vertical stretch by the factor of 2, followed by reflection through the y-axis. Construct its standard matrix in two ways:

    (a) By determining the ultimate destinations of the basic vectors $\mathbf{e}_1$ and $\mathbf{e}_2$.

    (b) By multiplying out the standard matrices of the individual transformations.

MATLAB Exercises.

1. Write a MATLAB function that would rotate the specified image through the specified angle by implementing the four-step procedure followed by interpolation as described in this section. Your function should work with any grayscale or color images. Test your function using a selection of images of your choice.

2. Similarly, write MATLAB functions that would implement other linear transformations, whose standard matrices you constructed in Exercise 10 of this section. Your function should accept the specified image and the transformation parameters (such as, for example, the shear factor) as inputs. Depending on the nature of the transformation, interpolation may or may not be necessary. Your function should work with any grayscale or color images. Test your function using a selection of images of your choice.

3. Prepare a selection of images suitable for rotations, reflections, shears, and other linear transformations.

4. Write a MATLAB program that will do the following:

   (a) Ask the user to specify which image is to be transformed.

   (b) Ask the user to specify which transformation needs to be performed.

   (c) Depending on which transformation was selected, ask the user to input the values for the appropriate transformation parameters (such as the angle of rotation, the shear factor, etc.)

   (d) Perform the specified transformation on the specified image.

   (e) Display both the original image and its transformed version.

   Test your program using several different grayscale and color images. Comment on the results.

5. Write a MATLAB program similar to the one you created in MATLAB Exercise 4, but this time, the user should have the opportunity to request a composition of two specified linear transformations. Compare the results of combining pairs of transformations in both orders.

6. Compile the results of the previous MATLAB exercises into a PowerPoint presentation.

4.3 Homogeneous Coordinates and Projective Transformations

At the end of Section 4.2, we asked an important practical question: is it possible to combine all the transformations performed during the first three steps of the image rotation procedure into just one step? The answer seemed to be no, but we were left with a vague indication of hope.

Is it possible to solve the equation $x^2 + 1 = 0$? Of course, not. Unless, of course, you introduce another dimension, called imaginary numbers. And then, lo and behold, the impossible becomes trivial.

Is it possible to travel from New York to London in six hours? Of course, not. Unless, of course, you introduce another dimension, called air travel at 37,000 feet above sea level. And then, lo and behold, the impossible becomes routine.

If we recall, the difficulty with combining the first three steps of the image-rotation procedure into one step lay in the annoying fact that translation is not a linear transformation. So, what extra dimension can we introduce to make it linear?

It turns out that all we have to do is move the origin of the coordinate system away from the center of the picture and place it where we are, one unit of length away from the computer screen. This way, any point with coordinates (x, y) will be assigned the coordinates $(x, y, 1)$, called *homogeneous coordinates.*

We first note that implementation of linear transformations in homogeneous coordinates does not require any major changes. A transformation T with the standard matrix

$$A = \begin{bmatrix} a_{11} & a_{12} \\ a_{21} & a_{22} \end{bmatrix}$$

with respect to the "regular" Cartesian coordinates will be given by the formula

$$T\left(\begin{bmatrix} x \\ y \\ 1 \end{bmatrix}\right) = \begin{bmatrix} a_{11} & a_{12} & 0 \\ a_{21} & a_{22} & 0 \\ 0 & 0 & 1 \end{bmatrix} \begin{bmatrix} x \\ y \\ 1 \end{bmatrix} \tag{4.12}$$

with respect to homogeneous coordinates, as the reader is encouraged to verify. Translating each point $(x, y, 1)$ to $(x+h, y+k, 1)$ is accomplished by the matrix transformation

$$T\left(\begin{bmatrix} x \\ y \\ 1 \end{bmatrix}\right) = \begin{bmatrix} 1 & 0 & h \\ 0 & 1 & k \\ 0 & 0 & 1 \end{bmatrix} \begin{bmatrix} x \\ y \\ 1 \end{bmatrix}, \tag{4.13}$$

as readers can easily check for themselves.

Our initial objective has been accomplished. We can now combine translations with linear operations and, therefore, perform all three steps of the image rotating procedure by means of just one matrix multiplication. But there is so much more we can get out of homogeneous coordinates!

Since 2×2 matrices only have four entries, and each entry can either be zero or nonzero, there are only $2^4 = 16$ types of linear transformations of the plane. On the other hand, 3×3 matrices operating on homogeneous coordinates have 9 entries and, therefore, they can provide for a much larger number of different types of transformations. We can anticipate how the introduction of homogeneous coordinates opens a whole world of new possibilities!

FIGURE 4.9: An image from a security camera.

For example, suppose that the image shown in Figure 4.9 came from our security camera, and we would like to obtain a nice frontal view of the vanity plate on the front bumper of the Toyota minivan. How can we digitally "turn" the picture to get a look from a different point of view? Can homogeneous coordinates be of help to us?

Geometrically, the problem comes down to designing the transformation T that would map the four corners (at (x_1, y_1), at (x_2, y_2), at (x_3, y_3), and at (x_4, y_4)) of the vanity plate to the four corners (at (x_1', y_1'), at (x_2', y_2'), at (x_3', y_3'), and at (x_4', y_4') of the rectangle that we would like the plate to be mapped to. Algebraically, we need to design the matrix

$$A = \begin{bmatrix} a_{11} & a_{12} & a_{13} \\ a_{21} & a_{22} & a_{23} \\ a_{31} & a_{32} & a_{33} \end{bmatrix}$$

for the transformation T, so that it can be expressed by

$$\lambda \begin{bmatrix} x_1' & x_2' & x_3' & x_4' \\ y_1' & y_2' & y_3' & y_4' \\ 1 & 1 & 1 & 1 \end{bmatrix} = \begin{bmatrix} a_{11} & a_{12} & a_{13} \\ a_{21} & a_{22} & a_{23} \\ a_{31} & a_{32} & a_{33} \end{bmatrix} \begin{bmatrix} x_1 & x_2 & x_3 & x_4 \\ y_1 & y_2 & y_3 & y_4 \\ 1 & 1 & 1 & 1 \end{bmatrix}. \tag{4.14}$$

Before proceeding any further, we have to explain the reason for introducing the scalar multiple λ. It is easy to see that the homogeneous coordinates $(x, y, 1)$ of the point (x, y) can be identified with the straight line L passing through the origin and the point $(x, y, 1)$. Since both $(x, y, 1)$ and $(\lambda x, \lambda y, \lambda)$ lie on L, they represent the same point (x, y).

How do we determine the unknown entries of the transformation matrix A? It is not easy, but we shall persevere. For each pair (x_k, y_k) and (x'_k, y'_k) of the corresponding points, (4.14) implies

$$\begin{aligned} \lambda x'_k &= a_{11}x_k + a_{12}y_k + a_{13} \\ \lambda y'_k &= a_{21}x_k + a_{22}y_k + a_{23} \\ \lambda &= a_{31}x_k + a_{32}y_k + a_{33}. \end{aligned} \tag{4.15}$$

Substituting the value for the scale factor λ in the first and second equations, we obtain the system

$$\begin{aligned} (a_{31}x_k + a_{32}y_k + a_{33})x'_k &= a_{11}x_k + a_{12}y_k + a_{13} \\ (a_{31}x_k + a_{32}y_k + a_{33})y'_k &= a_{21}x_k + a_{22}y_k + a_{23} \end{aligned}$$

which, after rearranging terms, becomes

$$\begin{aligned} -a_{11}x_k - a_{12}y_k - a_{13} + a_{31}x_k x'_k + a_{32}y_k x'_k + a_{33}x'_k &= 0 \\ -a_{21}x_k - a_{22}y_k - a_{23} + a_{31}x_k y'_k + a_{32}y_k y'_k + a_{33}y'_k &= 0. \end{aligned}$$

The problem of determining the entries of the transformation matrix A has thus been reduced to that of solving the homogeneous linear system

$$H\mathbf{a} = \mathbf{0}, \tag{4.16}$$

where

$$H = \begin{bmatrix} -x_1 & -y_1 & -1 & 0 & 0 & 0 & x_1x'_1 & y_1x'_1 & x'_1 \\ \vdots & & & \vdots & & \vdots & & & \vdots \\ -x_4 & -y_4 & -1 & 0 & 0 & 0 & x_4x'_4 & y_4x'_4 & x'_4 \\ 0 & 0 & 0 & -x_1 & -y_1 & -1 & x_1y'_1 & y_1y'_1 & y'_1 \\ \vdots & & & \vdots & & \vdots & & & \vdots \\ 0 & 0 & 0 & -x_4 & -y_4 & -1 & x_4y'_4 & y_4y'_4 & y'_4 \end{bmatrix} \tag{4.17}$$

and

$$\mathbf{a} = \begin{bmatrix} a_{11} & a_{12} & a_{13} & a_{21} & a_{22} & a_{23} & a_{31} & a_{32} & a_{33} \end{bmatrix}^T.$$

This system has nine variables but only eight equations, which is a very good thing, because we would like it to have a free variable and, hence, a nontrivial solution. Due to the size of the matrix, we do not attempt to row-reduce it by hand and recommend using a computer algebra system to complete the calculation of the entries a_{jk} of the transform matrix A. The numerical value of the scale parameter λ can next be obtained using the formula (4.15).

Just as with the implementation of the rotation transformation in the previous section, the change-of-perspective transformation is not surjective (onto), when its domain and range consist of pairs of integers (as opposed to real numbers). As a result, it is very likely that some of the pixels in the rotated image will not be assigned any values by the transformation. Such pixels must be interpolated using the neighboring pixels that **have** been assigned new values.

The implementation of the "change of perspective" transformation is similar to the rotation transform procedure discussed in the previous section. We outline it here for the reader's convenience.

A Step-by-Step Procedure for Implementation of the Change-of-Perspective Transformation (to Be Applied to Every Pixel).

1. Calculate the Cartesian coordinates (x, y) and define the homogeneous coordinates $(x, y, 1)$ for the pixel with the indices (m, n) with the help of the conversion formulas

$$x = n - \frac{N}{2} \quad \text{and} \quad y = \frac{M}{2} - m.$$

2. Apply the change-of-perspective transform to calculate the "new" homogeneous coordinates $(x'/\lambda, y'/\lambda, 1)$ that the pixel at (m, n) needs to be moved to.

3. Convert the "new" homogeneous coordinates $(x'/\lambda, y'/\lambda, 1)$ into the "new" indices (m', n') using the conversion formulas

$$m' = \frac{M}{2} - \frac{y'}{\lambda} \quad \text{and} \quad n' = \frac{x'}{\lambda} + \frac{N}{2}.$$

   If necessary, we will have to round m' and n' to the nearest integer.

4. The value of the pixel with the "former" indices (m, n) can now be assigned to the pixel with the "new" indices (m', n').

Since all the calculations are performed in homogeneous coordinates, all the transformations involved in the procedure can be performed with the help of matrix multiplication. All the transformation matrices can be multiplied together and, therefore, the first three steps can be combined and implemented with just one matrix multiplication.

Just as in the case of the rotation transformation that we studied in the previous section, even after we have applied the step-by-step procedure outlined above to every pixel of the image, our job is not fully completed, because the transformed image might have "holes" it it. Once again, this is due to the fact that the change-of-perspective transformation is not surjective (onto), when its domain and range consist of pairs of integers (as opposed to real numbers). Therefore, just as in the previous section, we have to interpolate the values of such pixels using the neighboring pixels that have been assigned new values.

Section 4.3 Exercises.

1. Verify that the transformation T with the standard matrix A is realized by the expression (4.12) in homogeneous coordinates.
2. Show that it does not seem possible to determine the entries of the matrix in (4.14) without introducing the scaling parameter λ.
3. Verify all the steps in the derivation of the system (4.16) with the coefficient matrix (4.17).
4. Combine the transformations in Exercise 10 – 13 of the previous section with the translation by $(x, y) \to (x + h, y + k)$ and find the matrices of the combined transformation in homogeneous coordinates.
5. Find general solutions for the homogeneous system $H\mathbf{x} = \mathbf{0}$ with the given coefficient matrices. Write the solutions in parametric vector form. Substitute a convenient value for the parameter and produce a fraction-free particular solution.

(a) $H = \begin{bmatrix} 2 & -1 & 5 \\ 4 & -2 & 10 \end{bmatrix}$, (d) $H = \begin{bmatrix} 1 & 1 & 8 \\ -2 & -2 & -16 \end{bmatrix}$,

(b) $H = \begin{bmatrix} 1 & 1 & 8 \\ 1 & -2 & 6 \\ 4 & -3 & 14 \end{bmatrix}$, (e) $H = \begin{bmatrix} 1 & 1 & 2 \\ -1 & -1 & 6 \\ 0 & 1 & 1 \end{bmatrix}$,

(c) $H = \begin{bmatrix} 1 & 1 & 8 \\ 1 & -2 & 6 \\ 4 & -3 & 14 \end{bmatrix}$, (f) $H = \begin{bmatrix} 1 & 2 & -12 \\ 2 & 1 & 12 \\ -1 & 1 & 0 \end{bmatrix}$.

Use a computer algebra system to check your answers.

MATLAB Exercises.

1. Revisit the functions you created in MATLAB Exercises 1 and 2 in the previous section. Streamline them by executing all the transformations with the help of just one matrix multiplication in homogeneous coordinates.
2. Create a MATLAB function that would obtain a workable particular solution for the system (4.16) with the coefficient matrix (4.17) given the four specified pairs of corresponding points.

3. Prepare a selection of images that have the "wrong perspective", that is, they may have been taken from the "wrong" point and you would like to digitally change that perspective point.

4. Write a MATLAB program that will do the following:

   (a) Ask the user to specify which image is to be transformed.

   (b) Ask the user to input the vertex coordinates (x_k, y_k) of the quadrilateral in the original image and the vertex coordinates (x'_k, y'_k) of the corresponding quadrilateral (rectangle) in the transformed image.

   (c) Apply the change-of-perspective transformation using the function you created in Problem 2. Follow up with interpolations of pixel values as necessary.

   (d) Display the transformed image.

   Test your program using several different grayscale and color images. Comment on the results.

5. Compile the results of the previous MATLAB exercises into a PowerPoint presentation.

Chapter 5

Convolution and Image Filtering

In this chapter, the reader will be exploring a variety of image-restoration transformations and edge-detection methods. In the process, the reader will discover that those seemingly disparate transformations have something in common mathematically – they are all effected by means of the operation of convolution. The middle part of the chapter is devoted to a discussion of the variants and basic properties of this new operation.

5.1 Image Blurring and Noise Reduction

In most situations, we prefer sharp images to blurry ones. Making an image look sharper is usually considered a desirable goal, and in Chapter 6, we will develop powerful image-sharpening techniques. Nevertheless, sometimes we choose to do the opposite: making an image appear somewhat blurry. Possible reasons for doing that are many, and they include, but are not limited to, the following:

- Protecting privacy

- Imparting a romantic ambiance to a landscape.
- Downplaying parts of an image in order to bring out the most important features
- Creating an illusion of a greater visual depth
- Imitation of the sfumato technique of allowing tones and colors to shade gradually into one another producing softened outlines or hazy forms
- Creating psychological effects that convey love, nostalgia, pain, and other emotions in art photography
- Imitating motion by applying motion blur to images of falling water
- Mitigating undesirable noise

Probably, the most famous example of the use of sfumato in Western art is Leonardo da Vinci's Mona Lisa (Figure 5.1). An example of background blurring with the purpose of putting the objects in the foreground into focus (a technique so often used in modern photography) is shown in Figure 5.2. In the exercises at the end of this section, the reader will be asked to find other examples of images that exhibit static or motion blur as integral and desirable features.

In this section, we will only be applying the blur to the entire image (as opposed to just selected parts of it). This is partly because methods of feature extraction are not discussed in sufficient detail within this book. However, the reader is encouraged to find images where the foreground and background can be easily separated and to experiment with a judicious application of various types of static and motion blur to selected parts of an image.

A somewhat naive and certainly the easiest way to blur an image is to replace each pixel with the average value of the pixels in its neighborhood. As an illustration of this approach, Figure 5.3 shows a grayscale version of a famous photo of Albert Einstein alongside its blurred version created by calculating the mean pixel value for the 3×3 neighborhood of each pixel. Formally, if we denote by A the $M \times N$ matrix of pixel values of the original image and by B – the matrix of its blurred version, then the entries of B are calculated as

$$B(m,n) = \frac{1}{9} \sum_{k=-1}^{1} \sum_{l=-1}^{1} A(m+k, m+l)$$

for all $1 < m < M$ and $1 < n < N$. Below is the MATLAB code used to generate Figure 5.3.

```
A=imread('einstein.tif');
A=double(A); B=A;
```

RE 5.1: Mona Lisa: an example of the use of sfumato.

FIGURE 5.2: An example of blurring the background.

```
);

-1
n,n)=(A(m-1,n-1)+A(m,n-1)+A(m+1,n-1)
+A(m-1,n)+A(m,n)+A(m+1,n)
+A(m-1,n+1)+A(m,n+1)+A(m+1,n+1))/9;



;
255]);
ginal Image','FontSize',18);
;
255]);
ed Image','FontSize',18);
```

with our usual policy in regards to MATLAB code examples, efficiency and style for the sake of transparency. The reader to improve upon this code in the exercises at the end of this probably worth pointing out the importance of converting the "double" by means of the command

$$>> \text{A=double(A);}$$

line of the code. Otherwise, the averages will not be calculated example, if we leave the image matrix A in the "uint8" class, ne pixels with the identical values of 128 will be 255 (since that value a "uint8"-class variable can take on) and, consequently, " will be calculated as $255/9 = 28$.

also point out our choice of starting the *for* loops at the index pposed to 1) and ending them at the index values $M - 1$ and M and N are the dimensions of the image A) in order to avoid oundary distortions.

cal, artistic, and atmospheric effects aside, probably the single reasons for applying a blur to an image is that doing so often tion of noise. In image processing, noise can be loosely defined andom variation of brightness or color, and it is often modeled ependent identically distributed random variables to the pixel ea behind blurring for the purpose of noise suppression is that a neighborhood of a pixel will have the effect of "canceling out" (provided that it has zero mean).

noise comes from many natural sources, the most common basic sed in signal and image processing is the so-called additive white e (AWGN), which is imitated by adding independent normal bles of zero mean and a specified standard deviation to the pixel

FIGURE 5.3: An example of blurring the image.

ler to create sample versions of images corrupted by AWGN, we use LAB function *normrnd* as in the code

```
rmrnd(0,sigma,[M,N]);
Noise;
```

to create a noisy version of the $M \times N$ image A (corrupted by the

› Gaussian noise with the standard deviation σ). In Figure 5.4(a), ı see the already familiar photo of Albert Einstein corrupted by f the standard deviation $\sigma = 30$. Figure 5.4(b) shows the effect 'ation by means of averaging over 3×3 neighborhoods of each examine it more closely.

nents come to mind when looking at Figure 5.4. The good news ethod of averaging over pixel neighborhoods appears reasonably moving most of the noise. The bad news is that noise mitigation cost of a significant loss of the image quality due to undesir- We can try to reduce this unwanted blurring by using *weighted* ›pposed to just arithmetic means) of pixel values. In that way, ›le to let the pixel at the center of its neighborhood make a bution to its "replacement" than the pixels on the periphery of ood.

5.4(c), we demonstrate the effect of replacing each pixel with the ge of its 3×3 neighborhood using the weights

$$\begin{array}{ccc} 2/25 & 3/25 & 2/25 \\ 3/25 & 5/25 & 3/25 \\ 2/25 & 3/25 & 2/25 \end{array} \qquad (5.1)$$

uniform weights

$$\begin{array}{ccc} 1/9 & 1/9 & 1/9 \\ 1/9 & 1/9 & 1/9 \\ 1/9 & 1/9 & 1/9. \end{array}$$

:amination of Figure 5.4(c) reveals that this "weighted blur" as effective in suppressing noise as simply calculating the arith- f pixel values, but its blurring effect is somewhat smoother and 5. Encouraged by this initial success with the use of weighted an feel confident in our ability to construct matrices of different ht distributions following the example of (5.1).

the reader might be wondering how we came up with the in the first place. To provide some guidance in the quest for r weights, we mention that there are several widely used fami- for the purpose of noise reduction. One of them is the Gaussian the weights are approximations of the Gaussian function of two

$$g(x, y) = \frac{1}{2\pi\sigma^2} e^{-\frac{x^2+y^2}{2\sigma^2}}, \qquad (5.2)$$

he integer values of x and y. The matrix (5.1), for example, can y choosing the value $\sigma = 1$, calculating the function g for the $x, y = -1, 0, 1$, and approximating the resulting values of g with fractions that add up to 1.

FIGURE 5.4: An example of noise mitigation.

We hope that the reader found the method of noise reduction in digital images described in this section straightforward and fairly uncomplicated. But there are still two areas of concern that need to be addressed.

1. First, there is the issue of terminology. Let us repeat the description of the noise-reduction method one more time in all of its annoying verboseness:

   > In order to reduce the level of noise in a digital image, *we replace each pixel with a weighted average of the pixels in its neighborhood using a specified matrix of weights.*

   That was quite a mouthful! We've got to be able to do better than that. There must be a way to phrase the same idea in a more concise language! We have arrived at a point where it seems painfully obvious that a new term us urgently needed. This new term is *convolution*. The above sentence can now be repeated in a form that is much more convenient and concise:

   > In order to reduce the level of noise in a digital image, we calculate its *convolution* with a specified matrix of weights.

   The matrix of weights used in this process is called a *filter*, and the action of a filter on an image by convolution is called *filtering*. With this terminology established, our subsequent discussion of image processing will become less convoluted – pun intended!

2. In order to be technically equipped to design new and better filters, to combine filters, to reverse the action of a filter, and to perform numerous other operations in image processing, it is necessary to undertake a more in-depth study of convolution. And that is what the next section will be devoted to.

Section 5.1 Exercises.

1. Find examples of images that exhibit static or motion blur as integral and desirable features.

2. Find several images that you believe could be enhanced by blurring due to acquiring a romantic ambiance or a new psychological effect.

3. Verify that the weights in (5.1) can be obtained by choosing $\sigma = 1$ in (5.2) and calculating $g(m, n)$ for the values $m, n = -1, 0, 1$ with subsequent rounding and simplification.

4. Use the Gaussian function (5.2) with different values of σ to design several 3×3 and 5×5 radially-symmetric filters similar to (5.1) for the purpose of mitigating noise in digital images.

5. Explain why it is important to make sure that the additive noise has zero mean if we want it to be "canceled out" by averaging over a pixel neighborhood.

MATLAB Exercises.

1. Select a few images to experiment with. Use the MATLAB function *normrnd* as shown in this section to create noisy versions of these images corrupted by additive white Gaussian noise of a specified standard deviation.

2. Write a MATLAB function that would blur the specified image by calculating its convolution with the specified filter. Your function should accept both the image and the filter as inputs. Make appropriate adjustments to avoid undesirable boundary effects, like a dark frame surrounding the image. Ideally, your function should work with any grayscale or color image.

3. Write a a MATLAB program that will do the following:

   (a) Use one or several of the filters designed in Exercise 4 of this section to reduce the noise in the noisy images that you found or created in MATLAB Exercise 1 by calling the MATLAB function you created in the previous problem.

   (b) Display the original images, the noisy versions, and the denoised images side-by-side. Comment on the degree of noise reduction.

4. Test your program using several different images (grayscale and color) and several different filters.

5. Compile the results of the previous exercises into a PowerPoint presentation.

5.2 Convolution: Definitions and Examples

At the end of the previous section, we defined the term *convolution* as a shorthand for describing the process of replacing each pixel by a weighted average of the pixels in its neighborhood. We could see for ourselves how the concept of convolution arose naturally in the context of image manipulations intended to create a blurring effect or to mitigate noise.

However, the role and usage of the operation of convolution is much broader. Convolution is ubiquitous in many areas of mathematics, science, and technology. It is of fundamental importance in such fields as algebra, number theory, functional analysis, probability theory, various areas of physics, signal, image, and data processing, acoustics, electrical engineering, analytical chemistry, and many others.

There is a multitude of different variants of convolutions: continuous and discrete, linear, circular and twisted, and so on. In this section, we will study the two types most closely associated with digital image processing: discrete linear and circular convolution of sequences and matrices.

5.2.1 Discrete Linear Convolution

We begin with the more straightforward variant - the discrete linear convolution of finite and infinite sequences. Here is our game plan: we begin with a formal definition, follow up with a few detailed examples, and proceed to establish general properties of this new and unfamiliar operation.

Definition 5.1 *Suppose that*

$$\boldsymbol{x} = (\cdots, x_{-2}, x_{-1}, x_0, x_1, x_2, x_3, \cdots)$$

and

$$\boldsymbol{y} = (\cdots, y_{-2}, y_{-1}, y_0, y_1, y_2, y_3, \cdots)$$

are two sequences (finite or infinite). Their ***linear convolution*** *is defined to be the sequence* $\boldsymbol{z} = \boldsymbol{x} * \boldsymbol{y}$ *given by*

$$z_n = \sum_k x_k y_{n-k} \tag{5.3}$$

with the sum calculated over all the values of the index k*, for which the product* $x_k y_{n-k}$ *is nonzero.*

Remark 1 on convergence: *Linear convolution is only defined for those pairs of sequences* $\boldsymbol{x}$ *and* $\boldsymbol{y}$ *for which the series in* (5.3) *converges (which, fortunately, includes all physically-realizable sequences). A discussion of general conditions that must be imposed on* $\boldsymbol{x}$ *and* $\boldsymbol{y}$ *in order to guarantee the series convergence is outside the scope of this text.*

Remark 2 on notation: *Throughout this text, we will be using* x_k *and* $x(k)$ *interchangeably. Both notations are equivalent and both refer to the* k*th element of the sequence* $\boldsymbol{x}$*. In any given situation, our choice of notation will be determined solely by considerations of typographical convenience.*

We recognize that Definition 5.1 might appear quite abstract and somewhat forbidding. Therefore, it is our first priority to remove the aura of inapproachability from it. We hope that a few concrete examples will be helpful in making the reader much more comfortable with this new construct.

Example 5.1 *As our very first example, we calculate the linear convolution* $\boldsymbol{z}$ *of the two finite sequences*

$$\boldsymbol{x} = (x_0, x_1, x_2, x_3) = (1, 2, 3, 4)$$

and

$$\boldsymbol{y} = (y_0, y_1, y_2) = (5, 6, 7)$$

of lengths 4 and 3 respectively. Applying the defining formula (5.3) *directly, we obtain*

$$\begin{aligned}
z_0 &= \sum_k x_k y_{-k} = x_0 y_0 = 1 \cdot 5 = 5; \\
z_1 &= \sum_k x_k y_{1-k} = x_0 y_1 + x_1 y_0 = 1 \cdot 6 + 2 \cdot 5 = 16; \\
z_2 &= \sum_k x_k y_{2-k} = x_0 y_2 + x_1 y_1 + x_2 y_0 \\
&= 1 \cdot 7 + 2 \cdot 6 + 3 \cdot 5 = 34; \\
z_3 &= \sum_k x_k y_{3-k} = x_1 y_2 + x_2 y_1 + x_3 y_0 \\
&= 2 \cdot 7 + 3 \cdot 6 + 4 \cdot 5 = 52; \\
z_4 &= \sum_k x_k y_{4-k} = x_2 y_2 + x_3 y_1 = 3 \cdot 7 + 4 \cdot 6 = 45; \\
z_5 &= \sum_k x_k y_{5-k} = x_3 y_2 = 4 \cdot 7 = 28.
\end{aligned}$$

This calculation can be checked with the help of the MATLAB function *conv* by executing the code

```
x=[1,2,3,4]; y=[5,6,7]; z=conv(x,y),
```

which produces the output

```
z =
  5 16 34 52 45 28
```

As with other arithmetic and algebraic operations, we would like to use graphs to demonstrate the convolution of two sequences. Figure 5.5 illustrates the result of the computation in Example 5.1 and contains the plots of both sequences and their linear convolution. The code used to generate Figure 5.5 is shown below.

```
x=[1,2,3,4]; y=[5,6,7];
z=conv(x,y);
subplot(1,3,1);
stem(1:4,x,'filled');
xlabel('k','FontSize',18);
ylabel('x_k','FontSize',18);
title('x=(1,2,3,4)','FontSize',18);
subplot(1,3,2);
stem(1:3,y,'filled');
xlabel('k','FontSize',18);
ylabel('y_k','FontSize',18);
title('y=(5,6,7)','FontSize',18);
subplot(1,3,3);
stem(1:6,z,'filled');
xlabel('k','FontSize',18);
ylabel('z_k','FontSize',18);
title('z=x*y','FontSize',18);
```

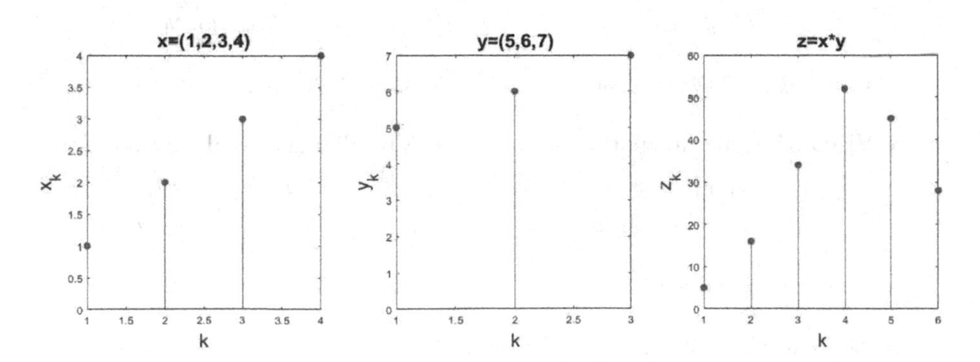

FIGURE 5.5: Plots of the two sequences and their linear convolution in Example 5.1.

We hope that the reader has found the calculations in Example 5.1 straightforward and easy to follow. Nevertheless, we do admit that using the formula (5.3) directly is a rather boring and tedious way to calculate convolutions. Another disadvantage of this method is that it lacks any intuitive insight or visual appeal. We, therefore, repeat the calculation using a different method (which we will dub the "flip-and-slide" method for reasons that will soon become apparent).

The "flip-and-slide" method consists of the following two steps:

1. Flip the sequence $\mathbf{y}$ using the position of y_0 as a pivot.
2. In order to calculate the value of $\mathbf{x} * \mathbf{y}$ at the point n, slide the flipped version of the sequence $\mathbf{y}$ along the sequence $\mathbf{x}$ precisely n positions to the right, so that y_0 is aligned with x_n. Calculate the "dot product" of the overlapping nonzero parts of the two sequences.

The diagrams below illustrate this method applied to the familiar sequences

$$\mathbf{x} = (x_0, x_1, x_2, x_3) = (1, 2, 3, 4) \quad \text{and} \quad \mathbf{y} = (y_0, y_1, y_2) = (5, 6, 7)$$

we have already worked with in Example 5.1.

- Visualizing the calculation of z_0:

			x_0	x_1	x_2	x_3		
	y_2	y_1	y_0					

$z_0 = x_0 y_0$.

- Visualizing the calculation of z_1:

			x_0	x_1	x_2	x_3			
		y_2	y_1	y_0					

$z_1 = x_0 y_1 + x_1 y_0$.

- Visualizing the calculation of z_2:

			x_0	x_1	x_2	x_3		
			y_2	y_1	y_0			

$z_2 = x_0 y_2 + x_1 y_1 + x_2 y_0$.

- Visualizing the calculation of z_3:

			x_0	x_1	x_2	x_3			
				y_2	y_1	y_0			

$z_3 = x_1 y_2 + x_2 y_1 + x_3 y_0$.

- Visualizing the calculation of z_4:

			x_0	x_1	x_2	x_3		
					y_2	y_1	y_0	

$z_4 = x_2 y_2 + x_3 y_1$.

- Visualizing the calculation of z_5:

			x_0	x_1	x_2	x_3			
						y_2	y_1	y_0	

$z_5 = x_3 y_2$.

In the exercises at the end of this section, the reader will be asked to practice calculating convolutions using both the defining formula (5.3) and the "flip-and-slide" method we have just described and illustrated.

We hope the reader has found the "flip-and-slide" method more visual, more intuitive, and easier to use. Nevertheless, it might seem very strange that convolution is defined in such a way as to require the "flipping" of the sequence $\mathbf{y}$. Would it not be easier to avoid that step altogether? After all, the notion of convolution was motivated by the task of blurring an image, and we never made any mention of flipping anything in that context!

This complaint is indeed completely justified. In fact, for all the applications discussed in this chapter, it would indeed have been sufficient to have a simplified definition of convolution (and some texts on digital image processing do make that choice). However, there are numerous good reasons why convolution is defined precisely the way it is. We will mention just two such reasons here:

1. The definition we adopted works perfectly in algebra in the context of multiplication of polynomials. Suppose that

$$p(x) = a_m x^m + a_{m-1} x^{m-1} + \cdots + a_1 x + a_0$$

   and

$$q(x) = b_n x^n + b_{n-1} x^{n-1} + \cdots + b_1 x + b_0$$

   are polynomials of degrees $m = M - 1$ and $n = N - 1$ respectively. Suppose also that

$$r(x) = p(x) \cdot q(x) = c_{m+n} x^{m+n} + c_{m+n-1} x^{m+n-1} + \cdots + c_1 x + c_0$$

   is their product. It turns out that the sequence $\mathbf{c}$ of the coefficients of the product $r(x)$ is always the convolution of the sequences $\mathbf{a}$ and $\mathbf{b}$ of the coefficients of the factors $p(x)$ and $q(x)$. For example, a direct calculation shows that the product of the two polynomials

$$p(x) = x^3 + 2x^2 + 3x + 4 \text{ and } q(x) = 5x^2 + 6x + 7$$

   equals

$$q(x) = 5x^5 + 16x^4 + 34x^3 + 52x^2 + 45x + 28,$$

   just as predicted by the result of Exercise 5.1.

2. As we will see in Chapter 6, frequency content of periodic signals is measured by the so-called Fourier coefficients. It turns out that the sequence of Fourier coefficients of a product of two periodic signals is the convolution of the Fourier coefficients of the individual signals being multiplied. Moreover, it will turn out that the frequency content of the convolution of two sequences is a product of frequency contents of individual sequences. Those are some of the reasons convolution plays a central role in frequency analysis of signals (and images).

Truth be told, when it comes to the type of applications we are discussing in this chapter, flipping or not flipping the second sequence rarely makes any significant difference. Nevertheless, we do believe it is important to gain experience working with the commonly accepted standard definition of convolution, because it has such a major presence in so many different fields.

5.2.2 Circular Convolution

Another variant of discrete convolution called *circular convolution* only applies to finite or periodic sequences, hence it is sometimes also called *periodic* convolution. It has strong similarities with its cousin, the linear convolution, and indeed the two are very closely related while remaining fundamentally different.

Definition 5.2 *Suppose that* $\boldsymbol{x} = (x_0, x_1, \cdots, x_{N-1})$ *and* $\boldsymbol{y} = (y_0, y_1, \cdots, y_{N-1})$ *are two finite sequences of the same length* N*. Their* circular convolution *is defined to be the finite sequence* $\boldsymbol{z} = \boldsymbol{x} * \boldsymbol{y}$ *of length* N *given by*

$$z_n = \sum_{k=0}^{N-1} x_k y_{n-k \ mod \ N}. \tag{5.4}$$

In order to better appreciate the similarities and differences between the two types of discrete convolution, our next example is intended to be almost identical to Example 5.1. The calculations are also very similar (both in concept and in detail), and we ask the reader to stay tuned to all the subtleties in using the defining formulas (5.3) and (5.4).

Example 5.2 *To illustrate Definition 5.2, we calculate the circular convolution* $\boldsymbol{z}$ *of the two finite sequences*

$$\boldsymbol{x} = (x_0, x_1, x_2, x_3) = (1, 2, 3, 4)$$

and

$$\boldsymbol{y} = (y_0, y_1, y_2, y_3) = (5, 6, 7, 8),$$

both of the same length $N = 4$*. Applying the defining formula (5.4) directly, we obtain*

$$\begin{aligned}
z_0 &= \sum_{k=0}^{3} x_k y_{-k \ mod \ 4} = x_0y_0 + x_1y_3 + x_2y_2 + x_3y_1 \\
&= 1 \cdot 5 + 2 \cdot 8 + 3 \cdot 7 + 4 \cdot 6 = 65, \\
z_1 &= \sum_{k=0}^{3} x_k y_{1-k \ mod \ 4} = x_0y_1 + x_1y_0 + x_2y_3 + x_3y_2 \\
&= 1 \cdot 6 + 2 \cdot 5 + 3 \cdot 8 + 4 \cdot 7 = 68, \\
z_2 &= \sum_{k=0}^{3} x_k y_{2-k \ mod \ 4} = x_0y_2 + x_1y_1 + x_2y_0 + x_3y_3 \\
&= 1 \cdot 7 + 2 \cdot 6 + 3 \cdot 5 + 4 \cdot 8 = 66, \ \ \textit{and}
\end{aligned}$$

$$\begin{aligned} z_3 &= \sum_{k=0}^{3} x_k y_{3-k \ mod\ 4} = x_0 y_3 + x_1 y_2 + x_2 y_1 + x_3 y_0 \\ &= 1 \cdot 8 + 2 \cdot 7 + 3 \cdot 6 + 4 \cdot 5 = 60. \end{aligned}$$

This calculation can also be checked with the help of the MATLAB function *cconv* (included in the Signal Processing Toolbox) by executing the code

```
x=[1,2,3,4]; y=[5,6,7,8]; z=cconv(x,y,4),
```

which produces the output

```
z =
    65 68 66 60
```

Following Example 5.2, we would like to emphasize three important differences between the linear and the circular convolution:

1. Unlike the formula
$$z_n = \sum_k x_k y_{n-k}$$
for the linear convolution, where the summation is taken over those values of the index k, for which the product $x_k y_{n-k}$ is nonzero and can thus have different numbers of terms for different n, the formula
$$z_n = \sum_{k=0}^{N-1} x_k y_{n-k \ \mathrm{mod}\ N}$$
for the circular convolution of two N-point sequences has the same number N of summation terms for all n.

2. When calculating the circular convolution of $\mathbf{x}$ and $\mathbf{y}$, once the sequence $\mathbf{y}$ "runs out of terms", it "wraps around" to the beginning, hence the term "circular". Linear convolutions do not have this "wrapping around" effect.

3. Whereas linear convolution is defined for sequences of any lengths (including infinite or doubly-infinite sequences), circular convolution is only defined for **finite** sequences of the **same** length.

The last point seems to be unequivocally prohibiting circular convolution of sequences of different lengths. But what if that is exactly what we need to calculate? There is a way to get around the prohibition, but it involves a trick called "zero-padding", that is, extending the shorter sequence to the required length by adding to it the necessary number of zero terms. For example, in order to calculate the circular convolution of $\mathbf{x} = (x_0, x_1, x_2, x_3) = (1, 2, 3, 4)$ and $\mathbf{y} = (y_0, y_1, y_2) = (5, 6, 7)$, we zero-pad $\mathbf{y}$ to

$$\mathbf{y}_{pad} = (y_0, y_1, y_2, y_3) = (5, 6, 7, 0)$$

and proceed to calculate the circular convolution of the sequences $\mathbf{x}$ and $\mathbf{y}_{pad}$ of the same length $N = 4$.

In fact, even though linear and circular convolutions are fundamentally different operations, they have more in common than meets the eye. Under certain conditions, they even turn out to be equivalent. For instance, it is not very difficult to show that the **linear** convolution of two sequences $\mathbf{x}$ and $\mathbf{y}$ of lengths M and N respectively equals the **circular** convolution of their extended versions $\mathbf{x}_{pad}$ and $\mathbf{y}_{pad}$, both zero-padded to the length $M + N - 1$.

In order to demonstrate the "flip-and-slide" method for circular convolution, we first need to introduce the concept of periodic extension.

Definition 5.3 *Given a finite sequence $\boldsymbol{x} = (x_0, x_1, \cdots, x_{N-1})$, its periodic extension $\boldsymbol{x}_p$ is defined by*

$$x_p(k) = x_{k \bmod N}$$

for all $k \in \mathbb{Z}$. Essentially, $\boldsymbol{x}_p$ is just a string of infinitely many copies of $\boldsymbol{x}$.

For example, the periodic extension of $\mathbf{x} = (x_0, x_1, x_2, x_3)$ is

$$\cdots, x_0, x_1, x_2, x_3, x_0, x_1, x_2, x_3, x_0, x_1, x_2, x_3, x_0, x_1, x_2, x_3, x_0, x_1, x_2, x_3, \cdots$$

With the help of this new concept, we can present the circular convolution of two N-point sequences $\mathbf{x}$ and $\mathbf{y}$ as the linear convolution of their periodic extensions $\mathbf{x}_p$ and $\mathbf{y}_p$ with the summation taken over just one period. Because of this, as we have already mentioned, circular convolution is sometimes also called *periodic* convolution. Formally, we can write

$$(\mathbf{x} * \mathbf{y})(n) = \sum_{k=0}^{N-1} \mathbf{x}_p(k)\mathbf{y}_p(n-k),$$

and this formula can be easily illustrated using the "flip-and-slide" method. We do so using the already familiar sequences

$$\mathbf{x} = (x_0, x_1, x_2, x_3) = (1, 2, 3, 4) \text{ and } \mathbf{y} = (y_0, y_1, y_2, y_3) = (5, 6, 7, 8)$$

that we saw in Example 5.2. For the convenience of the reader, we shade one period over which summation is taken.

- Visualizing the calculation of z_0:

$\cdots$	x_0	x_1	x_2	x_3	x_0	x_1	x_2	x_3	x_0	x_1	x_2	x_3	x_0	x_1	x_2	x_3	x_0	x_1	x_2	x_3	$\cdots$
$\cdots$	y_0	y_3	y_2	y_1	y_0	y_3	y_2	y_1	y_0	y_3	y_2	y_1	y_0	y_3	y_2	y_1	y_0	y_3	y_2	y_1	$\cdots$

$$z_0 = x_0 y_2 + x_1 y_3 + x_2 y_2 + x_3 y_1 = 1 \cdot 5 + 2 \cdot 8 + 3 \cdot 7 + 4 \cdot 6 = 65.$$

- Visualizing the calculation of z_1:

⋯	x_0	x_1	x_2	x_3	x_0	x_1	x_2	x_3	x_0	x_1	x_2	x_3	x_0	x_1	x_2	x_3	x_0	x_1	x_2	x_3	⋯
⋯	y_3	y_1	y_0	y_2	y_1	y_0	y_3	y_2	y_1	y_0	y_3	y_2	y_1	y_0	y_3	y_2	y_1	y_0	y_3	y_2	⋯

$z_1 = x_0y_1 + x_1y_0 + x_2y_3 + x_3y_2 = 1 \cdot 6 + 2 \cdot 5 + 3 \cdot 8 + 4 \cdot 7 = 68.$

- Visualizing the calculation of z_2:

⋯	x_0	x_1	x_2	x_3	x_0	x_1	x_2	x_3	x_0	x_1	x_2	x_3	x_0	x_1	x_2	x_3	x_0	x_1	x_2	x_3	⋯
⋯	y_2	y_1	y_0	y_3	y_2	y_1	y_0	y_3	y_2	y_1	y_0	y_3	y_2	y_1	y_0	y_3	y_2	y_1	y_0	y_3	⋯

$z_2 = x_0y_2 + x_1y_1 + x_2y_0 + x_3y_3 = 1 \cdot 7 + 2 \cdot 6 + 3 \cdot 5 + 4 \cdot 8 = 66.$

- Visualizing the calculation of z_3:

⋯	x_0	x_1	x_2	x_3	x_0	x_1	x_2	x_3	x_0	x_1	x_2	x_3	x_0	x_1	x_2	x_3	x_0	x_1	x_2	x_3	⋯
⋯	y_3	y_2	y_1	y_0	y_3	y_2	y_1	y_0	y_3	y_2	y_1	y_0	y_3	y_2	y_1	y_0	y_3	y_2	y_1	y_0	⋯

$z_3 = x_0y_3 + x_1y_2 + x_2y_1 + x_3y_0 = 1 \cdot 8 + 2 \cdot 7 + 3 \cdot 6 + 4 \cdot 5 = 60.$

5.2.3 Algebraic Properties of Convolution

In the last two subsections, we have learned to calculate linear and circular convolutions using both the definition and the more visual "flip-and-slide" method. Nevertheless, it is possible that some of the readers still feel that convolution is a relatively unfamiliar operation, with which they have very little experience. Therefore, it might prove helpful to investigate some of its basic properties and to establish whatever similarities we can find between convolution and other arithmetic operations, particularly multiplication. For instance, both linear and circular convolution are similar to multiplication because they are

- commutative in the sense that

  $$\mathbf{x} * \mathbf{y} = \mathbf{y} * \mathbf{x}$$

  for all sequences $\mathbf{x}$ and $\mathbf{y}$ for which convolution is defined.

- associative in the sense that

  $$(\mathbf{x} * \mathbf{y}) * \mathbf{z} = \mathbf{x} * (\mathbf{y} * \mathbf{z})$$

  for all sequences $\mathbf{x}$, $\mathbf{y}$, and $\mathbf{z}$ for which convolution is defined.

- distributive with respect to addition in the sense that

  $$(\mathbf{x} + \mathbf{y}) * \mathbf{z} = \mathbf{x} * \mathbf{z} + \mathbf{y} * \mathbf{z}$$

  for all sequences $\mathbf{x}$, $\mathbf{y}$, and $\mathbf{z}$ for which both operations of addition and convolution are defined.

It is fairly straightforward (although somewhat tedious) to verify those three properties, and we leave it to exercise. How about other algebraic properties, such as the existence of identity and inverses? When it comes to multiplication, we all know that there is a very special number 1 with the property that

$$a \cdot 1 = 1 \cdot a = a$$

for all real (or complex) numbers a. We call the number 1 the *multiplicative identity*. We would certainly be justified in asking whether there exists a convolutional identity, and the answer is in the affirmative. Consider the sequence $\boldsymbol{\delta}$ called the *unit impulse* and defined by

$$\delta(k) = \begin{cases} 1, & k = 0 \\ 0, & \text{otherwise.} \end{cases} \tag{5.5}$$

It is easy to see (particularly with the help of the "flip-and-slide" method) that under both the linear and the circular convolution,

$$\mathbf{x} * \boldsymbol{\delta} = \boldsymbol{\delta} * \mathbf{x} = \mathbf{x}$$

for any sequence $\mathbf{x}$.

Having lived with multiplication since elementary school, we take it for granted that every nonzero real or complex number a has a multiplicative inverse a^{-1} with the property that

$$a \cdot a^{-1} = a^{-1} \cdot a = 1.$$

Would the same property hold for the convolution? In other words, given a nonzero sequence $\mathbf{x}$, would there always exist another sequence denoted by $\mathbf{x}^{-1}$ so that

$$\mathbf{x} * \mathbf{x}^{-1} = \mathbf{x}^{-1} * \mathbf{x} = \boldsymbol{\delta},$$

or is it not always the case?

The truth is that in general, sequences do not have convolutional inverses, although some do. The only way to determine whether a given sequence is invertible and the only way to produce an easy formula for its convolutional inverse is to use methods of frequency analysis, which we will discuss in some detail in Chapter 6.

5.2.4 Convolution as a Linear Transformation

Let us suppose that we are given the sequence $\mathbf{h} = (h_0, h_1, \ldots, h_m)$. We can recast it in its column-vector form $\mathbf{h} = \begin{bmatrix} h_0 & h_1 & \cdots & h_m \end{bmatrix}^T$ and use linear convolution to define the mapping $T_h : \mathbb{R}^n \to \mathbb{R}^{n+m-1}$ by

$$T_h(\mathbf{x}) = \mathbf{x} * \mathbf{h}. \tag{5.6}$$

Alternatively, we can also use circular convolution and the same expression (5.6) to define the mapping T_h. In either case, the distributive property

$$(\mathbf{x}+\mathbf{y}) * \mathbf{h} = \mathbf{x} * \mathbf{h} + \mathbf{y} * \mathbf{h}$$

mentioned in the previous subsection implies that the mapping T is a linear transformation. Consequently, according to what we learned in Section 4.2, it has the form

$$T_h(\mathbf{x}) = H\mathbf{x},$$

where H is the standard matrix of T given by

$$H = \begin{bmatrix} T_h(\mathbf{e}_1) & T_h(\mathbf{e}_2) & \cdots & T_h(\mathbf{e}_n) \end{bmatrix},$$

with $\mathbf{e}_1$, $\mathbf{e}_2$, ..., and $\mathbf{e}_n$ being the basic vectors

$$\mathbf{e}_1 = \begin{bmatrix} 1 \\ 0 \\ \vdots \\ 0 \end{bmatrix}, \quad \mathbf{e}_2 = \begin{bmatrix} 0 \\ 1 \\ \vdots \\ 0 \end{bmatrix}, \quad \ldots, \quad \mathbf{e}_n = \begin{bmatrix} 0 \\ 0 \\ \vdots \\ 1 \end{bmatrix}.$$

The matrix H constructed in this manner is called *the convolution matrix* of $\mathbf{h}$. Let us see how all of this plays out in the following example:

Example 5.3 *Let $\boldsymbol{h} = (h_0, h_1) = (1, -1)$ and define $T_h : \mathbb{R}^4 \to \mathbb{R}^5$ by **linear** convolution as in (5.6). Then in the column-vector form,*

$$T_h(\boldsymbol{e}_1) = \begin{bmatrix} 1 \\ -1 \\ 0 \\ 0 \\ 0 \end{bmatrix}, \quad T_h(\boldsymbol{e}_2) = \begin{bmatrix} 0 \\ 1 \\ -1 \\ 0 \\ 0 \end{bmatrix}, \quad T_h(\boldsymbol{e}_3) = \begin{bmatrix} 0 \\ 0 \\ 1 \\ -1 \\ 0 \end{bmatrix}, \quad T_h(\boldsymbol{e}_4) = \begin{bmatrix} 0 \\ 0 \\ 0 \\ 1 \\ -1 \end{bmatrix}$$

and, consequently, the linear-convolution matrix H of $\boldsymbol{h}$ is

$$H = \begin{bmatrix} 1 & 0 & 0 & 0 \\ -1 & 1 & 0 & 0 \\ 0 & -1 & 1 & 0 \\ 0 & 0 & -1 & 1 \\ 0 & 0 & 0 & -1 \end{bmatrix},$$

as the reader can easily verify.

*Alternatively, if $\boldsymbol{h} = (h_0, h_1, h_2, h_3) = (1, -1, 0, 0)$ and we define $T_h : \mathbb{R}^4 \to \mathbb{R}^4$ as in (5.6) by **circular** convolution, then*

$$T_h(\boldsymbol{e}_1) = \begin{bmatrix} 1 \\ -1 \\ 0 \\ 0 \end{bmatrix}, \quad T_h(\boldsymbol{e}_2) = \begin{bmatrix} 0 \\ 1 \\ -1 \\ 0 \end{bmatrix}, \quad T_h(\boldsymbol{e}_3) = \begin{bmatrix} 0 \\ 0 \\ 1 \\ -1 \end{bmatrix}, \quad T_h(\boldsymbol{e}_4) = \begin{bmatrix} -1 \\ 0 \\ 0 \\ 1 \end{bmatrix}$$

and, consequently, the circular-convolution matrix H of $\mathbf{h}$ is

$$H = \begin{bmatrix} 1 & 0 & 0 & -1 \\ -1 & 1 & 0 & 0 \\ 0 & -1 & 1 & 0 \\ 0 & 0 & -1 & 1 \end{bmatrix}.$$

In the exercises at the end of this section, the reader will be asked to generalize this example. We will be making extensive use of convolution matrices of filters in Chapter 7.

5.2.5 Convolution in Two Dimensions

It was precisely the technique of filtering digital images (represented by two-dimensional matrices) that motivated our introduction of the term *convolution* all the way back at the end of Section 5.1. We have come a long way, and having studied both linear and circular convolutions of sequences, we are now prepared to state an official definition of the convolution of two matrices. In doing so, we will get closer to the main theme of this text – operations on digital images.

Definition 5.4 *Suppose that A and B are two (possibly infinite) matrices. Their* ***linear convolution*** *(sometimes called the convolutional product) is defined to be the matrix $C = A * B$ given by*

$$C(m,n) = \sum_k \sum_l A(k,l)B(m-k, n-l) \tag{5.7}$$

with the sum calculated over all the values of the indices k and l for which the product $A(k,l)B(m-k,n-l)$ is nonzero.

Remark 1 on convergence: *Just like linear convolution of sequences, linear convolution of matrices is only defined for those pairs of matrices A and B for which the series in* (5.7) *converges (which, fortunately, includes all physically realizable matrices). A discussion of general conditions that must be imposed on A and B in order to guarantee the series convergence is outside the scope of this text.*

Remark 2 on notation: *Throughout this text, we will be using a_{mn}, $a_{m,n}$ and $A(m,n)$ interchangeably. All three notations are equivalent, and they all refer to the element of the matrix A located at the intersection of the mth row and nth column. In any given situation, our choice of notation will be determined solely by considerations of typographical convenience.*

Remark 3 on indices: *Although it is customary to start the count of matrix rows and columns at one, the convolution formula* (5.7) *is set up to begin the count at zero. In applications, one usually uses a shift or zero-padding to reconcile the difference.*

To illustrate Definition 5.4, we calculate the linear convolution C of a 3×3 matrix

$$A = \begin{bmatrix} a_{11} & a_{12} & a_{13} \\ a_{21} & a_{22} & a_{23} \\ a_{31} & a_{32} & a_{33} \end{bmatrix} = \begin{bmatrix} 1 & 2 & 3 \\ 4 & 5 & 6 \\ 7 & 8 & 9 \end{bmatrix}$$

with a 2×2 matrix

$$B = \begin{bmatrix} b_{11} & b_{12} \\ b_{21} & b_{22} \end{bmatrix} = \begin{bmatrix} 1 & -1 \\ -2 & 3 \end{bmatrix}.$$

Applying the defining formula (5.7) directly, we obtain several selected entries of C as follows:

$$\begin{aligned}
c_{m1} &= c_{1n} = 0 \quad \text{for all} \quad m, n; \\
c_{mn} &= 0 \quad \text{whenever} \quad m > 5 \quad \text{or} \quad n > 5; \\
c_{22} &= \sum_k \sum_l a_{kl} b_{2-k,2-l} = a_{11}b_{11} = 1 \cdot 1 = 1; \\
c_{23} &= \sum_k \sum_l a_{kl} b_{2-k,3-l} = a_{11}b_{12} + a_{12}b_{11} \\
&= 1 \cdot (-1) + 2 \cdot 1 = 1; \\
c_{24} &= \sum_k \sum_l a_{kl} b_{2-k,4-l} = a_{12}b_{12} + a_{13}b_{11} \\
&= 2 \cdot (-1) + 3 \cdot 1 = 1; \\
c_{25} &= \sum_k \sum_l a_{kl} b_{2-k,5-l} = a_{13}b_{12} = 3 \cdot (-1) = -3; \\
\cdots \\
c_{33} &= \sum_k \sum_l a_{kl} b_{3-k,3-l} = a_{11}b_{22} + a_{12}b_{21} + a_{21}b_{12} + a_{22}b_{11} \\
&= 1 \cdot 3 + 2 \cdot (-2) + 4 \cdot (-1) + 5 \cdot 1 = 0; \\
\cdots \\
c_{55} &= \sum_k \sum_l a_{kl} b_{5-k,5-l} = a_{33}b_{22} = 3 \cdot 9 = 27.
\end{aligned}$$

We are confident that the readers can complete the computation on their own and obtain the answer

$$\begin{bmatrix} 1 & 2 & 3 \\ 4 & 5 & 6 \\ 7 & 8 & 9 \end{bmatrix} * \begin{bmatrix} 1 & -1 \\ -2 & 3 \end{bmatrix} = \begin{bmatrix} 1 & 1 & 1 & -3 \\ 2 & 0 & 1 & 3 \\ -1 & 3 & 4 & 9 \\ -14 & 5 & 6 & 27 \end{bmatrix},$$

with the first row and the first columns of zeros removed as mere annoying artifacts resulting from the indexing conventions. This result can be checked using the MATLAB function *conv2* by executing the code

```
A=[1,2,3;4,5,6;7,8,9];
B=[1,-1;-2,3];
C=conv2(A,B),
```

which produces the output

```
C =
     1     1     1    -3
     2     0     1     3
    -1     3     4     9
   -14     5     6    27
```

We can make an observation that the dimensions of A, B, and C are related in a way similar to that for finite sequences. It would appear that the number of rows of the convolutional product of two matrices equals the sum of the numbers of rows of individual matrices minus 1, with an identical relationship for the columns. In the exercises following this section, the reader will be asked to verify this assertion.

The definition of circular convolution for matrices is also completely analogous to its counterpart for sequences:

Definition 5.5 *Suppose that A and B are two $M \times N$ matrices. Their* ***circular convolution*** *is defined to be the $M \times N$ matrix $C = A * B$ given by*

$$C(m,n) = \sum_{k=1}^{M} \sum_{l=1}^{N} A(k,l)B(m-k \ mod \ M, \quad n-l \ mod \ N). \tag{5.8}$$

Remark on indices: *Because it is customary to begin counting the rows and columns of a matrix at 1 (as opposed to zero), the summation in (5.5) also begins at 1. This does not affect the final result because the only thing that is important is that the summation is performed over the entire range of indices.*

To illustrate this definition, we calculate the circular convolution C of the matrices

$$A = \begin{bmatrix} a_{11} & a_{12} \\ a_{21} & a_{22} \end{bmatrix} = \begin{bmatrix} 1 & 2 \\ 3 & -4 \end{bmatrix}$$

and

$$B = \begin{bmatrix} b_{11} & b_{12} \\ b_{21} & b_{22} \end{bmatrix} = \begin{bmatrix} 1 & -1 \\ 2 & 3 \end{bmatrix}.$$

Applying the defining formula (5.8) directly, we get:

$$\begin{aligned}
c_{11} &= \sum_{k=1}^{2}\sum_{l=1}^{2} A(k,l)B(1-k \bmod 2, \quad 1-l \bmod 2) \\
&= a_{11}b_{22} + a_{12}b_{21} + a_{21}b_{12} + a_{22}b_{11} \\
&= 1\cdot 3 + 2\cdot 2 + 3\cdot(-1) + (-4)\cdot 1 = 0; \\
c_{12} &= \sum_{k=1}^{2}\sum_{l=1}^{2} A(k,l)B(1-k \bmod 2, \quad 2-l \bmod 2) \\
&= a_{11}b_{21} + a_{12}b_{22} + a_{21}b_{11} + a_{22}b_{12} \\
&= 1\cdot 2 + 2\cdot 3 + 3\cdot 1 + (-4)\cdot(-1) = 15; \\
c_{21} &= \sum_{k=1}^{2}\sum_{l=1}^{2} A(k,l)B(2-k \bmod 2, \quad 1-l \bmod 2) \\
&= a_{11}b_{12} + a_{12}b_{11} + a_{21}b_{22} + a_{22}b_{21} \\
&= 1\cdot(-1) + 2\cdot 1 + 3\cdot 3 + (-4)\cdot 2 = 2; \\
c_{22} &= \sum_{k=1}^{2}\sum_{l=1}^{2} A(k,l)B(2-k \bmod 2, \quad 2-l \bmod 2) \\
&= a_{11}b_{11} + a_{12}b_{12} + a_{21}b_{21} + a_{22}b_{22} \\
&= 1\cdot 1 + 2\cdot(-1) + 3\cdot 2 + (-4)\cdot 3 = -7,
\end{aligned}$$

so the answer is

$$\begin{bmatrix} 1 & 2 \\ 3 & -4 \end{bmatrix} * \begin{bmatrix} 1 & -1 \\ 2 & 3 \end{bmatrix} = \begin{bmatrix} 0 & 15 \\ 2 & -7 \end{bmatrix}.$$

In practice, it is the linear convolution of matrices that is used most of the time. MATLAB does not even seem to have an easily accessible function for calculating the two-dimensional circular convolution, and one usually obtains it in a roundabout way by creating a periodic extension for one of the matrices.

We leave it to the reader to develop an analogue to the "flip-and-slide" method of calculating convolutional products of matrices. Probably, the easiest way is to write the flipped version of the matrix B on a semi-transparent tape and slide it along the matrix A calculating the dot products or the overlap.

The algebraic properties of two-dimensional convolution are identical to those we have established for the convolution of sequences. The 2-dimensional convolutional identity is provided by the matrix $\boldsymbol{\delta}$ defined by

$$\delta(m,n) = \begin{cases} 1, & m = n = 0 \\ 0, & \text{otherwise.} \end{cases}$$

We leave it to exercises to verify that

$$A * \boldsymbol{\delta} = \boldsymbol{\delta} * A = A$$

for all matrices A.

The important takeaway from this section is that the operation of two-dimensional convolution is doing exactly what we originally expected it to do when we first introduced the term "convolution" in the closing paragraph of Section 5.1: it replaces each entry of the matrix A with a linear combination of the values of A with the coefficients of the linear combination coming from the matrix B.

Section 5.2 Exercises.

1. Calculate both the linear and the circular convolution of the following pairs of finite sequences. If one of the sequences is shorter than the other one, extend it as necessary by means of zero-padding. Unless indicated otherwise, the sequence indices begin at 0.

   (a) $\mathbf{x} = \{1, 2, 1, -1\}$ and $\mathbf{y} = \{1, 2, 3, 1\}$;

   (b) $\mathbf{x} = \{1, 2, 4\}$ and $\mathbf{y} = \{1, 1, 1, 1\}$;

   (c) $\mathbf{x} = \{1, 2, 4\}$ and $\mathbf{y} = \{0, 0, 1, 1, 1\}$;

   (d) $\mathbf{x} = \{1, 2, 4\}$ and $\mathbf{y} = \mathbf{x} = \{1, 2, 4\}$;

   (e) $\mathbf{x} = \{1, 2, 4\}$ and $\mathbf{y} = \{y_{-2}, y_{-1}, y_0, y_1, y_2\} = \{1, 1, 0, 1, 1\}$;

   (f) A finite geometric sequence

   $$x(k) = \begin{cases} a^k, & 0 \leq k < N \\ 0, & \text{otherwise,} \end{cases}$$

   and the finite unit step sequence

   $$u(k) = \begin{cases} 1, & 0 \leq k < M \\ 0, & \text{otherwise.} \end{cases}$$

   Plot graphs of both sequences and their convolution for $M = N = 10$, $a = \frac{1}{2}$, $a = -\frac{1}{2}$, $a = 2$, and $a = -2$. Experiment with several other combinations of values of M, N, and a.

   (g) Two finite geometric sequences

   $$x(k) = \begin{cases} a^k, & 0 \leq k < M \\ 0, & \text{otherwise,} \end{cases} \quad \text{and} \quad y(k) = \begin{cases} b^k, & 0 \leq k < N \\ 0, & \text{otherwise.} \end{cases}$$

   Plot graphs of both sequences and their convolution for $M = N = 10$, $a = \frac{1}{2}$ and $b = \frac{1}{4}$. Experiment with several other combinations of values of M, N, a, and b.

Check your answers with the help of MATLAB functions *conv* and *cconv.*

2. Calculate the linear convolution of the infinite geometric sequence

$$x(k) = \begin{cases} a^k, & k \geq 0 \\ 0, & \text{otherwise} \end{cases}$$

with $|a| < 1$ and the infinite unit step sequence

$$u(k) = \begin{cases} 1, & k \geq 0 \\ 0, & \text{otherwise.} \end{cases}$$

Plot graphs of both sequences and their convolution for $a = \frac{1}{2}$, $a = -\frac{1}{2}$, and $a = \frac{1}{4}$. Experiment with several other values of a.

3. Calculate the linear convolution of two infinite geometric sequences

$$x(k) = \begin{cases} a^k, & k \geq 0 \\ 0, & \text{otherwise} \end{cases} \quad \text{and} \quad y(k) = \begin{cases} b^k, & k \geq 0 \\ 0, & \text{otherwise} \end{cases}$$

with $|a| < 1$ and $|b| < 1$. Plot graphs of both sequences and their convolution for $a = \frac{1}{2}$ and $b = \frac{1}{4}$. Experiment with several other combinations of values of a and b.

4. Construct the convolution matrix for the mapping T_h defined by the formula

$$T_h(\mathbf{x}) = \mathbf{x} * \mathbf{h}$$

for the following filters $\mathbf{h}$. Do the problem twice – for the linear convolution and for the circular convolution.

(a) $\mathbf{h} = (h_0, h_1) = (1, 1)$,
(b) $\mathbf{h} = (h_0, h_1) = (-1, 1)$,
(c) $\mathbf{h} = (h_0, h_1, h_2) = (1, 2, 3)$,
(d) $\mathbf{h} = (h_{-1}, h_0, h_1) = (1, 2, 3)$,
(e) $\mathbf{h} = (h_0, h_1, h_2) = (3, 2, 1)$,
(f) $\mathbf{h} = (h_{-1}, h_0, h_1) = (3, 2, 1)$,
(g) $\mathbf{h} = (h_0, h_1, h_2) = (1, 2, 1)$,
(h) $\mathbf{h} = (h_{-1}, h_0, h_1) = (1, 2, 1)$.

5. Given a sequence $\mathbf{h} = (h_0, h_1, \ldots, h_m)$, define the mapping T_h acting on vectors in $\mathbf{x} \in \mathbb{R}^n$ by $T_h(\mathbf{x}) = \mathbf{x} * \mathbf{h}$ using

(a) linear convolution,
(b) circular convolution,

Determine the co-domain of T_h and derive the general formulas for its convolution matrix.

6. Calculate the linear and circular convolution of the following pair of matrices:

(a) $\begin{bmatrix} 3 & 2 \\ 8 & 5 \end{bmatrix}$ and $\begin{bmatrix} 2 & 4 \\ -4 & -6 \end{bmatrix}$;

(b) $\begin{bmatrix} 3 & 2 \\ 8 & -12 \end{bmatrix}$ and $\begin{bmatrix} 2 & 4 \\ -4 & 8 \end{bmatrix}$;

(c) $\begin{bmatrix} 2 & -3 & 4 \\ 1 & 0 & -2 \\ -3 & 1 & 4 \end{bmatrix}$ and $\begin{bmatrix} 3 & 2 \\ 8 & 5 \end{bmatrix}$;

(d) $\begin{bmatrix} 1 & -2 & 1 \\ 0 & 1 & 2 \\ -5 & -2 & 1 \end{bmatrix}$ and $\begin{bmatrix} 2 & 4 \\ -4 & -6 \end{bmatrix}$;

(e) $\begin{bmatrix} 0 & 3 & -3 \\ -1 & 2 & 2 \\ 4 & 1 & 1 \end{bmatrix}$ and $\begin{bmatrix} 3 & 2 \\ 8 & -12 \end{bmatrix}$;

(f) $\begin{bmatrix} 1 & 0 & 1 \\ 1 & 0 & 2 \\ -5 & 1 & 1 \end{bmatrix}$ and $\begin{bmatrix} 2 & 4 \\ -4 & 8 \end{bmatrix}$.

Check your answers using the MATLAB function *conv*. Plot both sequences and their convolution using the MATLAB function *stem* as shown in Example 5.1 of Subsection 5.2.1.

7. Show that both linear and circular convolutions are commutative, which means that

$$\mathbf{x} * \mathbf{y} = \mathbf{y} * \mathbf{x}$$

for all sequences $\mathbf{x}$ and $\mathbf{y}$ for which convolution is defined.

8. Show that both linear and circular convolutions enjoy the distributive property, which means that

$$(\mathbf{x} + \mathbf{y}) * \mathbf{z} = \mathbf{x} * \mathbf{z} + \mathbf{y} * \mathbf{z}$$

for all sequences $\mathbf{x}$, $\mathbf{y}$, and $\mathbf{z}$ for which both operations of addition and convolution are defined.

9. Explain why the distributive property you just proved implies that the mapping defined by

$$T_h(\mathbf{x}) = \mathbf{x} * \mathbf{h}$$

for a given sequence $\mathbf{h} = (h_0, h_1, \ldots, h_m)$ is a linear transformation.

10. Verify all the calculations in Example 5.3 of Subsection 5.2.4.

11. Show that both linear and circular convolutions are associative, which means that

$$(\mathbf{x} * \mathbf{y}) * \mathbf{z} = \mathbf{x} * (\mathbf{y} * \mathbf{z})$$

for all sequences $\mathbf{x}$, $\mathbf{y}$, and $\mathbf{z}$ for which convolution is defined.

12. Show that the unit impulse $\boldsymbol{\delta}$ defined by (5.5) is the convolutional identity in the sense that
$$\mathbf{x} * \boldsymbol{\delta} = \boldsymbol{\delta} * \mathbf{x} = \mathbf{x}$$
for any sequence $\mathbf{x}$.

13. Suppose that two sequences $\mathbf{x}$ and $\mathbf{y}$ have lengths M and N respectively. Show that their linear convolution $\mathbf{z} = \mathbf{x} * \mathbf{y}$ has length $M + N - 1$.

14. Suppose that two sequences $\mathbf{x}$ and $\mathbf{y}$ have lengths M and N respectively. Show that their linear convolution $\mathbf{z} = \mathbf{x} * \mathbf{y}$ equals the **circular** convolution of their extended versions $\mathbf{x}_{pad}$ and $\mathbf{y}_{pad}$, both zero-padded to the length $M + N - 1$.

15. Show that the coefficients of the product of two polynomials can be determined by calculating the linear convolution of the coefficients of the factor polynomials being multiplied.

16. Verify that the product of two polynomials
$$p(x) = x^3 + 2x^2 + 3x + 4 \text{ and } q(x) = 5x^2 + 6x + 7$$
is
$$q(x) = 5x^5 + 16x^4 + 34x^3 + 52x^2 + 45x + 28,$$
and explain why the result was predicted by Exercise 5.1.

17. Find the product of the following polynomials both directly and by calculating the convolution of their coefficients:

    (a) $p(x) = x + 1 \quad \text{and} \quad q(x) = p(x) = x + 1;$
    (b) $p(x) = x^3 + 3x^2 + 3x + 1 \quad \text{and} \quad q(x) = x^2 + 2x + 1;$
    (c) $p(x) = x^4 + x^3 + x^2 + x + 1 \quad \text{and} \quad q(x) = x - 1;$
    (d) $p(x) = x^4 + 5x^3 - 3x^2 + 7x + 1 \quad \text{and} \quad q(x) = 3x^2 - 2x - 1.$

    Check your answers using the appropriate MATLAB functions.

18. Show that the linear convolution of an $M_1 \times N_1$ matrix with an $M_2 \times N_2$ matrix has $M_1 + M_2 - 1$ rows and $N_1 + N_2 - 1$ columns.

19. Explain how one can calculate circular convolution of two matrices by taking the linear convolution of one matrix with the periodic extension of the other one.

MATLAB Exercises.

1. Write a MATLAB function that would calculate linear convolution of two finite sequences $\mathbf{x}$ and $\mathbf{y}$. Your function should accept the two sequences as inputs. Test your function using several different pairs of sequences and make sure it gives the same answers as the MATLAB function *conv*.

2. Write a MATLAB function that would calculate circular convolution of two finite sequences **x** and **y**. If the sequences have different lengths, the shorter one must be extended by means of zero-padding to the length of the longer one. Your function should accept the two sequences as inputs. Test your function using several different pairs of sequences (of identical and different lengths) and make sure it gives the same answers as the MATLAB function *cconv*.

3. Write a MATLAB function that would calculate the linear-convolution matrix for a specified sequence **h**.

4. Write a MATLAB function that would calculate the circular-convolution matrix for a specified sequence **h**.

5. Write a MATLAB program to do the following:

   (a) Ask the user whether the linear or the circular convolution is desired;

   (b) Ask the user to input the two sequences **x** and **y** whose convolution needs to be calculated.

   (c) Calculate the appropriate convolution using the functions you created in Problems 1 and 2.

   (d) Plot graphs of the sequences **x**, **y**, and their convolution $\mathbf{x} * \mathbf{y}$.

6. Test your program using several different pairs of sequences. Compile the results into a PowerPoint presentation.

5.3 Edge Detection

Kids love coloring. One of the most reliable ways to keep kids occupied is to place coloring paper and crayons in front of them, and they will be busy for hours. And not just kids. Painters and illustrators often start a new project with a pencil sketch and later proceed with turning a drawing into a painting.

Sometimes, however, the exact opposite needs to be done - something tantamount to turning a painting into a drawing. *Edge detection* is a procedure for finding the boundaries of objects within images. It is used in image processing for the purposes of image segmentation, feature detection, and feature extraction. Edge detection has numerous applications in various areas such as

- computer and machine vision,
- video surveillance,

- satellite imaging,
- medicine,
- geology,
- glaciology,

and many others. Numerous sophisticated methods of edge detection have been developed, and anything resembling a comprehensive discussion is beyond the scope of this book. We will introduce a few simple and basic techniques based on concepts that belong in a standard college math curriculum.

5.3.1 Partial Derivatives and the Gradient Edge Detector

Edge detection methods work by detecting sharp and abrupt changes in light intensity. The tried and tested instrument for measuring change is the derivative. If we denote by $f(x, y)$ the light intensity at the point (x, y), then the rate of change of light intensity in the direction of the x-axis at any given point (a, b) is the partial derivative

$$f_x(a, b) = \lim_{h \to 0} \frac{f(a+h, b) - f(a, b)}{h} \tag{5.9}$$

provided, of course, that f is differentiable at that point. When we are dealing with digital images, the closest h can get to zero is $h = 1$. Therefore, the partial derivative (5.9) can best be approximated by

$$f_x(a, b) \approx f(a+1, b) - f(a, b). \tag{5.10}$$

Similarly, the rate of change of light intensity in the direction of the y-axis at any given point (a, b) is the partial derivative

$$f_y(a, b) = \lim_{h \to 0} \frac{f(a, b+h) - f(a, b)}{h}, \tag{5.11}$$

which can be approximated by

$$f_y(a, b) \approx f(a, b+1) - f(a, b) \tag{5.12}$$

in order to be usable in a discrete setting.

We recall that a grayscale digital image A can be thought of as a function of two integer-valued variables m and n. The value of $A = A(m, n)$ represents the light intensity at the pixel location (m, n), where the first index m stands for the vertical position of the pixel, and the second index n stands for its horizontal position. Therefore, (5.10) and (5.12) imply that the rate of change in light intensity in the vertical direction at the pixel location (m, n) can be approximated by

$$G_1(m, n) = A(m+1, n) - A(m, n), \tag{5.13}$$

and the rate of change in the horizontal direction by

$$G_2(m,n) = A(m,n+1) - A(m,n), \tag{5.14}$$

where the subscripts 1 and 2 were chosen to indicate the first and second index respectively. However, we find the subscripts x and y more familiar, and we usually expect the x-axis to be oriented horizontally, and the y-axis to be oriented vertically. We, therefore, rename the rates of change

$$G_x(m,n) = G_2(m,n) = A(m,n+1) - A(m,n) \tag{5.15}$$

and

$$G_y(m,n) = G_1(m,n) = A(m+1,n) - A(m,n) \tag{5.16}$$

in accordance with our expectations regarding the orientation of the coordinate axes.

We also recall from our study of multivariate calculus that the *gradient vector* of a differentiable function f of two variables x and y at the point (a,b) is defined by

$$\nabla f(a,b) = f_x(a,b)\vec{i} + f_y(a,b)\vec{j}.$$

The magnitude

$$||\nabla f(a,b)|| = \sqrt{f_x(a,b)^2 + f_y(a,b)^2}$$

of the gradient vector is the maximum rate of change of f at the point (a,b), and the direction ϕ of the gradient vector, which can be determined by means of

$$\tan(\phi) = \frac{f_y(a,b)}{f_x(a,b)},$$

(provided, of course, that $f_x(a,b) \neq 0$) is the *direction* of the maximum rate of increase of f at the point (a,b). In the exercises at the end of this section, the reader will have an opportunity to review these concepts from multivariate calculus and refresh their understanding of the properties of the gradient vector.

When considered in the context of digital images, the magnitude of the gradient vector of light intensity at the point (m,n) can be interpreted as the overall "strength" of the edge at that point. In view of (5.15) and (5.16), this "strength" can be approximated by

$$\begin{aligned} ||G(m,n)|| &= \sqrt{G_x(m,n)^2 + G_y(m,n)^2} \\ &= \sqrt{|A(m,n+1) - A(m,n)|^2 + |A(m+1,n) - A(m,n)|^2}. \end{aligned}$$

Likewise, the direction of the gradient vector at the point (m,n) can be interpreted as the orientation ϕ of the edge passing through that point and can be approximated by

$$\tan(\phi) \approx \frac{G_y(m,n)}{G_x(m,n)} = \frac{A(m+1,n) - A(m,n)}{A(m,n+1) - A(m,n)}, \tag{5.17}$$

FIGURE 5.6: Illustration of the gradient edge detector.

provided that $G_x(m, n) \neq 0$.

It is time to put all those theoretical considerations to the test. Figure 5.6(a) shows a view of the facade of the Pashkov House - the main building of the Russian State Library in Moscow. In the other parts of Figure 5.6, we can see the plots of the absolute values of the approximations G_x and G_y of the partial derivatives, and, ultimately, the overall magnitude of the gradient vector illustrating the process of edge detection using the gradient method.

An examination of Figure 5.6 reveals that the plot of $|G_x|$ contains most of the vertical lines, and the plot of $|G_y|$ – most of the horizontal lines. This is not surprising because vertical edges tend to be comprised of the points with a large rate of change in pixel value in the horizontal direction and, likewise, horizontal lines consist of points with a large rate of change in the vertical direction.

Below is the code used to generate Figure 5.6, which we include here for the reference in anticipation of the MATLAB exercises at the end of this section.

```
A=imread('Library.jpg');
A=rgb2gray(A);
[M, N]=size(A);
G=zeros(M,N);
Gx=G; Gy=G;
for m=1:M-1
    for n=1:N-1
        Gx(m,n)=A(m,n+1)-A(m,n);
        Gy(m,n)=A(m+1,n)-A(m,n);
        G(m,n)=sqrt(Gx(m,n)∧ 2+Gy(m,n)∧ 2);
    end
end
subplot(2,2,1); imshow(A);
title('(a) Original Image','Fontsize',10);
subplot(2,2,2); imshow(abs(Gx),[]);
title('(b) The x-derivative, Gx','Fontsize',10);
subplot(2,2,3); imshow(abs(Gy),[]);
title('(c) The y-derivative, Gy','Fontsize',10);
subplot(2,2,4); imshow(abs(G),[]);
title('(d) The Magnitude of the Gradient','Fontsize',10);
```

5.3.2 Directional Derivatives and the Roberts Cross Operator

Not all images are dominated by horizontal and vertical lines. Oftentimes, it is diagonal lines that are more prevalent, and it is precisely for that reason that one of the oldest edge detectors ever devised – the *Roberts cross* operator – is designed to maximize its response to the edges running at 45° with the coordinate axes.

We recall from Multivariate Calculus that the derivative of a differential function f of two variables x and y at the point (a, b) in the direction of the vector $\vec{u} = u_1\vec{i} + u_2\vec{j}$ is defined by

$$f_{\vec{u}}(a,b) = \lim_{h\to 0} \frac{f(a+hu_1, b+hu_2) - f(a,b)}{h||\vec{u}||}, \tag{5.18}$$

where $\vec{u}$ is usually taken to be of unit length to simplify notation. Since the smallest value that h can attain in the context of digital images is $h = 1$, we can approximate the directional derivative by

$$f_{\vec{u}}(a,b) \approx \frac{1}{||\vec{u}||}\big[f(a+u_1, b+u_2) - f(a,b)\big]. \tag{5.19}$$

The two vectors at 45° with the coordinate axes that we need are $\vec{u} = \vec{i} + \vec{j}$

and $\vec{v} = \vec{i} - \vec{j}$ (both of magnitude $\sqrt{2}$). The derivatives in the directions of those vectors at the point (a, b) can be approximated by

$$f_{\vec{u}}(a,b) \approx \frac{\sqrt{2}}{2}\left[f(a+1,b+1) - f(a,b)\right]$$

and by

$$f_{\vec{v}}(a,b) \approx \frac{\sqrt{2}}{2}\left[f(a+1,b-1) - f(a,b)\right],$$

which, for a digital image A, results in

$$G_1(m,n) = A(m+1,n+1) - A(m,n) \tag{5.20}$$

and

$$G_2(m,n) = A(m+1,n) - A(m,n+1) \tag{5.21}$$

after some adjustment and scaling. Formulas (5.20) and (5.21) describe the action of the Roberts cross operator on a digital image. Similar to the gradient edge detector, the "strength" of the edge at the point (m, n) is measured by

$$||G(m,n)|| = \sqrt{G_1(m,n)^2 + G_2(m,n)^2} \tag{5.22}$$

with $G_1(m, n)$ and $G_2(m, n)$ given by (5.20) and (5.21) respectively.

Figure 5.7(a) shows an image of the Pyramids of Giza, which is dominated by diagonal lines. We can compare the results of edge detection using the gradient operator (Figure 5.7(b)) and the Roberts cross operator (Figure 5.7(c)) and compare the performance of both operators when applied to an image of this kind. A significant advantage of Roberts cross edge detection seems apparent, particularly with respect to the upper-right outlines of the pyramids.

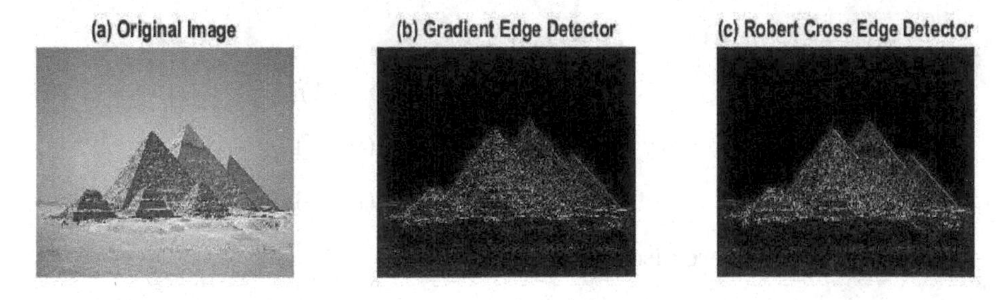

FIGURE 5.7: Illustration of the Roberts cross edge detector.

5.3.3 The Prewitt and Sobel Edge Detectors

In the last two subsections, we used the notions of partial and directional derivatives to develop two rudimentary edge detectors. How can we further improve and develop our methods of edge detection? Two possible areas of improvement immediately come to mind:

- we would like to make the edges thicker and more prominent;
- we would like our edge detector to be less susceptible to image noise.

It turns out that both goals can be accomplished using similar strategies, which can intuitively be described as "anticipating" an approaching edge one step ahead, letting the memory of an edge "linger" for one step longer, and feeling for edges that are slightly off to the side of our "main path".

We next put these ideas into practice. When calculating "horizontal" differences at a point (m, n), we could add together the differences in pixel values immediately "before" and "after" that point, specifically, by calculating

$$\begin{aligned} \Delta &= [A(m,n+1) - A(m,n)] + [A(m,n) - A(m,n-1)] \\ &= A(m,n+1) - A(m,n-1). \end{aligned} \tag{5.23}$$

We could do the same thing one pixel row above and one pixel row below the point (m, n) to calculate

$$\begin{aligned} \Delta_a &= [A(m-1,n+1) - A(m-1,n)] + [A(m-1,n) - A(m-1,n-1)] \\ &= A(m-1,n+1) - A(m-1,n-1) \end{aligned} \tag{5.24}$$

and

$$\begin{aligned} \Delta_b &= [A(m+1,n+1) - A(m+1,n)] + [A(m+1,n) - A(m+1,n-1)] \\ &= A(m+1,n+1) - A(m+1,n-1) \end{aligned} \tag{5.25}$$

respectively. Adding equations (5.23), (5.24), and (5.25) together, we define

$$\begin{aligned} G_x(m,n) &= \Delta + \Delta_a + \Delta_b \\ &= A(m,n+1) - A(m,n-1) \qquad (5.26) \\ &+ A(m-1,n+1) - A(m-1,n-1) \\ &+ A(m+1,n+1) - A(m+1,n-1). \end{aligned}$$

In a similar manner, we also define

$$\begin{aligned} G_y(m,n) &= A(m+1,n) - A(m-1,n) \qquad (5.27) \\ &+ A(m+1,n-1) - A(m-1,n-1) \\ &+ A(m+1,n+1) - A(m-1,n+1) \end{aligned}$$

and calculate the "strength" of the edge at the point (m, n) by

$$||G(m,n)|| = \sqrt{G_x(m,n)^2 + G_y(m,n)^2} \tag{5.28}$$

with $G_x(m,n)$ and $G_y(m,n)$ given by (5.26) and (5.27) respectively. The direction ϕ of the edge at the point (m,n) can be estimated by

$$\tan(\phi) = \frac{G_y(m,n)}{G_x(m,n)},$$

just as in the case of the gradient edge detector. Together, the formulas (5.26), (5.27), and (5.28) constitute the *Prewitt edge detector*.

With the formulas getting long and ugly the way (5.26) and (5.27) do, we might start looking for a more efficient notation. Fortunately, thanks to Section 5.2, we have exactly the type of notation and terminology that is needed. All we are really doing when calculating G_x and G_y is taking the convolution of the given image with the filter matrices

$$P_x = \begin{bmatrix} 1 & 0 & -1 \\ 1 & 0 & -1 \\ 1 & 0 & -1 \end{bmatrix} \quad \text{and} \quad P_y = \begin{bmatrix} 1 & 1 & 1 \\ 0 & 0 & 0 \\ -1 & -1 & -1 \end{bmatrix} \tag{5.29}$$

respectively. In its most concise form, the action of the Prewitt edge detector on the image A can be expressed as

$$G_x = A * P_x \quad \text{and} \quad G_y = A * P_y, \tag{5.30}$$

and we can appreciate how much space this notation has saved us already and how much time and work it will save us in the future (when we undertake the project of implementing edge detectors in MATLAB).

Going over the derivation of the Prewitt edge detector, particularly over the considerations that led to the formulas (5.26) and (5.27), we can argue that the row containing the point (m,n) probably ought to be making a greater contribution to the "strength" of the edge at (m,n) than the rows above and below. Similarly, the column containing the point (m,n) should be making a greater contribution to the "strength" of the edge than the columns to the left and to the right of it. These considerations lead to what is known as the *Sobel edge detector*, defined by the filter matrices

$$S_x = \begin{bmatrix} 1 & 0 & -1 \\ 2 & 0 & -2 \\ 1 & 0 & -1 \end{bmatrix} \quad \text{and} \quad S_y = \begin{bmatrix} 1 & 2 & 1 \\ 0 & 0 & 0 \\ -1 & -2 & -1 \end{bmatrix}, \tag{5.31}$$

and whose filtering action on the image A can be concisely described by

$$G_x = A * S_x \quad \text{and} \quad G_y = A * S_y,$$

where S_x and S_y are given by (5.31). The direction ϕ of the edge at the point (m,n) is given by

$$\tan(\phi) = \frac{G_y(m,n)}{G_x(m,n)}$$

just as in the cases of the gradient and Pruitt edge detectors.

(a) Original Image

(b) Gradient Edge Detector

(c) Prewitt Edge Detector

(d) Sobel Edge Detector

FIGURE 5.8: Comparison of the gradient, Prewitt, and Sobel edge detectors.

Figure 5.8 compares the performance of the gradient, Prewitt, and Sobel edge detectors when applied to the photo of the facade of the Russian State Library we have already seen in Subsection 5.3.1. The difference in quality of the edges produced by the three detectors is nothing short of remarkable. Especially obvious is the overwhelming advantage of the Prewitt filter over the naive direct use of just the gradient vector. More subtle but still quite perceptible is the improvement in the sharpness of the edges as we moved from the Prewitt to the Sobel edge detector.

5.3.4 Laplacian Edge Detection

All four of the edge detecting methods we have studied (the gradient, Roberts cross, Prewitt, and Sobel) measured the rate of change in pixel values separately in the directions of two chosen orthogonal axes. As a result, all these methods use two filters and require two convolutions. The reader might be wondering whether it is possible to produce the edges with just one filter (and, hence, just one convolution to compute)?

The answer is yes, and the mathematical instrument we can use for the purpose is called the *Laplacian*, which is defined for smooth differentiable functions by

$$\nabla^2 f(x,y) = f_{xx}(x,y) + f_{yy}(x,y). \tag{5.32}$$

Originally discovered in the context of the study of celestial mechanics, the Laplacian has since found a prominent place in almost all fields of natural sciences. For example, it is used in differential equations that describe such physical phenomena as electric and gravitational potentials, in the diffusion equation for heat and fluid flow, and in quantum mechanics. It is remarkable (but hardly surprising) that the Laplacian also has applications in image processing.

Our first task is to derive an approximation of the Laplacian in the discrete setting. By definition, the second-order partial derivative of a smooth differentiable function f with respect to x is

$$f_{xx}(x,y) = \lim_{h\to 0} \frac{f_x(x+h,y) - f_x(x,y)}{h}. \tag{5.33}$$

Since the closest h can get to zero in the discrete setting is $h = 1$ or $h = -1$, we substitute $h = -1$ in (5.33) and get

$$f_{xx}(x,y) \approx \frac{f_x(x-1,y) - f_x(x,y)}{-1} = f_x(x,y) - f_x(x-1,y).$$

Substituting the approximation (5.10) of the partial derivative into this estimate yields

$$\begin{aligned} f_{xx}(x,y) &\approx f(x+1,y) - f(x,y) - [f(x,y) - f(x-1,y)] \\ &= f(x+1,y) - 2f(x,y) + f(x-1,y). \end{aligned} \tag{5.34}$$

In a similar manner, we obtain the approximation

$$f_{yy}(x,y) \approx f(x,y-1) - 2f(x,y) + f(x,y+1) \tag{5.35}$$

for the second-order partial derivative of f with respect to y. Adding the equations (5.34) and (5.35) together, we obtain the desired approximation

$$\begin{aligned} \nabla^2 f(x,y) &\approx f(x-1,y) + f(x,y-1) - 4f(x,y) \\ &+ f(x+1,y) + f(x,y+1) \end{aligned} \tag{5.36}$$

for the Laplacian defined by (5.32).

We can expect the discrete approximation of the Laplacian operator in the form (5.36) to be sensitive to the changes in the directions of the coordinate axes. Therefore, we expect it to be effective in detecting horizontal and vertical edges. However, as we discovered while experimenting with the Roberts cross operator, it is also important to take the diagonal edges into account.

Thus, if instead of calculating the second-order partial derivatives in the directions of the coordinate axes, we calculated the second-order **directional** derivatives in the directions of the vectors

$$\vec{u} = \vec{i} + \vec{j} \quad \text{and} \quad \vec{v} = \vec{i} - \vec{j},$$

we would have obtained (after dropping the nuisance factor $\frac{\sqrt{2}}{2}$ the way we did when deriving the Roberts cross operator) the estimate

$$\begin{aligned} \nabla^2 f(x,y) &\approx f(x-1,y-1) + f(x-1,y+1) - 4f(x,y) \\ &+ f(x+1,y-1) + f(x+1,y+1). \end{aligned} \tag{5.37}$$

Adding the two estimates (5.36) and (5.37) together (and factoring our -1 for aesthetic reasons), we obtain the Laplacian filter matrix

$$L = \begin{bmatrix} -1 & -1 & -1 \\ -1 & 8 & -1 \\ -1 & -1 & -1 \end{bmatrix}. \tag{5.38}$$

With the help of the matrix form (5.38), the action of the Laplacian edge detector on the image A can be expressed by

$$G = A * L,$$

and, as we hoped, it only involves one convolution (as opposed to two convolutions required by all the methods we have developed in previous subsections).

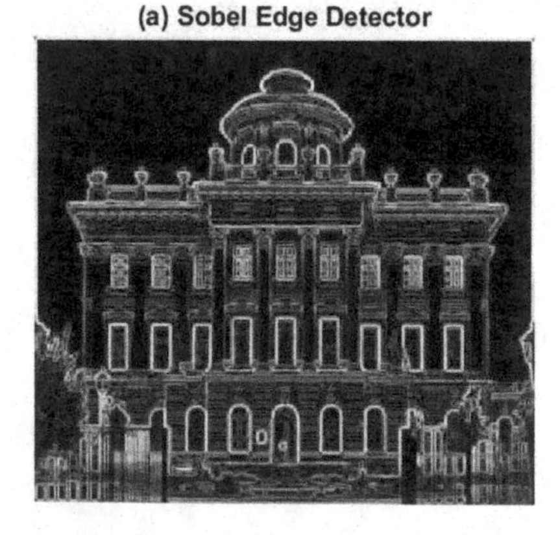

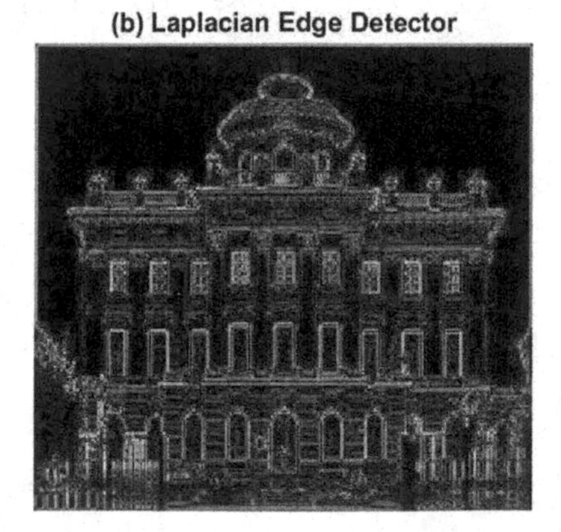

FIGURE 5.9: Comparison of Sobel and Laplacian edge detectors.

Figure 5.9 compares the performance of the Sobel and the Laplacian edge detectors. As we could have expected, the two filters have different areas of strength and weakness. In the MATLAB exercises at the end of this section, the reader will be asked to compare all of the edge detectors we have discussed by applying them to a selection of images of their choice.

5.3.5 Edge Detection in Noisy Images

All images contain a certain degree of noise, and applying an edge detector to a noisy image directly does not always produce desirable outcomes. The images in the left columns of Figures 5.10 and 5.11 show the results of applying the gradient, Sobel, and Laplacian edge detectors directly to the photo of the Russian State Library corrupted by additive white Gaussian noise with the standard deviation $\sigma = 30$. We can see that some edge detectors are more vulnerable to noise than others. Particularly susceptible are the gradient and Laplacian operators.

A reasonable way to remedy the situation would be to attempt a reduction of the noise level using methods of Section 5.1 prior to calculating edges. The images in the right column of Figures 5.10 and 5.11 show the result of applying edge detectors in series with the approximate Gaussian filter

$$W = \begin{bmatrix} 2/25 & 3/25 & 2/25 \\ 3/25 & 5/25 & 3/25 \\ 2/25 & 3/25 & 2/25 \end{bmatrix}$$

described in Section 5.1. We can see the dramatic improvement, particularly on the part of the gradient and Laplacian edge detectors.

It would seem that this improvement comes at the cost of increased computational complexity, because the process

$$A \to A * W \to (A * W) * L$$

requires two convolutions (each involving a large-size matrix). But there is a trick! In Section 5.2, we learned that convolution is associative! Therefore, we do not have to perform calculations in this inefficient order. Instead, we use the order

$$A \to A * (W * L)$$

and complete the process of filtering the image with just one convolution. For this specific approximate realization of the Gaussian filter, the approximation of the so-called "Laplacian of Gaussian" filter matrix is

$$LoG = W * L = \frac{1}{25} \begin{bmatrix} -1 & -1 & -1 \\ -1 & 8 & -1 \\ -1 & -1 & -1 \end{bmatrix} * \begin{bmatrix} 2 & 3 & 2 \\ 3 & 5 & 3 \\ 2 & 3 & 2 \end{bmatrix},$$

and the reader is encouraged to verify (by hand and with the help of the

FIGURE 5.10: Edge detection with and without Gaussian denoising - the gradient operator.

MATLAB function *conv2*) that this linear convolutional product equals

$$LoG = \frac{1}{25}\begin{bmatrix} -2 & -5 & -7 & -5 & -2 \\ -5 & 5 & 9 & 5 & -5 \\ -7 & 9 & 20 & 9 & -7 \\ -5 & 5 & 9 & 5 & -5 \\ -2 & -5 & -7 & -5 & -2 \end{bmatrix}. \tag{5.39}$$

An appropriate Laplacian of Gaussian matrix can always be prepared in advance, and the combination of noise reduction and edge detection can still be performed in one pass! We can appreciate how the right terminology and

FIGURE 5.11: Edge detection with and without Gaussian denoising - Sobel and Laplacian filters.

deeper understanding of abstract algebraic properties of convolution have helped us reduce computational complexity of digital image filtering.

5.3.6 Boolean Convolution and Edge Dilation

We have studied several different methods of edge detection, and the reader has applied those methods to a selection of images. But what if the results are not to our complete and full satisfaction? What if the edges are still too thin? Is there a way to further dilate them?

The section on edge detection would not be complete without at least a brief mention of one other type of convolution – the Boolean convolution. As the name implies, Boolean convolution applies to binary images , that is, to the images whose pixels take on one of the two discrete logical values (for example, 0 and 1, with 0 representing the dark "background" and 1 representing the bright "foreground").

For the purposes of this text, we define the **linear Boolean convolution** of the $M \times N$ binary image A with the so-called "structural element" H (which is just another binary matrix) by

$$(A \oplus H)(m,n) = \bigvee_{k,l} A(k,l) \wedge H(m-k, n-l), \tag{5.40}$$

where the symbol $\vee$ represents the "logical sum" (disjunction) and the symbol $\wedge$ represents the "logical product" (conjunction). Similarly, the **circular Boolean convolution** of A with H can be defined by

$$(A \oplus H)(m,n) = \bigvee_{k,l} A(k,l) \wedge H(m-k \mod M, \quad n-l \mod N). \tag{5.41}$$

We leave it to the reader to establish the algebraic properties of the Boolean convolution.

To help us visualize the use of the Boolean convolution for binary image dilation, we use the tried-and-tested "flip-and-slide" method. We flip the structural element H and move its center to the position (m,n). If any of the "foreground" (having the value 1) pixels of the flipped and shifted structural element overlap with any of the foreground elements of the image A, then the image pixel in the position (m,n) is assigned the value 1; otherwise it retains its current value.

Edge dilation is not the only possible application of the Boolean convolution. In general, it has the effect of broadening the narrow features of a binary image, "rounding" the boundaries if its regions, and joining separate regions. Of critical importance is the choice of the structural element H, as different structural elements have dramatically different effects on specific shapes and structures of binary images.

In the MATLAB exercises at the end of this section, the reader will be asked to implement the Boolean-convolution, to use it for the purpose of improving the visual quality of the edges and to experiment with a variety of structural elements.

Section 5.3 Exercises.

1. Use the limit definition to find derivatives of the following functions:

   (a) $f(x) = 5x^2$,
   (b) $f(x) = 2x^3$,
   (c) $f(x) = \pi/x$,
   (d) $f(x) = 7/x^2$,
   (e) $f(x) = \sqrt{3x}$,
   (f) $f(x) = 1/\sqrt{x}$.

2. Estimate the derivative of the function $f(x) = x^x$ at the points $a = 1$ and $a = 2$.

3. Use the limit definition to find both partial derivatives of the following functions:

   (a) $f(x) = 5x^2y$,
   (b) $f(x) = 2x^3/y$,
   (c) $f(x) = 7y/x^2$,
   (d) $f(x) = y\sqrt{3x} - 1/\sqrt{x}$.

4. Calculate the magnitude and the direction of the gradient vector of the given function of two variables at the specified point.

   (a) $f(x, y) = x^2y + 2xy^3$, at $(2, 3)$,
   (b) $f(x, y) = \ln(x^2 + xy)$ at $(4, 1)$,
   (c) $f(x, y) = 1/(x^2 + y^2)$ at $(1, 2)$,
   (d) $f(x, y) = e^{2x} \sin(3y) - e^{3x} \cos(5y)$ at $(0, 0)$.

5. Prove that the directional derivative of a differentiable function f at the point (a, b) in the direction of the unit vector $\mathbf{u} = u_1\vec{i} + u_2\vec{j}$ can be calculated using the formula

$$f_{\mathbf{u}}(a, b) = f_x(a, b)u_1 + f_y(a, b)u_2.$$

6. Use the result of Exercise 5 to find the directional derivatives of the functions in Exercise 4 in the directions of the vectors $\vec{u} = \vec{i} + \vec{j}$ and $\vec{v} = \vec{i} - \vec{j}$.

7. Suppose that you are standing above the point $(3, 1)$ on the surface described by the function

$$f(x, y) = 100 - x^2 - 2y^2.$$

   (a) What is the direction of the steepest ascent?
   (b) What is the direction of the steepest descent?
   (c) If you start to walk in the direction of the steepest descent, what is the slope of your path?

8. Repeat Exercise 7 with the following functions and points:

   (a) $f(x) = x^4 - x^2y + 2y^2 + 10$ above the point $(1, 1)$,
   (b) $f(x) = y + 1/x^3$ above the point $(1, 4)$.

9. Derive the formulas (5.20) and (5.21) that describe the action of the Roberts cross operator.

10. Derive the formulas (5.27) for the Prewitt edge detector.

11. Derive the formula (5.31) for the Sobel edge detector.

12. Derive the approximation (5.35) for the second-order partial derivative with respect to y.

13. Derive the approximation (5.37) for the “diagonal” Laplacian and verify the formula (5.38) for the Laplacian filter matrix.

14. Verify the result of computation (5.39) of the combined matrix of a Gaussian filter followed by the Laplacian edge detector. Calculate several other “Laplacian of Gaussian” filter matrices using your results in Exercise 4 at the end of Section 5.1.

15. Prove that the Boolean convolution defined by the formulas (5.40) and (5.41) is commutative, associative, and distributive with respect to disjunction.

MATLAB Exercises.

1. Prepare a selection of images that you would like to turn into drawings. An architectural ensemble or a decorative pattern could be two possible themes. If necessary, you can use the MATLAB *rgb2gray* function to convert a color image to grayscale.

2. Write MATLAB functions that would create convolution matrices for implementing the gradient, Roberts, Prewitt, Sobel, and Laplacian edge detectors.

3. Write a MATLAB program that would do the following:

   (a) Display the original image.

   (b) Ask the user which edge detector is to be used.

   (c) Call the appropriate function you created in the previous problem to perform the edge detection using the selected method. You can either use the MATLAB function *conv2* or the 2-D convolution function you created in MATLAB exercises in Section 5.2.

   (d) Display the “drawing” obtained by using the selected edge detector.

   Test your program using several different images (grayscale and color) and different edge detectors. Compare the results.

4. Identify some of the most prominent edges in the most successful “drawings” you created in the previous problem. Identify a few points the edges pass through by examining the image matrices. Determine the orientation of the selected edges using the formula (5.17).

5. Create and display noisy versions of your images by adding white Gaussian noise of the specified standard deviation.

6. Call the functions you implemented in MATLAB Exercise 2 to perform edge detection on the noisy images.

7. Next, use any of the approximate Gaussian filter matrices you created in Exercise 4 at the end of Section 5.1 to perform image noise reduction prior to edge detection. For Laplacian edge detection, feel free to use the filter matrices you created in Exercise 14. Compare the results obtained with and without preliminary noise reduction.

8. Write a MATLAB function that would produce a binary image from a grayscale image as follows: set those pixels whose values are greater than the specified threshold to 1 and set those pixels whose values are below the threshold to 0. Your function should accept the image and the value of the threshold as inputs.

9. Test the MATLAB function you created in the previous exercise using a selection of suitable grayscale images (such as the "drawings" created in MATLAB Exercise 3). While several methods of automatic threshold selection have been developed, most notably, the Otsu's method, we ask the reader to select the threshold manually by examining the image histogram and identifying obvious gaps in pixel values.

10. Write MATLAB functions that would implement the linear and circular Boolean convolution defined by the formulas (5.40) and (5.40). Apply this function to the digital images obtained from the "drawings" created in MATLAB Exercise 3. Experiment with different structural elements and compare the resulting edge-dilating effect. As a starting point, you can use the structural elements

$$\text{(a) } H = \begin{bmatrix} 0 & 0 & 0 \\ 0 & 1 & 0 \\ 0 & 0 & 0 \end{bmatrix}, \qquad \text{(d) } H = \begin{bmatrix} 0 & 1 & 0 \\ 1 & 1 & 1 \\ 0 & 1 & 0 \end{bmatrix},$$

$$\text{(b) } H = \begin{bmatrix} 0 & 0 & 0 \\ 1 & 1 & 1 \\ 0 & 0 & 0 \end{bmatrix}, \qquad \text{(e) } H = \begin{bmatrix} 1 & 0 & 1 \\ 0 & 1 & 0 \\ 1 & 0 & 1 \end{bmatrix},$$

$$\text{(c) } H = \begin{bmatrix} 0 & 1 & 0 \\ 0 & 1 & 0 \\ 0 & 1 & 0 \end{bmatrix}, \qquad \text{(f) } H = \begin{bmatrix} 1 & 1 & 1 \\ 1 & 1 & 1 \\ 1 & 1 & 1 \end{bmatrix}.$$

11. Compile the results of the previous MATLAB exercises into a PowerPoint presentation.

5.4 Chapter Summary

In this chapter, we have studied three seemingly different applied topics: image blurring, noise reduction, and edge detection. All three topics were united by one underlying concept, namely that of *convolution*, which was first introduced as a shorthand term but has since acquired a life of its own keeping us busy for the entire middle section of this chapter.

The deeper insight into convolution we have acquired has already delivered some benefits. Our notation has grown considerably cleaner, and our language has grown more concise. Our understanding of algebraic properties of convolution has helped reduce computational complexity of image filtering and led the way to creating unified matrices for entire cascades of operators. But we can still guess what the reader is probably thinking: are the benefits worth the effort? Why did we spend so much time learning about convolution if, at least up to this point, the benefits have been mostly a matter of style, language, and improvements in computational speed that most of us aren't even going to notice?

This is a completely valid question, and we will answer it more fully in Chapter 6. At this point, we will only reiterate the point made in the closing paragraph of Section 5.1: it is the task of *reversing* the effects of *unwanted* filtering that renders a deeper understanding of convolution and its connection with frequency analysis truly indispensable.

Chapter 6

Analysis and Processing in the Frequency Domain

In this chapter, the reader will discover that reversing undesirable blurring can be accomplished by studying and manipulating the image frequency content. To facilitate this process, we develop the foundations of frequency analysis, beginning with Fourier coefficients and ending with discrete Fourier transforms in two dimensions and their applications to digital images. Also included in this section is a review of complex numbers and complex-valued functions.

6.1 Introduction

In everyday language, the word "frequency" is usually associated with radio and with audible sounds. We are used to our AM/FM car radio and happily scan the radio frequencies for our favorite music and news channels. We sometimes talk about the kids' ability to hear the high frequency sounds that are no longer audible for most adults. The contexts in which we mention the word "frequency" also come from many different areas of social and natural sciences. It might come as a surprise, however, that frequency analysis and frequency manipulations also play a very important role in image processing.

Probably the easiest way to see the connection between frequency analysis of signals (such as sounds and radio waves) and frequency analysis of images is to interpret high-frequency components as those that correspond to rapid changes on a small temporal scale and, similarly, to interpret the low-frequency components as those corresponding to slow changes over longer time intervals. Armed with this understanding, we can interpret small-scale details in an image as its high-frequency components and large-scale features of an image as its low-frequency components.

Why would we be concerned about separating the small-scale details from the large-scale ones? It turns out that many of the distortions introduced into images by the imperfections of camera lenses and by the way in which we take our photos (such as static blur, motion blur, and many other types) are frequency-specific in the sense that they affect certain frequency ranges and do not affect others. Thus, the task of image restoration often comes down to separating the image into its frequency components, "repairing" the affected ones, and reconstructing the (hopefully) improved version of the image from those "repaired" components.

FIGURE 6.1: Restoration of a blurred photo of Emanuel Lasker.

Some examples of images that can be improved using the techniques just described are shown in Figures 6.1 and 6.2. For instance, Figure 6.1(a) shows the effects of a static blur on the photo of Emanuel Lasker (1868-1941), who was chess world champion for twenty-seven years from 1894 until 1921, a strong mathematician, one of the co-discoverers of a famous theorem in abstract algebra, a philosopher, and a playwright. Alongside, in Figure 6.1(b), the blurring effect is largely removed from the photo by means of methods of frequency analysis. Figure 6.2(a) shows a photo of the Parthenon that may have been taken from a moving vehicle, and in Figure 6.2(b) we can see the result of reversing the motion blur (also using methods of frequency analysis).

We hope that these examples provide the reader with a strong motivation for the hard work we must undertake in order to develop the mathematical machinery necessary for this type of image restoration.

FIGURE 6.2: Restoration of a photo of the Parthenon affected by a strong motion blur.

Our game plan for this chapter is as follows: we will begin with the least complicated and most straightforward area of frequency analysis – the one devoted to studying frequency content of continuous periodic signals. We will subsequently expand our investigation to include finite and periodically infinite discrete signals. We will conclude the chapter by applying our results and observations to digital images (which can be thought of as two-dimensional signals). As the ultimate high point of the chapter we will demonstrate how one can restore images like the ones shown in Figures 6.1(a) and 6.2(a).

6.2 Frequency Analysis of Continuous Periodic Signals

In this first experience with frequency analysis, we will learn how to analyze the frequency content of continuous periodic signals (or functions) because a study of this specific class of signals requires the least amount of background in abstract mathematics. We shall begin with the simplest case of 1-periodic functions and will subsequently generalize our findings to functions of any given period T. Even though we will be using the term "continuous" throughout this section, the class of functions under consideration will also include *piecewise-continuous* functions.

The development of frequency analysis in this section will be guided by intuition and educated guesses. However, in the following Section 6.3, we will undertake an excursion into linear algebra, which will help us place our results on a firmer mathematical foundation and will also help us see how abstract concepts and methods can greatly simplify and streamline the development of frequency analysis.

Throughout this section we will use the terms "signal" and "function" (or "signal" and "sequence" in the discrete settings) almost interchangeably. The units of frequency are cycles per unit of time for continuous functions (or cycles per period for discrete sequences), and they will usually be omitted.

6.2.1 Trigonometric Fourier Coefficients of 1-Periodic Signals

As mentioned in the foreword to this section, we begin our exploration by developing methods of frequency analysis for 1-periodic functions (that is, functions of period $T = 1$). As a reminder, a function $x(t)$ is called **1-periodic** if the condition

$$x(t+1) = x(t)$$

holds for all $t \in \mathbb{R}$. The simplest examples of such functions are the sine waves defined by

$$s(t) = A\sin(n \cdot 2\pi t)$$

and the cosine waves defined by

$$c(t) = A\cos(n \cdot 2\pi t)$$

with the amplitude A and frequency n (measured in revolutions per unit of time).

It is important to note that all the results derived in this section for periodic functions also apply to functions defined on the semi-open interval $[0, 1)$. Indeed, for any such function $x(t)$, we can define its periodic extension $x_p(t)$ by means of

$$x_p(t) = x(t-n)$$

whenever $n \leq t < n+1$. It is evident that the domain of $x_p(t)$ is the entire real line, that $x_p(t)$ is 1-periodic, and that $x_p(t)$ agrees with $x(t)$ on the interval $[0, 1)$. If, in addition, the function $x(t)$ has the property that $x(0) = x(1)$, its periodic extension $x_p(t)$ agrees with $x(t)$ on the closed interval $[0, 1]$.

The principal problem explored in this section can be stated as follows: how does one break up an arbitrary 1-periodic signal into elementary constituent elements that are easy to analyze and manipulate? This idea is as old as scientific exploration itself. Throughout the entire history of science, researchers have sought to discover the basic building blocks of nature. The existence of atoms, for example, was conjectured as early as in the 5th century B.C.E. on the basis of purely metaphysical reasoning. Mathematicians too have often looked for ways to express complex mathematical structures as superpositions of simpler objects. It is this guiding principle that motivates the following bold assumption as the starting point in the development of frequency analysis:

A Bold Assumption:

Any physically realizable signal with the period $T = 1$ consists of sine and cosine waves as its constituent components and can, therefore, be expressed in the form

$$x(t) = A_0 + \sum_{n=1}^{\infty} A_n \cos(n \cdot 2\pi t) + \sum_{n=1}^{\infty} B_n \sin(n \cdot 2\pi t), \tag{6.1}$$

with the precise meaning of series convergence to be worked out.

This assumption may seem rather surprising at first, but it begins to appear more plausible once we think of a musical analogy. Consider the example of a musical chord played on a pipe organ; such chords consist of the main tones (which correspond to the notes being played) and of numerous overtones characteristic of that particular instrument.

In general, we feel comfortable with the idea that most (if not all) musical sounds "consist" of various tones and overtones. We are confident that ultimately any piece of music represents complex superpositions of pure pitches (and this confidence is the very reason for the existence of our musical notation). In order to represent the musical sounds to be produced, we indicate the keys (or strings) that need to be engaged and the instruments to be played, all of which correspond to certain sets of tones and overtones. The latter can be mathematically modeled by the sine and cosine functions of appropriate frequencies and amplitudes.

Numerous other examples of frequency analysis can be provided by the fields of radio- or acoustic communications.

In its essence, frequency analysis of a received signal comes down to determining the frequencies and amplitudes of the specific sine and cosine terms the signal consists of and expressing it in the form (6.1). We will approach this task very slowly and carefully: we will first consider several examples of the simplest possible kind and will gradually extend our findings to the more general type of signals.

Example 6.1 *Suppose that we are expecting to receive the signal described by the formula*

$$m(t) = A\sin(5 \cdot 2\pi t), \tag{6.2}$$

that is, a sine wave with the frequency of 5Hz and with the amplitude denoted by A. Suppose that we did receive a signal $x(t)$, but we are not certain what it is and would like to make sure that the received signal is indeed a sine wave, that its frequency does indeed equal 5Hz, and that its amplitude is indeed A. How would we approach the task if the only available instruments in our mathematical toolbox were trigonometric functions, multiplication, and integration?

The standard approach of frequency analysis works as follows: we multiply the received signal $x(t)$ by the basic trigonometric function of exactly the same type and frequency, that is, by $\sin(5 \cdot 2\pi t)$, integrate the product over one period, that is, over the interval $[0, 1]$ (or, alternatively, over the interval $[-1/2, 1/2]$, if that is more convenient), and denote the result by b_5. There are three cases to consider:

Case 1. If we did indeed receive the "correct" signal $m(t)$, that is, if we have $x(t) = A\sin(5 \cdot 2\pi t)$, the result would be

$$\begin{aligned} b_5 &= \int_0^1 x(t)\sin(5 \cdot 2\pi t)dt \qquad (6.3) \\ &= \int_0^1 A\sin^2(5 \cdot 2\pi t)dt \\ &= \frac{A}{2}\int_0^1 [1 - \cos(10 \cdot 2\pi t)]dt \\ &= \frac{A}{2}, \end{aligned}$$

where we used the trigonometric identity

$$\sin^2 \alpha = \frac{1}{2}[1 - \cos(2\alpha)] \tag{6.4}$$

in the penultimate step of the calculation.

Case 2. What if we had received the "wrong" signal? For example, what if the received signal had a "wrong" frequency, say 6Hz instead of 5Hz? In that case, the received signal would be described by the formula $x(t) = A\sin(6 \cdot 2\pi t)$, and the same process of multiplication of the received signal by $\sin(5 \cdot 2\pi t)$ and integration over one period yields

$$b_5 = \int_0^1 x(t)\sin(5 \cdot 2\pi t)dt \tag{6.5}$$

$$= \int_0^1 A\sin(6\cdot 2\pi t)\sin(5\cdot 2\pi t)dt$$
$$= \frac{A}{2}\int_0^1 \big[\cos(2\pi t) - \cos(11\cdot 2\pi t)\big]dt$$
$$= 0,$$

where we used the trigonometric identity

$$\sin\alpha\sin\beta = \frac{1}{2}\big[\cos(\alpha-\beta) - \cos(\alpha+\beta)\big] \tag{6.6}$$

in the penultimate step of the calculation.

Case 3. Finally, how would we go about verifying that it is the sine function, as opposed to the cosine that brings us the expected frequency 5Hz? Had we received $x(t) = A\cos(5\cdot 2\pi t)$ (instead of the expected sine wave $m(t)$), multiplied the received signal by $\sin(5\cdot 2\pi t)$ and integrated the product over the unit interval, we would again get

$$\begin{aligned} b_5 &= \int_0^1 x(t)\sin(5\cdot 2\pi t)dt \qquad (6.7)\\ &= \int_0^1 A\cos(5\cdot 2\pi t)\sin(5\cdot 2\pi t)dt\\ &= \frac{A}{2}\int_0^1 \sin(10\cdot 2\pi t)dt\\ &= 0, \end{aligned}$$

where we used the trigonometric identity

$$\sin\alpha\cos\alpha = \frac{1}{2}\sin(2\alpha) \tag{6.8}$$

in the penultimate step of the calculation.

Based on the calculations (6.3), (6.5), and (6.7), we can make the following educated guess regarding detection and analysis of pure sine waves of a specified frequency:

In order to confirm the type of the trigonometric function, to verify the frequency n, and to measure the amplitude A in a received signal of the form

$$x(t) = A\sin(n \cdot 2\pi t),$$

all we have to do is calculate the quantity

$$b_n = \int_0^1 x(t)\sin(n \cdot 2\pi t)dt,$$

and the result, if nonzero, will give us *half* the amplitude A. If $b_n = 0$, then either the frequency n is not present in the received signal, or the received signal is not in the form of the sine wave.

With the task set before us in Example 6.1 successfully completed, we can confidently take the same approach to detect and analyze the other type of elementary trigonometric signals - pure cosine waves.

Example 6.2 *In Example 6.1 and the subsequent discussion, we were expecting to receive the frequency 5Hz in the form of a pure sine wave. What if, instead, we were expecting to receive the signal described by the formula*

$$m(t) = A\cos(5 \cdot 2\pi t), \tag{6.9}$$

that is, a pure cosine wave with the frequency of 5Hz and with the amplitude A? Again, as in Example 6.1, we would like to verify the frequency, to measure the amplitude, and to confirm the type of the trigonometric function we are receiving. How would we go about this task?

Just as we did in Example 6.1, we consider the following three cases:

Case 1. We first assume that the received signal $x(t)$ indeed equals $m(t)$ and is, therefore, given by the formula (6.9). This time, we multiply the received signal by the cosine function of exactly the same frequency, that is, by $\cos(5 \cdot 2\pi t)$, integrate the product over the interval $[0, 1]$, and denote the result by a_5. This is accomplished by calculating

$$\begin{aligned} a_5 &= \int_0^1 x(t)\cos(5 \cdot 2\pi t)dt & (6.10) \\ &= \int_0^1 A\cos^2(5 \cdot 2\pi t)dt \\ &= \frac{A}{2}\int_0^1 \left[1 + \cos(10 \cdot 2\pi t)\right]dt \\ &= \frac{A}{2}, \end{aligned}$$

where we used the trigonometric identity

$$\cos^2\alpha = \frac{1}{2}\big[1 + \cos(2\alpha)\big] \tag{6.11}$$

in the penultimate step of the calculation.

Case 2. What if the received signal $x(t)$ had the "wrong" frequency, say 6Hz as opposed to the expected 5Hz? In that case,

$$\begin{aligned} a_5 &= \int_0^1 x(t)\cos(5\cdot 2\pi t)dt \\ &= \int_0^1 A\cos(6\cdot 2\pi t)\cos(5\cdot 2\pi t)dt \\ &= \frac{A}{2}\int_0^1 \big[\cos(2\pi t) + \cos(11\cdot 2\pi t)\big]dt \quad = \quad 0, \end{aligned} \tag{6.12}$$

where we used the trigonometric identity

$$\cos\alpha\cos\beta = \frac{1}{2}\big[\cos(\alpha-\beta) + \cos(\alpha+\beta)\big] \tag{6.13}$$

in the penultimate step of the calculation.

Case 3. Finally, if the expected frequency was received in the form of the sine function instead of the cosine function, we would get

$$\begin{aligned} a_5 &= \int_0^1 x(t)\cos(5\cdot 2\pi t)dt \\ &= \int_0^1 A\sin(5\cdot 2\pi t)\cos(5\cdot 2\pi t)dt \\ &= \frac{A}{2}\int_0^1 \sin(10\cdot 2\pi t)dt \\ &= 0, \end{aligned} \tag{6.14}$$

where we used the trigonomentic identity (6.8) in the penultimate step of the calculation.

Based on the calculations (6.10), (6.12), and (6.14), we can make the following educated guess regarding detection and analysis of pure cosine waves of a specified frequency:

In order to confirm the type of the trigonometric function, to verify the frequency n and to measure the amplitude A in a received signal of the form

$$x(t) = A\cos(n \cdot 2\pi t),$$

all we have to do is calculate the quantity

$$a_n = \int_0^1 x(t)\cos(n \cdot 2\pi t)dt,$$

and the result, if nonzero, will give us *half* the amplitude A. If $a_n = 0$, then either the frequency n is not present in the received signal, or the received signal is not in the form of the cosine wave.

Here an attentive reader might point out that we almost never receive pure sine waves or pure cosine waves without any phase shift or time shift. This is a valid point, and, for the sake of completeness, we must indeed consider one more example:

Example 6.3 *What if, instead, we receive the signal described by the formula*

$$x(t) = A\cos(n \cdot 2\pi t - \phi), \tag{6.15}$$

that is, a periodic signal with the frequency of n *and with the amplitude* A*, but this time in the form of the* cosine *wave with a phase shift* ϕ*? Again, as in the two preceding examples, we would like to verify the frequency, to measure the amplitude, and to confirm the type of the trigonometric function we are receiving. This time, we would also like to measure the phase shift* ϕ*. How would we go about this task?*

We begin by using the trigonometric identity

$$\cos(\alpha - \beta) = \cos(\alpha)\cos(\beta) + \sin(\alpha)\sin(\beta) \tag{6.16}$$

to rewrite $x(t)$ in the form

$$x(t) = A\cos(\phi)\cos(n \cdot 2\pi t) + A\sin(\phi)\sin(n \cdot 2\pi t)$$

and calculate

$$a_n = \frac{A\cos(\phi)}{2} \quad \text{and} \quad b_n = \frac{A\sin(\phi)}{2}$$

using the same techniques as the ones we used in the previous two examples. From here, we can easily determine both the amplitude A and the phase shift ϕ using the formulas

$$\tan(\phi) = \frac{b_n}{a_n} \quad \text{and} \quad A^2 = 4(a_n^2 + b_n^2), \tag{6.17}$$

as readers can easily verify on their own.

Having developed the rudimentary tools of frequency analysis in Examples 6.1, 6.2, and 6.3, where the simplest possible periodic functions were considered, we now move on to the general case of an arbitrary physically realizable signal $x(t)$, which, according to the assumption (6.1), can be expressed as

$$x(t) = A_0 + \sum_{n=1}^{\infty} A_n \cos(n \cdot 2\pi t) + \sum_{n=1}^{\infty} B_n \sin(n \cdot 2\pi t)$$

with the precise set of the components and the amplitudes A_n and B_n to be determined. Thanks to all the work we have done in the Examples 6.1 and 6.2, it is evident that

$$\begin{aligned} a_n &= \int_0^1 x(t)\cos(n \cdot 2\pi t)dt \qquad (6.18) \\ &= \int_0^1 \Big[\sum_{n=0}^{\infty} A_n \cos(n \cdot 2\pi t) + \sum_{n=1}^{\infty} B_n \sin(n \cdot 2\pi t)\Big] \cos(n \cdot 2\pi t)dt \\ &= A_n/2, \end{aligned}$$

which implies $A_n = 2a_n$. Thus, in order to determine whether the component $A_n \cos(n \cdot 2\pi t)$ with the given frequency n is present in $x(t)$ and to measure its contribution, all we need to do is calculate a_n. Similarly,

$$\begin{aligned} b_n &= \int_0^1 x(t)\sin(n \cdot 2\pi t)dt \qquad (6.19) \\ &= \int_0^1 \Big[\sum_{n=0}^{\infty} A_n \cos(n \cdot 2\pi t) + \sum_{n=1}^{\infty} B_n \sin(n \cdot 2\pi t)\Big] \sin(n \cdot 2\pi t)dt \\ &= B_n/2, \end{aligned}$$

which in turn yields $B_n = 2b_n$. Thus, in order to determine whether the component $B_n \sin(n \cdot 2\pi t)$ with the given frequency n is present in $x(t)$ and to measure its contribution, all we need to do is calculate b_n.

It is also evident that in order to measure the constant term A_0, all we need to do is calculate

$$a_0 = \int_0^1 x(t)dt$$

and set $A_0 = a_0$. We have arrived at the following definition:

Definition 6.1 *For a 1-periodic function $x(t)$, the nth* ***trigonometric Fourier coefficients*** *a_n and b_n are defined by*

$$a_n = \int_0^1 x(t)\cos(n \cdot 2\pi t)dt, \qquad (6.20)$$

$$b_n = \int_0^1 x(t)\sin(n \cdot 2\pi t)dt, \tag{6.21}$$

and

$$a_0 = \int_0^1 x(t)dt \tag{6.22}$$

provided that the integrals exist.

The issue of integrability rarely (if ever) arises for physically realizable signals. It is evident from the discussion that preceded Definition 6.1 that the Fourier coefficients a_n and b_n measure the contribution of the frequency n to the signal $x(x)$ delivered in the form of the cosine and sine functions respectively. Therefore, the following notion of the Fourier series expansion appears as an immediate and natural continuation of that discussion.

Definition 6.2 *A physically realizable 1-periodic function $x(t)$ can be expressed in the form of its* ***trigonometric Fourier series expansion***

$$x(t) = a_0 + 2\sum_{n=1}^{\infty} a_n \cos(n \cdot 2\pi t) + 2\sum_{n=1}^{\infty} b_n \sin(n \cdot 2\pi t), \tag{6.23}$$

with the coefficients a_n and b_n defined by the formulas (6.20), (6.21), *and* (6.22).

Unfortunately, the precise mathematical meaning of convergence of series like (6.23) is beyond the scope of this book. However, we will make an attempt to explain it as rigorously as possible in Section 6.3.

The intuitive meaning of Fourier coefficients is rooted in the way they measure the contribution of all the integer frequencies to the signal $x(t)$. The pair of terms

$$h_n(t) = a_n \cos(n \cdot 2\pi t) + b_n \sin(n \cdot 2\pi t) \tag{6.24}$$

is called **the nth harmonic** of $x(t)$, the quantity $\sqrt{a_n^2 + b_n^2}$ is called **the amplitude of the nth harmonic**, and its square,

$$E_n = a_n^2 + b_n^2, \tag{6.25}$$

represents the amount of the energy of the signal that is contained in its nth harmonic. The most common way to represent the distribution of the signal energy across the frequencies is by means of the so-called **energy spectrum**, which is the graph of E_n as a function of n.

The next several examples are intended to further illustrate the concept of trigonometric Fourier coefficients:

Example 6.4 *In the simplest situations (when the signal $x(t)$ is a trigonometric polynomial, that is, a finite superposition of sine and cosine functions),* *we do not even need to resort to integration. For instance, it is evident that the Fourier coefficients of*

$$\begin{aligned} x(t) &= 3 + 11\cos(2\pi t) - 2\cos(3\cdot 2\pi t) \\ &+ 5\sin(2\cdot 2\pi t) - 7\sin(4\cdot 2\pi t) + 4\sin(5\cdot 2\pi t) \end{aligned}$$

are

$$a_0 = 3, \quad a_1 = 11/2, \quad a_3 = -1$$

and

$$b_2 = 5/2, \quad b_4 = -7/2, \quad b_5 = 2,$$

which follows directly from their intuitive meaning. The graph of this signal is shown in Figure 6.3 alongside its energy spectrum.

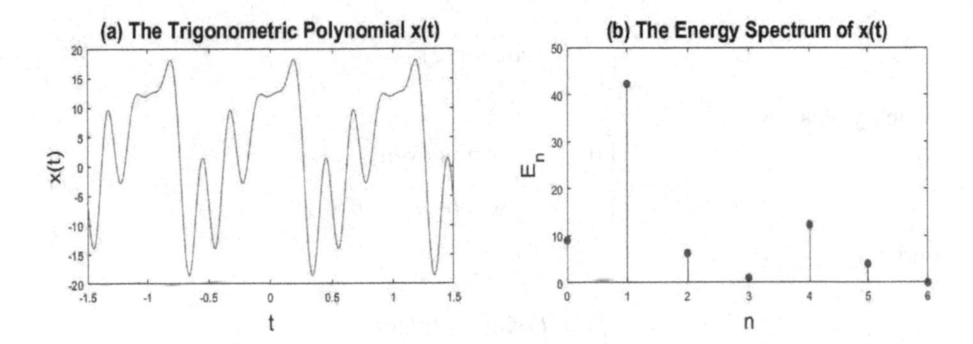

FIGURE 6.3: The trigonometric polynomial of Example 6.4 and its energy spectrum.

We started this section with the assumption that any physically realizable signal is a superposition of sine and cosine functions, just as any musical sound is a superposition of pure pitches. To illustrate this point and also to make the defining formulas of the Fourier coefficients and Fourier series expansions seem less forbidding to the reader, we next offer the following example of a square-wave function:

Example 6.5 *The 1-periodic square-wave function is defined by*

$$x(t) = \begin{cases} 1, & n - 1/4 \leq t \leq n + 1/4 \\ 0, & \textit{otherwise} \end{cases} \tag{6.26}$$

for all $n \in \mathbb{Z}$. We first observe that

$$a_0 = \int_0^1 x(t)dt$$

$$= \int_{-1/4}^{1/4} 1dt$$
$$= 1/2$$

and next calculate the trigonometric Fourier coefficients of $x(t)$ for all $n \in \mathbb{N}$. Using (6.20) and (6.21), we obtain

$$\begin{aligned} a_n &= \int_0^1 x(t)\cos(n \cdot 2\pi t)dt \\ &= \int_{-1/4}^{1/4} \cos(n \cdot 2\pi t)dt \\ &= \frac{1}{n \cdot 2\pi} \sin(n \cdot 2\pi t)\Big|_{-1/4}^{1/4} \\ &= \frac{1}{n\pi} \sin(n\pi/2), \end{aligned}$$

which yields

$$a_n = \begin{cases} 0, & \text{if } n \text{ is even} \\ \frac{(-1)^k}{\pi n}, & \text{where } n = 2k+1 \end{cases} \tag{6.27}$$

and

$$\begin{aligned} b_n &= \int_0^1 x(t)\sin(n \cdot 2\pi t)dt \\ &= \int_{-1/4}^{1/4} \sin(n \cdot 2\pi t)dt \\ &= -\frac{1}{n \cdot 2\pi} \cos(n \cdot 2\pi t)\Big|_{-1/4}^{1/4} = 0, \end{aligned}$$

which we could have expected from the very beginning because integrals of odd functions over symmetric intervals always equal zero.

The square-wave function $x(t)$ defined by (6.26) can therefore be expanded into its trigonometric Fourier series as

$$\begin{aligned} x(t) &= \frac{1}{2} + \frac{2}{\pi}\cos(2\pi t) - \frac{2}{3\pi}\cos(3 \cdot 2\pi t) \\ &+ \frac{2}{5\pi}\cos(5 \cdot 2\pi t) - \frac{2}{7\pi}\cos(7 \cdot 2\pi t) + \ldots \end{aligned} \tag{6.28}$$

In practice, we can only add up *finitely* many terms of the series expansion (6.28) for an approximate representation of $x(t)$. Such finite expressions that use terms with frequencies up to n are called **nth-degree Fourier polynomials**. Figure 6.4 illustrates the approximation of several periods of the square-wave function considered in Example 6.5 by means of the third-degree, ninth-degree, and fifteenth-degree Fourier polynomials. The graph of the first fifteen trigonometric Fourier coefficients of the square-wave functions is shown in Figure 6.5.

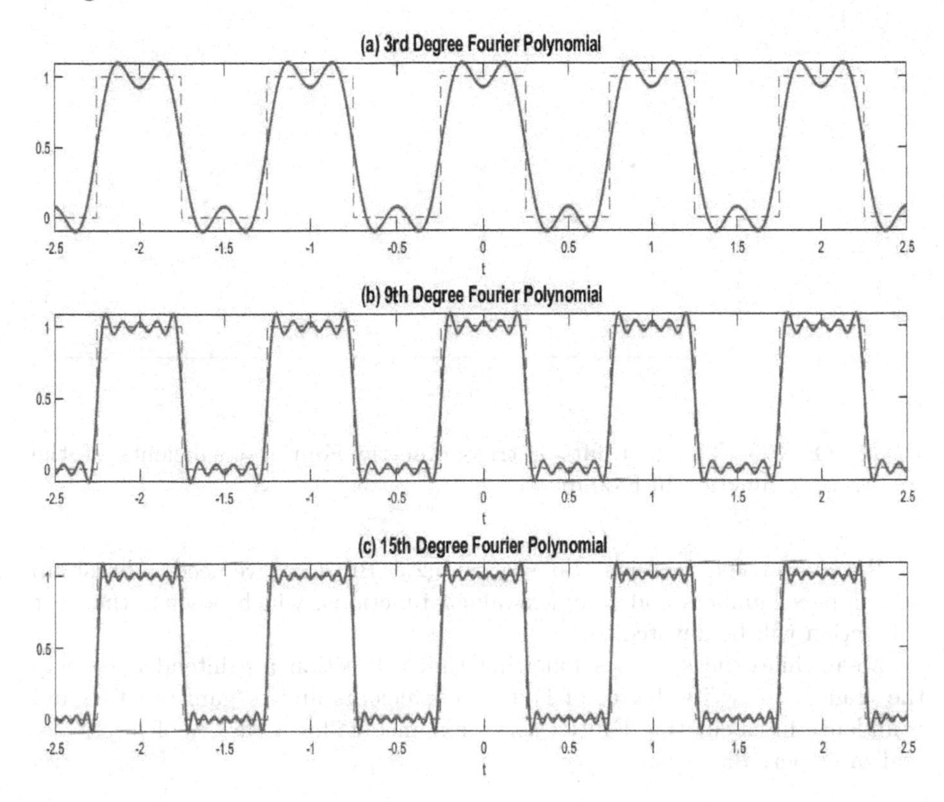

FIGURE 6.4: Approximations of the square-wave function in Example 6.5 by Fourier polynomials.

Example 6.5 demonstrates convincingly that a periodic signal that does not even seem to be of trigonometric nature can nevertheless be expressed as a superposition of sine and cosine waves. By doing so, it provides even more evidence for the plausibility of our Bold Assumption (6.1). For further evidence, the reader will be asked to calculate the trigonometric Fourier coefficients and to construct trigonometric Fourier series expansions for several other functions in the exercises at the end of this section.

We have made a very significant accomplishment having discovered nothing less than the building blocks of nature (at least in the context of continuous 1-periodic functions). But here is a troubling thought: why are we treating the same frequency n differently depending on which trigonometric function – sine or cosine – is bringing it to us? After all, isn't it the frequency itself (and not the carrier) that is of interest?

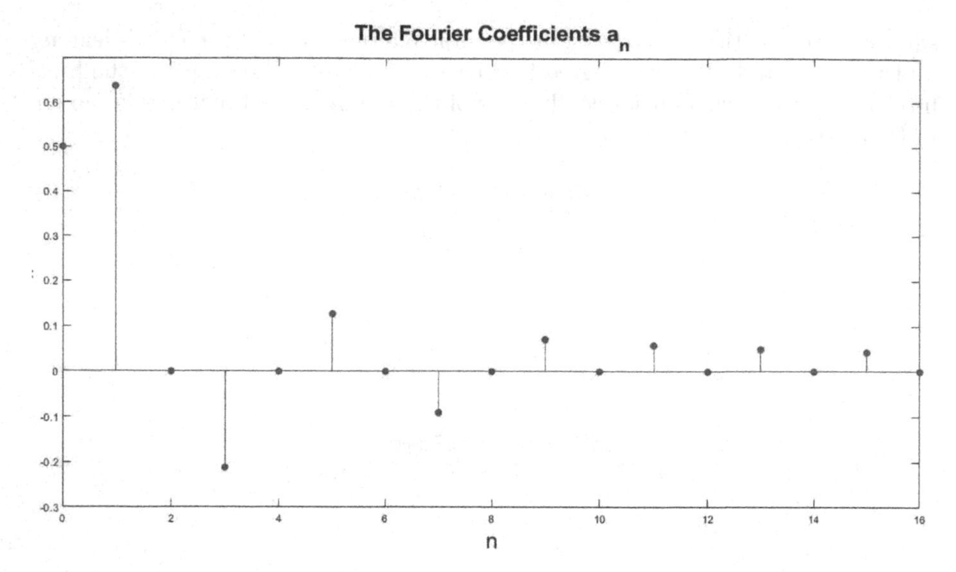

FIGURE 6.5: The first fifteen trigonometric Fourier coefficients of the square-wave function in Example 6.5.

We will settle this issue in Subsection 6.2.3. But first, we need to brush up on complex numbers and complex-valued functions, which is what the next subsection will be devoted to.

Meanwhile, the exercises following this subsection are intended to help the reader review the basics of Fourier coefficients and to gain practice and confidence in calculating Fourier series expansions for a variety of physically realizable periodic signals.

Subsections 6.2.1 Exercises.

1. Perform the integration and verify the results in the calculations (6.3), (6.5), and (6.7).

2. Perform the integration and verify the results in the calculations (6.10), (6.12), and (6.14).

3. Repeat the calculations in Examples 6.1 and 6.2 of this section, but this time, consider an arbitrary frequency n as opposed to the specific frequency $n = 5$ used in the text.

4. How would you go about attempting to prove the trigonometric identities (6.4), (6.6), (6.8), (6.11), (6.13), and (6.16)?

5. Verify the integration in the calculations (6.18) and (6.19).

6. Verify the formula (6.17) for the phase and the amplitude of a generalized cosine wave.

7. Substitute $t = 0$ into the Fourier series expansion (6.28) of the square-wave function (6.26) to prove **Leibnitz summation formula**

$$1 - \frac{1}{3} + \frac{1}{5} - \frac{1}{7} + \frac{1}{9} - \frac{1}{11} + ... = \frac{\pi}{4}. \tag{6.29}$$

8. Calculate the trigonometric Fourier coefficients a_n and b_n for square-pulse functions defined by

   (a)

$$x(t) = \begin{cases} 1, & n - 1/8 \leq t \leq n + 1/8 \\ 0, & \text{otherwise,} \end{cases} \tag{6.30}$$

   (b)

$$y(t) = \begin{cases} 1, & n \leq t \leq n + 1/8 \\ 0, & \text{otherwise,} \end{cases} \tag{6.31}$$

   for all $n \in \mathbb{Z}$. Use these Fourier coefficients to construct the Fourier series expansions of $x(t)$ and $y(t)$. Compare your answers to the formulas derived in Example 6.5 for the square-wave function.

9. Calculate the trigonometric Fourier coefficients a_n and b_n for a triangular-wave function defined by

$$x(t) = \begin{cases} 1 + 2(t - n), & n - 1/2 \leq t \leq n \\ 1 - 2(t - n), & n \leq t \leq n + 1/2 \end{cases} \tag{6.32}$$

   for all $n \in \mathbb{Z}$. Use these Fourier coefficients to construct the Fourier series expansion of $x(t)$.

10. Calculate the trigonometric Fourier coefficients a_n and b_n for the function defined by

$$y(t) = \begin{cases} -1, & n - 1/4 \leq t \leq n \\ 1, & n \leq t \leq n + 1/4 \\ 0, & \text{elsewhere,} \end{cases} \tag{6.33}$$

    for all $n \in \mathbb{Z}$. Use these Fourier coefficients to construct the Fourier series expansion of $y(t)$.

11. Calculate the trigonometric Fourier coefficients a_n and b_n for the saw-tooth function defined by

$$r_1(t) = 2(t - n), \qquad n - 1/2 \leq t < n + 1/2 \tag{6.34}$$

and the ramp-shape function defined by

$$r_2(t) = \begin{cases} 2(t-n), & n \le t \le n+1/2 \\ 0, & \text{otherwise,} \end{cases} \tag{6.35}$$

for all $n \in \mathbb{Z}$. Use these Fourier coefficients to construct the Fourier series expansions of $r_1(t)$ and $r_2(t)$.

12. Calculate the trigonometric Fourier coefficients a_n and b_n for the function defined by

$$p(t) = 4(t-n)^2, \qquad n - 1/2 \le t \le n + 1/2 \tag{6.36}$$

for all $n \in \mathbb{Z}$. Use these Fourier coefficients to construct the Fourier series expansion of $p(t)$.

13. Calculate the trigonometric Fourier coefficients a_n and b_n for the function defined by

$$s(t) = \begin{cases} 4(t-n)^2, & n \le t \le n+1 \\ 0, & \text{otherwise} \end{cases} \tag{6.37}$$

for all $n \in \mathbb{Z}$. Use these Fourier coefficients to construct the Fourier series expansion of $s(t)$.

14. Use the Fourier series expansion of the periodic second-power function (6.37) to prove the identities

(a)

$$\sum_{n=1}^{\infty} \frac{1}{n^2} = 1 + \frac{1}{2^2} + \frac{1}{3^2} + \frac{1}{4^2} + \frac{1}{5^2} + \dots = \frac{\pi^2}{6}, \tag{6.38}$$

(b)

$$\sum_{n=1}^{\infty} \frac{(-1)^{n+1}}{n^2} = 1 - \frac{1}{2^2} + \frac{1}{3^2} - \frac{1}{4^2} + \dots = \frac{\pi^2}{12}, \tag{6.39}$$

(c)

$$\sum_{n \text{ odd}} \frac{1}{n^2} = 1 + \frac{1}{3^2} + \frac{1}{5^2} + \frac{1}{7^2} + \frac{1}{9^2} + \dots = \frac{\pi^2}{8} \tag{6.40}$$

by substituting appropriate values for t.

15. Calculate the trigonometric Fourier coefficients a_n and b_n for the functions defined by

(a)

$$q_1(t) = (t-n)^3, \qquad n - 1/2 \le t < n + 1/2 \tag{6.41}$$

(b)

$$q_2(t) = \begin{cases} (t-n)^3, & n \leq t \leq n + 1/2 \\ 0, & \text{otherwise}, \end{cases} \tag{6.42}$$

for all $n \in \mathbb{Z}$. Use these Fourier coefficients to construct the Fourier series expansion of $q_1(t)$ and $q_2(t)$.

16. Use the Fourier series expansion of the third-power function (6.41) to prove the identity

$$1 - \frac{1}{3^3} + \frac{1}{5^3} - \frac{1}{7^3} + \frac{1}{9^3} - \frac{1}{11^3} + ... = \frac{\pi^3}{32} \tag{6.43}$$

by substituting an appropriate value for t.

MATLAB Exercises.

1. Write MATLAB functions to create graphs of the Fourier polynomial approximations of degree n for the following 1-periodic signals:
   (a) The square-wave function defined by (6.26).
   (b) The triangular-wave function defined by (6.32).
   (c) The step function defined by (6.33).
   (d) The ramp-shape functions defined by (6.34) and (6.35).
   (e) The two-branch periodic parabola defined by (6.36).
   (f) The one-branch periodic parabola defined by (6.37).
   (g) The periodic third-power function defined by (6.41).

   Your MATLAB functions should accept the value of n as a parameter and plot both the function to be approximated and the approximating nth degree Fourier polynomial on the same set of axes.
2. Test your functions using several different values of n. Comment on the quality of approximation.
3. Prepare a PowerPoint presentation to illustrate your results.

6.2.2 A Refresher on Complex Numbers

Although there is a multitude of excellent algebraic and geometric reasons convincingly justifying the existence of a number whose square equals -1, a detailed discussion of the origins and history of the imaginary unit i is outside the scope of this book. We just begin with the assumption that there is a number denoted by i with the property that

$$i^2 = -1$$

and define a *complex number* to be an expression of the form

$$z = a + bi,$$

where both a and b are real numbers. The number a is called the real part of z and is denoted by $Re(z)$, and the number b is called the imaginary part of z, and is denoted by $Im(z)$. Two complex numbers $z_1 = a_1 + b_1 i$ and $z_2 = a_2 + b_2 i$ are equal if and only if both $a_1 = a_2$ and $b_1 = b_2$.

A complex number $z = a + bi$ can be identified with the point (a, b) in the two-dimensional plane $\mathbb{R}^2$ or, equivalently, with the vector $\vec{v} = \begin{bmatrix} a \\ b \end{bmatrix}$. The operations of addition and subtraction of complex numbers can, therefore, be defined accordingly by

$$(a_1 + b_1 i) \pm (a_2 + b_2 i) = (a_1 \pm a_2) + (b_1 \pm b_2)i.$$

Unlike vectors in $\mathbb{R}^2$, complex numbers can also be multiplied and divided. From the defining property $i^2 = -1$, it follows that the product of two complex numbers

$$(a + bi)(c + di) = (ac - bd) + (bc + ad)i, \tag{6.44}$$

and the quotient of two complex numbers

$$\frac{a + bi}{c + di} = \frac{(a + bi)(c - di)}{(c + di)(c - di)} = \frac{ac + bd}{c^2 + d^2} + \frac{bc - ad}{c^2 + d^2} i \tag{6.45}$$

provided, of course, that the denominator is nonzero. For example,

$$(1 + 2i)(3 + 4i) = (1 \cdot 3 - 2 \cdot 4) + (1 \cdot 4 + 2 \cdot 3)i = -5 + 10i$$

and

$$\frac{1 + 2i}{3 + 4i} = \frac{(1 + 2i)(3 - 4i)}{(3 + 4i)(3 - 4i)} = \frac{1 \cdot 3 + 2 \cdot 4}{3^2 + 4^2} + \frac{2 \cdot 3 - 1 \cdot 4}{3^2 + 4^2} i = \frac{11}{25} + \frac{2}{25} i.$$

Given a complex number $z = a + bi$, its complex conjugate $\bar{z}$ is defined by

$$\bar{z} = a - bi,$$

and its absolute value $|z|$ is defined by

$$|z| = \sqrt{a^2 + b^2},$$

the latter being in complete agreement with the interpretation of complex numbers as vector in $\mathbb{R}^2$. It is also important to point out that

$$|z|^2 = z \cdot \bar{z}. \tag{6.46}$$

As we have already mentioned, complex numbers are identified with points in $\mathbb{R}^2$, which can be described either by their Cartesian coordinates (x, y) or by their polar coordinates (r, θ) with the conversion formulas

$$x = r\cos\theta, \qquad y = r\sin\theta$$

and

$$r^2 = x^2 + y^2, \qquad \tan\theta = \frac{y}{x}.$$

As a consequence, complex numbers can be written either in the Cartesian form as

$$z = x + iy$$

or in the polar form as

$$z = r(\cos\theta + i\sin\theta),$$

which is illustrated in Figure 6.6.

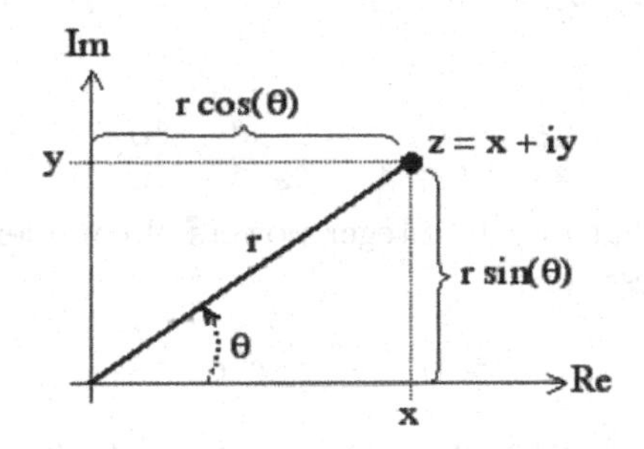

FIGURE 6.6: The Cartesian and polar forms of a complex number.

Certain elementary functions of a complex variable (such as integer powers and polynomials) are defined exactly the same way as their real-variable counterparts. To properly extend the definitions of other functions, some degree of imagination and resourcefulness is called for. For example, in order to extend the definition of the exponential function to the entire complex plane, we take advantage of its Maclaurin series expansion to define

$$e^z = 1 + z + \frac{z^2}{2!} + \frac{z^3}{3!} + \frac{z^4}{4!} + \frac{z^5}{5!} + \frac{z^6}{6!} + \frac{z^7}{7!} + \cdots \tag{6.47}$$

In a similar manner, Maclaurin series allow us to define trigonometric functions of a complex variable by means of

$$\cos(z) = 1 - \frac{z^2}{2!} + \frac{z^4}{4!} - \frac{z^6}{6!} + \cdots \tag{6.48}$$

and

$$\sin(z) = z - \frac{z^3}{3!} + \frac{z^5}{5!} - \frac{z^7}{7!} + \cdots \tag{6.49}$$

Substituting $z = i\theta$ into (6.47) and $z = \theta$ into (6.48) and (6.49) yields *Euler's equation*

$$e^{i\theta} = \cos\theta + i\sin\theta, \tag{6.50}$$

which allows us to express the polar form of a complex number in a particularly concise way as

$$z = r(\cos\theta + i\sin\theta) = re^{i\theta}. \tag{6.51}$$

Some of the frequently encountered complex numbers in this complex exponential notation take the form

$$1 = e^{2\pi i}, \quad i = e^{(\pi/2)i}, \quad -1 = e^{\pi i}, \quad \text{and} \quad -i = e^{-(\pi/2)i}.$$

The complex exponential notation allows us to greatly simplify expressions for products and quotients of complex numbers. Specifically, given $z_1 = r_1e^{i\theta_1}$ and $z_2 = r_2e^{i\theta_2}$, standard rules of working with exponents imply that

$$z_1 \cdot z_2 = r_1e^{i\theta_1}r_2e^{i\theta_2} = r_1r_2e^{i(\theta_1+\theta_2)} \tag{6.52}$$

and

$$\frac{z_1}{z_2} = \frac{r_1e^{i\theta_1}}{r_2e^{i\theta_2}} = \frac{r_1}{r_2}e^{i(\theta_1-\theta_2)}, \tag{6.53}$$

provided, of course, that $r_2 \neq 0$. Integer powers of complex numbers can also be conveniently expressed as

$$z^n = \left(re^{i\theta}\right)^n = r^ne^{in\theta},$$

which greatly simplifies calculations as compared to the algebraic notation that involves Cartesian coordinates.

Adding together both sides of the equations

$$e^{i\theta} = \cos\theta + i\sin\theta$$

and

$$e^{-i\theta} = \cos\theta - i\sin\theta,$$

allows us to express the cosine function as

$$\cos\theta = \frac{1}{2}\left(e^{i\theta} + e^{-i\theta}\right), \tag{6.54}$$

and subtracting the second equation from the first one yields a similar expression

$$\sin\theta = \frac{1}{2i}\left(e^{i\theta} - e^{-i\theta}\right) \tag{6.55}$$

for the sine function. The complex expressions (6.54) and (6.55) are tremendously useful in all areas of frequency analysis, as readers will see for themselves in subsequent sections. They often allow us to do away with cumbersome trigonometric manipulations altogether.

When studying trigonometry as part of our pre-calculus sequence, we rarely get to work on the proofs of trigonometric identities, and there is a very good reason for that: those proofs are often difficult. However, with the help of the complex expressions (6.54) and (6.55), they can be reduced to simple (albeit somewhat tedious) algebraic manipulations. For example, to prove the Pythagorean identity

$$\cos^2\theta + \sin^2\theta = 1,$$

all we need to do is simplify

$$\begin{aligned}\cos^2\theta + \sin^2\theta &= \left[\frac{1}{2}\left(e^{i\theta} + e^{-i\theta}\right)\right]^2 + \left[\frac{1}{2i}\left(e^{i\theta} - e^{-i\theta}\right)\right]^2 \\ &= \frac{1}{4}\left[\left(e^{2i\theta} + 2 + e^{-2i\theta}\right) - \left(e^{2i\theta} - 2 + e^{-2i\theta}\right)\right] \\ &= 1,\end{aligned}$$

which is much shorter and conceptually simpler than the standard geometry-based proofs provided in pre-calculus texts.

Raising both sides of Euler's equation (6.50) to the nth power, we obtain what is known as *De Moivre's formula*

$$\left(\cos\theta + i\sin\theta\right)^n = \cos(n\theta) + i\sin(n\theta), \tag{6.56}$$

which is also useful in establishing certain types of trigonometric identities. For example, substituting $n = 2$ in De Moivre's formula gives us

$$\cos^2\theta + 2i\cos\theta\sin\theta - \sin^2\theta = \cos(2\theta) + i\sin(2\theta),$$

and setting the real and imaginary parts on both sides of this equation equal to each other yields the well-known double-angle identities

$$\cos(2\theta) = \cos^2\theta - \sin^2\theta \tag{6.57}$$

and

$$\sin(2\theta) = 2\cos\theta\sin\theta$$

without any tedious and cumbersome trigonometric manipulations.

In the exercises that follow, the reader will be asked to practice complex-number arithmetic, to verify all the formulas and identities mentioned in this subsection, and to use Euler's equation and De Moivre's formulas to prove a few of the trigonometric identities typically covered without proof in standard pre-calculus courses. Upon completing those exercises, the reader will be fully ready to proceed with the study of frequency analysis of signals and images.

Subsection 6.2.2 Exercises.

1. Calculate the sums, differences, products, quotients, complex conjugates, and absolute values of the following complex numbers:

(a) $z = 1 + i$ and $w = 1 - i$,
(b) $z = 1 + i$ and $w = -1 + i$,
(c) $z = 1 + i$ and $w = z$,
(d) $z = 1 + i$ and $w = -z$,
(e) $z = 3 + 2i$ and $w = 5 - i$,
(f) $z = 8 + 5i$ and $w = -1 - i$.

Use MATLAB to check your answers.

2. Convert the following numbers into their polar and complex exponential forms. Proceed to calculate their products and quotients in the complex exponential form.

(a) $z = 1 + i$ and $w = 1 - i$,
(b) $z = 1 + i$ and $w = -1 + i$,
(c) $z = 1 + i$ and $w = -1 - i$,
(d) $z = 1 + i$ and $w = -z$.

3. Use the complex exponential form of the following complex numbers to calculate z^2, z^3, and z^4. For each part of the exercise, try to come up with a general formula for z^n.

(a) $z = i$,
(b) $z = -i$,
(c) $z = 1 + i$,
(d) $z = 1 - i$
(e) $z = -1 + i$,
(f) $w = -1 - i$.

Use MATLAB to check your answers.

4. Verify the calculations in (6.44), (6.45), and (6.46).

5. Verify that the commutative, associative, and distributive properties hold for the complex numbers.

6. Verify that 0 and 1 are the additive and multiplicative identities for the complex numbers.

7. Show that there are no "divisors of zero" among the complex numbers. In other words, prove that if $z_1 z_2 = 0$, then $z_1 = 0$ or $z_2 = 0$.

8. Illustrate and prove the following two forms of the "triangle inequality" which state that for any two complex numbers z and w,

(a) $|z + w| \leq |z| + |w|$,
(b) $\big||z| - |w|\big| \leq |z - w|$.

9. Use mathematical induction to prove the generalized triangle inequality in the form

$$\left|\sum_{k=1}^{n} z_k\right| \leq \sum_{k=1}^{n} |z_k|.$$

10. Use Maclaurin series expansions of the exponential and trigonometric functions to verify Euler's equation (6.50) and its consequences (6.54) and (6.55).

11. Verify De Moivre's formula (6.56).

12. Show that the multiplicative inverse of the complex number z is

$$z^{-1} = \frac{\bar{z}}{|z|^2}.$$

In particular, show that if $|z| = 1$, then $z^{-1} = \bar{z}$.

13. In Subsection 6.2.1, we used the power-reduction formulas

$$\cos^2 \alpha = \frac{1}{2}\big[1 + \cos(2\alpha)\big] \quad \text{and} \quad \sin^2 \alpha = \frac{1}{2}\big[1 - \cos(2\alpha)\big].$$

Prove these identities using

(a) The double-angle identity (6.57),

(b) The complex expressions (6.54) and (6.55) for the cosine and sine functions.

14. Use De Moivre's formula (6.56) to prove the following trigonometric identities:

(a) $\cos(3\theta) = \cos^3 \theta - 3\cos\theta \sin^2 \theta$,

(b) $\sin(3\theta) = 3\cos^2 \theta \sin\theta - \sin^3 \theta$.

15. Derive similar formulas for $\cos(4\theta)$ and $\sin(4\theta)$ using

(a) De Moivre's formula (6.56),

(b) The double-angle identities (6.57).

16. Use the complex expressions (6.54) and (6.55) for the cosine and sine functions to prove the following four sum-and-difference trigonometric identities:

(a) $\sin(\alpha + \beta) = \sin\alpha \cos\beta + \cos\alpha \sin\beta$,

(b) $\sin(\alpha - \beta) = \sin\alpha \cos\beta - \cos\alpha \sin\beta$,

(c) $\cos(\alpha + \beta) = \cos\alpha \cos\beta - \sin\alpha \sin\beta$,

(d) $\cos(\alpha - \beta) = \cos\alpha \cos\beta + \sin\alpha \sin\beta$.

17. Use the complex expressions (6.54) and (6.55) for the cosine and sine functions to prove the following three product-to-sum trigonometric identities, which we have already used in Subsection 6.2.1:

(a) $\sin\alpha \sin\beta = \frac{1}{2}\big[\cos(\alpha - \beta) - \cos(\alpha + \beta)\big]$,

(b) $\cos\alpha\cos\beta = \frac{1}{2}\left[\cos(\alpha-\beta)+\cos(\alpha+\beta)\right]$,

(c) $\sin\alpha\cos\beta = \frac{1}{2}\left[\sin(\alpha-\beta)+\sin(\alpha+\beta)\right]$.

18. Although not commonly used in frequency analysis, the sum-to-product formulas are, nevertheless, extremely important for solving trigonometric equations that arise in applications of differential calculus to optimization. Prove the following four identities using either the substitution $u = (\alpha+\beta)/2$ and $v = (\alpha+\beta)/2$ and the product-to-sum identities established in the previous exercise, or the complex expressions (6.54) and (6.55):

    (a) $\cos\alpha + \cos\beta = 2\cos\frac{\alpha+\beta}{2}\cos\frac{\alpha-\beta}{2}$,

    (b) $\cos\alpha - \cos\beta = -2\sin\frac{\alpha+\beta}{2}\sin\frac{\alpha-\beta}{2}$,

    (c) $\sin\alpha + \sin\beta = 2\sin\frac{\alpha+\beta}{2}\cos\frac{\alpha-\beta}{2}$,

    (d) $\sin\alpha - \sin\beta = 2\cos\frac{\alpha+\beta}{2}\sin\frac{\alpha-\beta}{2}$.

    Explain why the identity (d) is an immediate consequence of the identity (c).

6.2.3 Complex Fourier Coefficients

In Subsection 6.2.1, we developed a method for resolving a given signal $x(t)$ into separate frequency components and measuring the individual contribution of those components. Our approach proved fruitful, culminating in the definition of Fourier coefficients, construction of Fourier series expansions, and, overall, in a method for representing general physically realizable signals as superpositions of sine and cosine waves. However, as has already been mentioned a few pages earlier, we might find it a bit strange and counter-intuitive that Fourier coefficients measure the contribution of a given frequency n to the signal $x(t)$ separately depending on whether that frequency comes in the form of the sine or the cosine function.

That does not feel completely satisfactory. After all, it is the *frequency* n that we are interested in and not necessarily the precise type of the trigonometric function that brings it to us. We might, therefore, wonder whether it is possible to come up with "unified" Fourier coefficients that would somehow combine the contributions of the related components $\sin(n\cdot 2\pi t)$ and $\cos(n\cdot 2\pi t)$ and thus carry complete information on the entire nth harmonic (6.24).

Fortunately, we do not have to look too far. With the help of Euler's equation

$$e^{i\theta} = \cos\theta + i\sin\theta$$

and its consequences, the complex expressions

$$\cos\theta = \frac{1}{2}\left(e^{i\theta} + e^{-i\theta}\right) \quad \text{and} \quad \sin\theta = \frac{1}{2i}\left(e^{i\theta} - e^{-i\theta}\right)$$

for the cosine and sine functions, we can rewrite the individual trigonometric components of the Fourier series expansion

$$x(t) = a_0 + 2\sum_{n=1}^{\infty} a_n \cos(n \cdot 2\pi t) + 2\sum_{n=1}^{\infty} b_n \sin(n \cdot 2\pi t)$$

as

$$a_n \cos(n \cdot 2\pi t) = \frac{a_n}{2}\left(e^{n \cdot 2\pi i t} + e^{-n \cdot 2\pi i t}\right)$$

and

$$b_n \sin(n \cdot 2\pi t) = \frac{b_n}{2i}\left(e^{n \cdot 2\pi i t} - e^{-n \cdot 2\pi i t}\right).$$

In that way, we can replace trigonometric expressions with complex exponential functions of the form $e^{n \cdot 2\pi i t}$ and $e^{-n \cdot 2\pi i t}$. It follows that the signal $x(t)$ can be expressed in the form

$$c_n = \sum_{n=-\infty}^{\infty} c_n e^{n \cdot 2\pi i t} \tag{6.58}$$

with the (complex) coefficients

$$c_n = a_n - b_n i \quad \text{and} \quad c_{-n} = a_n + b_n i. \tag{6.59}$$

The complex exponentials can, therefore, be viewed as the constituent frequency components of $x(t)$, and we must learn to detect them and measure their contribution to $x(t)$ directly. To do so, we will employ a method analogous to the one used in Subsection 6.2.1. In the process, we are going to discover that it is much easier to develop frequency analysis with complex exponentials than to do the same with trigonometric functions.

Example 6.6 *Suppose that we are expecting to receive the frequency component described by the formula*

$$m(t) = Ae^{n \cdot 2\pi i t} \tag{6.60}$$

and would like to make sure that the received signal $x(t)$ is indeed a complex exponential function of the form (6.60), that its frequency does indeed equal n, and that its amplitude is indeed A. Again, similar to the conditions imposed in the examples of Subsection 6.2.1, the only available instruments in our mathematical toolbox are complex exponential functions, multiplication, and integration.

Guided by the approach we followed in Subsection 6.2.1, we multiply the received signal $x(t)$ by $e^{-n \cdot 2\pi i t}$, integrate the product over one period (that is, over the interval $[0, 1]$), and denote the result by c_n. As we are working with complex exponentials (as opposed to trigonometric functions), there are two (as opposed to three) cases to consider:

Case 1. If we received the "correct" signal, that is, if $x(t) = m(t) = Ae^{n\cdot 2\pi it}$, the result would be

$$\begin{aligned} c_n &= \int_0^1 x(t)e^{-n\cdot 2\pi it}dt \qquad (6.61) \\ &= \int_0^1 Ae^{n\cdot 2\pi it} \cdot e^{-n\cdot 2\pi it}dt \\ &= \int_0^1 Adt = A, \end{aligned}$$

with $c_n = A$ being the precise measure of the contribution of the component $e^{n\cdot 2\pi it}$ to the received signal $x(t)$.

Case 2. What if the received signal had the "wrong" frequency $k \neq n$? In that case, the same process of multiplication of $x(t) = Ae^{k\cdot 2\pi it}$ by $e^{-n\cdot 2\pi it}$ and integration over the interval $[0, 1]$ yields

$$\begin{aligned} c_n &= \int_0^1 x(t)e^{-n\cdot 2\pi it}dt \qquad (6.62) \\ &= \int_0^1 Ae^{k\cdot 2\pi it} \cdot e^{-n\cdot 2\pi it}dt \\ &= \int_0^1 Ae^{(k-n)\cdot 2\pi it}dt \\ &= \frac{A}{(k-n)\cdot 2\pi i}e^{(k-n)\cdot 2\pi it}\Big|_0^1 = 0. \end{aligned}$$

Based on this example, we can draw the following conclusion:

> In order to confirm the frequency n and to measure the amplitude A in a received signal
>
> $$x(t) = Ae^{n\cdot 2\pi it},$$
>
> all we have to do is calculate the quantity
>
> $$c_n = \int_0^1 x(t)e^{-n\cdot 2\pi it}dt,$$
>
> and the result, if nonzero, will give us precisely the amplitude A.
> If $c_n = 0$, then the frequency n is not present in the received signal.

This observation leads to the following definition:

Definition 6.3 *For a physically realizable 1-periodic function $x(t)$, the nth* ***complex Fourier coefficients*** *c_n are defined by the formula*

$$c_n = \int_0^1 x(t)e^{-n\cdot 2\pi it}dt, \tag{6.63}$$

provided that the integral exists.

As in the case of trigonometric Fourier coefficients, the issue of integrability of complex Fourier coefficients rarely (if ever) arises for physically realizable signals.

It is evident from the discussion that preceded Definition 6.3 that the Fourier coefficients c_n measure the contribution of the frequency n to the signal $x(t)$, which renders natural the concept of the complex Fourier series expansion:

Definition 6.4 *A physically realizable 1- periodic function $x(t)$ can be expressed in the form of its* ***(complex) Fourier series expansion***

$$x(t) = \sum_{n=-\infty}^{\infty} c_n e^{n\cdot 2\pi it} \tag{6.64}$$

with the coefficients c_n determined by the formula (6.63).

Unfortunately, the precise meaning of series convergence in (6.64) is beyond the scope of this book. However, as we have already mentioned in the comments following the definition of trigonometric Fourier series expansion, an effort will be made in a subsequent section to get as close as possible to a rigorous explanation.

Remark on notation. When two or more different signals are under consideration, and, as a consequence, there is a possibility for ambiguity, the alternative notation

$$X(n) \equiv c_n \quad \text{or} \quad \hat{x}_n \equiv c_n$$

is often used for complex Fourier coefficients of $x(t)$.

We hope that the reader has been impressed by the relative simplicity of the derivation of all the relevant formulas for complex Fourier coefficients, especially when compared to the amount of work we had to do when using trigonometric identities in order to derive similar formulas for the counterpart trigonometric Fourier coefficients. Nevertheless, the reader might still be wondering how complex coefficients of the form (6.63) and complex expressions of the form (6.64) can possibly be used to express real-valued physically realizable signals. To dispel any remaining doubts, we revisit the square-wave function

$$x(t) = \begin{cases} 1, & n - 1/4 \le t \le n + 1/4 \\ 0, & \text{otherwise} \end{cases}$$

already discussed in Example 6.5. We observe that

$$\begin{aligned} c_0 &= \int_0^1 x(t)dt \\ &= \int_{-1/4}^{1/4} 1dt \\ &= 1/2 \end{aligned}$$

and rely on the defining formula (6.63) to determine that the (complex) Fourier coefficients of $x(t)$ are

$$\begin{aligned} c_n &= \int_0^1 x(t)e^{-n\cdot 2\pi it}dt \\ &= \int_{-1/4}^{1/4} e^{-n\cdot 2\pi it}dt \\ &= \frac{1}{-n\cdot 2\pi i}e^{-n\cdot 2\pi it}\Big|_{-1/4}^{1/4} \\ &= \frac{1}{n\pi}\sin(n\pi/2), \end{aligned}$$

where we used the identity

$$\sin\theta = \frac{1}{2i}\left(e^{i\theta} - e^{-i\theta}\right)$$

in the last step of the calculation. Thus, we can see that the Fourier coefficients of $x(t)$ are all real-valued and are given by the formula

$$c_n = \begin{cases} \frac{(-1)^k}{n\pi}, & n = 2k+1 \\ 0, & n = 2k \end{cases}$$

for all $n \in \mathbb{Z}$. Therefore, the formula (6.64) tells us that the Fourier series expansion of $x(t)$ is

$$\begin{aligned} x(t) &= \frac{1}{2} + \frac{1}{\pi}e^{2\pi it} + \frac{1}{\pi}e^{-2\pi it} - \frac{1}{3\pi}e^{3\cdot 2\pi it} - \frac{1}{3\pi}e^{-3\cdot 2\pi it} \\ &+ \frac{1}{5\pi}e^{5\cdot 2\pi it} - \frac{1}{5\pi}e^{-5\cdot 2\pi it} - \frac{1}{7\pi}e^{7\cdot 2\pi it} - \frac{1}{7\pi}e^{-7\cdot 2\pi it} + \ldots \\ &= \frac{1}{2} + \frac{2}{\pi}\cos(2\pi t) - \frac{2}{3\pi}\cos(3\cdot 2\pi t) \\ &+ \frac{2}{5\pi}\cos(5\cdot 2\pi t) - \frac{2}{7\pi}\cos(7\cdot 2\pi t) + \ldots \end{aligned}$$

in complete agreement with (6.28) in Example 6.5.

Subsection 6.2.3 Exercises.

1. Perform the integration and verify the results in the calculations (6.61) and (6.62).

2. Verify the relationship (6.59) between the real and complex Fourier coefficients.

3. Calculate the complex Fourier coefficients c_n for square-pulse functions $x(t)$ and $y(t)$ defined by formulas (6.30) and (6.31). Use these Fourier coefficients to construct the Fourier series expansions for $x(t)$ and $y(t)$. Compare your answers to the ones obtained for Exercise 8 following Subsection 6.2.1.

4. Calculate the complex Fourier coefficients c_n for the triangular-wave function $x(t)$ defined by (6.32). Use these Fourier coefficients to construct the Fourier series expansion of $x(t)$. Compare your answers to the ones obtained for Exercise 9 following Subsection 6.2.1.

5. Calculate the complex Fourier coefficients c_n for the function $y(t)$ defined by (6.33). Use these Fourier coefficients to construct the Fourier series expansion of $y(t)$. Compare your answers to the ones obtained for Exercise 10 following Subsection 6.2.1.

6. Calculate the complex Fourier coefficients for the saw-tooth function $r_1(t)$ and the ramp-shape function $r_2(t)$ defined by (6.34) and (6.35) respectively. Use these Fourier coefficients to construct the Fourier series expansions of $r_1(t)$ and $r_2(t)$. Compare your answers to the ones obtained for Exercise 11 following Subsection 6.2.1.

7. Calculate the complex Fourier coefficients c_n for the function $p(t)$ defined by (6.36). Use these Fourier coefficients to construct the Fourier series expansion of $p(t)$. Compare your answers to the ones obtained for Exercise 12 following Subsection 6.2.1.

8. Calculate the complex Fourier coefficients c_n for the function $s(t)$ defined by (6.37). Use these Fourier coefficients to construct the Fourier series expansion of $s(t)$. Compare your answers to the ones obtained for Exercise 13 following Subsection 6.2.1.

9. Calculate the complex Fourier coefficients c_n for the functions $q_1(t)$ and $q_2(t)$ defined by (6.41) and (6.42) respectively. Use these Fourier coefficients to construct the Fourier series expansion of $q_1(t)$ and $q_2(t)$. Compare your answers to the ones obtained for Exercise 13 following Subsection 6.2.1.

10. Comment on the symmetry properties of the complex Fourier coefficients in Exercises 3 through 9.

6.2.4 Properties of Fourier Coefficients

We have already noted the relative simplicity of using complex Fourier coefficients, which stems from the fact that they measure the contribution of a given frequency irrespective of the type of the trigonometric function that brings it to us. In this subsection, the reader is going to see that it is also easier to formulate and prove the relevant mathematical and physical properties of complex Fourier coefficients as compared to the corresponding properties of their trigonometric counterparts.

To save space during the subsequent discussion, we begin with the following general assumption and convention:

> All signals considered in this subsection will be real-valued and physically realizable. All of them will be assumed to be 1-periodic.
>
> Whenever we mention Fourier coefficients or Fourier series expansions without specifying the type, it is the complex versions that are meant, as opposed to the trigonometric ones.

As a starting point of the discussion devoted to the basic properties of Fourier coefficients, we are going to inquire about conditions under which we can expect them to be symmetric. We could not help but notice that in the examples discussed in Subsection 6.2.1, the Fourier coefficients did turn out to be symmetric in the sense that $c_{-n} = c_n$ for all $n \in \mathbb{Z}$. Is this always going to be the case? Let us consider an arbitrary real-valued physically realizable 1-periodic signal $x(t)$ and calculate the Fourier coefficient

$$\begin{aligned} c_{-n} &= \int_0^1 x(t)e^{-(-n)\cdot 2\pi it}dt \\ &= \overline{\int_0^1 x(t)e^{-n\cdot 2\pi it}dt} = \bar{c}_n, \end{aligned}$$

where we used the fact that the complex conjugate of $e^{-i\theta}$ is $e^{i\theta}$ for any $\theta \in \mathbb{R}$. Thus, we can conclude that complex Fourier coefficients of real-valued signals are always *complex conjugate*-symmetric. It follows that if Fourier coefficients also happen to be real-valued, then they are symmetric.

Which brings us to the next topic: since we are living in the "real" world (where all the quantities we come in practical contact with are real-valued), we are naturally conditioned to be more comfortable dealing with real numbers than with complex ones. Therefore, we might be curious when and under what conditions complex Fourier coefficients turn out to be real-valued.

To answer this question, we first observe that for the cosine function

$$x(t) = A\cos(n \cdot 2\pi t) = \frac{A}{2}\left(e^{n\cdot 2\pi it} + e^{-n\cdot 2\pi it}\right)$$

the Fourier coefficients are

$$c_n = c_{-n} = \frac{A}{2},$$

and they are indeed real-valued (and symmetric). On the other hand, for the sine function

$$x(t) = A\sin(n \cdot 2\pi t) = \frac{A}{2i}\left(e^{n\cdot 2\pi it} - e^{-n\cdot 2\pi it}\right)$$

the Fourier coefficients are

$$c_n = -c_{-n} = \frac{A}{2i},$$

and they are purely imaginary (and complex conjugate-symmetric). We might suspect that these properties have something to do with the fact that the cosine function is even and that the sine function is odd.

This is indeed the case. To prove it, we suppose that $x(t)$ is *even* and real-valued. Then $x(-t) = x(t)$, and, therefore, for any $n \in \mathbb{Z}$, the complex conjugate of its complex Fourier coefficient c_n is

$$\begin{aligned}
\bar{c}_n &= \overline{\int_0^1 x(t)e^{-n\cdot 2\pi it}dt} \\
&= \int_0^1 x(t)e^{n\cdot 2\pi it}dt \\
&= \int_0^1 x(u)e^{-n\cdot 2\pi iu}du = c_n,
\end{aligned}$$

where the substitution $u = -t$ and the assumption that $x(t)$ is even were used in the last step. Since the only numbers that equal their complex conjugates are the real numbers and since $\bar{c}_n = c_n$, we can conclude that the Fourier coefficients c_n of $x(t)$ are real-valued. Similarly, if $x(t)$ is *odd* and real-valued, then $x(-t) = -x(t)$ for all $t \in \mathbb{R}$, and the complex conjugates of its complex Fourier coefficients are

$$\begin{aligned}
\bar{c}_n &= \overline{\int_0^1 x(t)e^{-n\cdot 2\pi it}dt} \\
&= \int_0^1 x(t)e^{n\cdot 2\pi it}dt \\
&= -\int_0^1 x(u)e^{-n\cdot 2\pi iu}du = -c_n,
\end{aligned}$$

where the substitution $u = -t$ and the assumption that $x(t)$ is odd were used in the last step. Since $\bar{c}_n = -c_n$, c_n must be purely imaginary. To summarize,

The Fourier coefficients of a real-valued signal are always (conjugate-) symmetric. The Fourier coefficients of an **even** real-valued signal are real and symmetric. The Fourier coefficients of an **odd** real-valued signal are purely imaginary.

We would like to encourage the reader to investigate the corresponding symmetry properties of the trigonometric Fourier coefficients a_n and b_n.

Next, we proceed to derive the time-shift and frequency-shift properties of Fourier coefficients. Suppose that a 1-periodic signal $x(t)$ gets delayed by τ units of time, and instead of $x(t)$ we end up receiving its delayed version $y(t) = x(t - \tau)$. What are the Fourier coefficients of $y(t)$?

Using the notation $Y(n)$ for the nth complex Fourier coefficient of $y(t)$, we calculate

$$\begin{aligned} Y(n) &= \int_0^1 y(t)e^{-n\cdot 2\pi it}dt \\ &= \int_0^1 x(t-\tau)e^{-n\cdot 2\pi it}dt \\ &= \int_0^1 x(u)e^{-n\cdot 2\pi i(u+\tau)}du \\ &= e^{-n\cdot 2\pi i\tau}\int_0^1 x(u)e^{-n\cdot 2\pi iu}du \\ &= e^{-n\cdot 2\pi i\tau}X(n), \end{aligned}$$

where we used the substitution $u = t - \tau$. We have derived the following "time-shift rule":

The delay of a signal by τ time units is equivalent to the multiplication of its Fourier coefficients by the factor of $e^{-n\cdot 2\pi i\tau}$. Formally, if

$$y(t) = x(t - \tau)$$

then

$$Y(n) = e^{-n\cdot 2\pi i\tau}X(n).$$

What if the given signal had somehow undergone a so-called frequency shift? In other words, what if it got multiplied by $e^{k\cdot 2\pi it}$? The nth Fourier coefficient of $y(t) = e^{k\cdot 2\pi it}x(t)$ would then turn out to be

$$Y(n) = \int_0^1 y(t)e^{-n\cdot 2\pi it}dt$$

$$\begin{aligned}
&= \int_0^1 e^{k\cdot 2\pi it} e^{-n\cdot 2\pi it} dt \\
&= \int_0^1 x(t) e^{-(n-k)\cdot 2\pi it} dt \\
&= X(n-k),
\end{aligned}$$

which explains the term "frequency shift" used here. We have thus established the following "frequency-shift rule":

> Multiplication of a signal by $e^{k\cdot 2\pi it}$ is equivalent to shifting its Fourier coefficients by k positions to the right. Formally, if
>
> $$y(t) = e^{k\cdot 2\pi it} x(t)$$
>
> then
>
> $$Y(n) = X(n-k).$$

In real life, we do not usually encounter complex exponentials, but we do commonly encounter trigonometric functions. If the signal $x(t)$ happens to get multiplied by $\cos(k \cdot 2\pi t)$, we can use the complex expression of the cosine function and calculate the Fourier coefficients of $y(t) = x(t)\cos(k \cdot 2\pi t)$ to be

$$\begin{aligned}
Y(n) &= \int_0^1 y(t) e^{-n\cdot 2\pi it} dt \\
&= \int_0^1 \cos(k \cdot 2\pi t) \cdot x(t) e^{-n\cdot 2\pi it} dt \\
&= \frac{1}{2} \int_0^1 \left(e^{k\cdot 2\pi it} + e^{-k\cdot 2\pi it}\right) \cdot x(t) e^{-n\cdot 2\pi it} dt \\
&= \frac{1}{2} \left[\int_0^1 x(t) e^{-(n-k)\cdot 2\pi it} dt + \int_0^1 x(t) e^{-(n+k)\cdot 2\pi it} dt\right] \\
&= \frac{1}{2} \Big[X(n-k) + X(n+k)\Big],
\end{aligned}$$

thus establishing the following "modulation property":

Multiplication of a signal by $\cos(k \cdot 2\pi t)$ is equivalent to averaging the two shifts of its Fourier coefficients by $-k$ and k. Formally, if

$$y(t) = \cos(k \cdot 2\pi t) \cdot x(t)$$

then

$$Y(n) = \frac{1}{2}\Big[X(n-k) + X(n+k)\Big].$$

Multiplication by $\cos(k \cdot 2\pi t)$ is used frequently in radio communications, particularly in various AM (amplitude modulation) techniques and is of great practical importance. In the exercises at the end of this section, the reader will be asked to derive an analogous rule for multiplying a signal by the sine function.

As important as the symmetry and shift properties are, the most interesting and consequential question is what happens to the Fourier coefficients when two signals get multiplied together. How does the frequency content of the product of signals relate to the frequency contents of the factors? To calculate the Fourier coefficients of $g(t) = x(t) \cdot y(t)$, we use the Fourier series expansion

$$x(t) = \sum_{k=-\infty}^{\infty} X(k)e^{2\pi ikt}$$

of $x(t)$ to calculate

$$\begin{aligned}
G(n) &= \int_0^1 x(t) \cdot y(t)e^{-n \cdot 2\pi it}dt \\
&= \int_0^1 \Big(\sum_{k=-\infty}^{\infty} X(k)e^{2\pi ikt}\Big) \cdot y(t)e^{-n \cdot 2\pi it}dt \\
&= \sum_{k=-\infty}^{\infty} X(k)\Big(\int_0^1 x(t)e^{-(n-k) \cdot 2\pi it}dt\Big) \\
&= \sum_{k=-\infty}^{\infty} X(k)Y(n-k) \\
&= \big(X * Y\big)(n),
\end{aligned}$$

where the interchange of the order of summation and integration can be justified for physically realizable signals using methods of advanced analysis. We have derived a result of fundamental importance:

Multiplication of periodic functions is equivalent to the convolution of their complex Fourier coefficients. Formally, if

$$g(t) = x(t) \cdot y(t)$$

then

$$G(n) = \big(X * Y\big)(n).$$

We first encountered the operation of convolution of two sequences in Chapter 5 in the context of calculating weighted averages of successive sequence elements and, by extension, in the context of calculating weighted averages of neighboring pixels. Subsequently, we encountered it once again in the context of edge detection. The ubiquitous nature and fundamental importance of convolution have been further made evident in this section as it appeared naturally in the context of Fourier coefficients of products of function.

The section on Fourier coefficients would not be complete without at least a brief mention of convolution of continuous signals. Given two 1-periodic physically realizable signals $x(t)$ and $y(t)$, their **continuous circular convolution** $(x * y)(t)$ is defined by

$$(x * y)(t) = \int_0^1 x(u)y(t-u)du, \tag{6.65}$$

provided that the integral exists, which it does for physically realizable signals.

Since this text is primarily concerned with *digital*, that is, discrete signals and images, we will not be discussing continuous convolution in any detail. However, for the sake of completeness, we must still ask the following question: what are the Fourier coefficients of the convolution of two signals? Substituting (6.65) into the definition of Fourier coefficients, we obtain

$$\begin{aligned}
\widehat{(x * y)}(n) &= \int_0^1 (x * y)(t)e^{-n\cdot 2\pi i t}dt \\
&= \int_0^1 \Big(\int_0^1 x(u)y(t-u)du\Big)e^{-n\cdot 2\pi i t}dt \\
&= \int_0^1 x(u)\Big(\int_0^1 y(t-u)e^{-n\cdot 2\pi i(t-u)}dt\Big)e^{-n\cdot 2\pi i u}du \\
&= X(n)\cdot Y(n),
\end{aligned}$$

where the interchange of the order of integration can be justified using methods of advanced analysis. Thus,

The continuous circular convolution of two signals is equivalent to the product of their (complex) Fourier coefficients. Formally, if

$$g(t) = (x * y)(t)$$

then

$$G(n) = X(n) \cdot Y(n).$$

We conclude this subsection with the observation of complete duality between the time domain of continuous signals and the frequency domain of their Fourier coefficients. As we have seen, time shifts of signals are equivalent to multiplying their Fourier coefficients by complex exponentials, whereas frequency modulation (multiplying signals by complex exponentials) is equivalent to frequency shifts (that is, shifts in Fourier coefficients). Products of signals in the time domain are equivalent to convolutions of their Fourier coefficients in the frequency domain and, vice versa, convolutions of signals in the time domain are equivalent to products of their Fourier coefficients. The list of such duality relationships can be continued, but it is, unfortunately, beyond the scope of this book. We hope that the reader feels motivated to undertake further study of frequency analysis.

Subsection 6.2.4 Exercises.

1. Suppose $x(t)$ and $y(t)$ are 1-periodic physically realizable signals and α is a constant (real or complex). Define $g(t) = x(t)+y(t)$ and $h(t) = \alpha \cdot x(t)$.

   (a) Show that $G(n) = X(n) + Y(n)$.

   (b) Show that $H(n) = \alpha X(n)$.

   Together, these two properties imply that calculating Fourier coefficients constitutes a linear operator.

2. Suppose that $x(t)$ is a 1-periodic physically realizable signal and that $y(t) = x'(t)$. Use integration by parts and the periodicity of $x(t)$ to prove that

   $$Y(n) = 2\pi i n X(n)$$

   for all $n \in \mathbb{Z}$. Extend this rule to the general case of T-periodic signals.

3. As in the previous problem, suppose that $x(t)$ is a 1-periodic physically realizable signal and that $y(t) = x'(t)$. Derive analogous formulas relating the trigonometric Fourier coefficients a_n and b_n of the functions $x(t)$ and $y(t)$. Extend this relation to the general case of T-periodic signals.

4. Suppose $x(t)$ is a 1-periodic physically realizable signal and $y(t) = x(t)\sin(k \cdot 2\pi t)$. Calculate the Fourier coefficients $Y(n)$ of $y(t)$.

5. Calculate the Fourier coefficients a_n and b_n as well as the complex Fourier coefficients c_n for a triangular-wave function defined by

$$x(t) = 2|t - n|, \quad n - 1/2 \leq t \leq n + 1/2$$

for all $n \in \mathbb{Z}$ in two different ways: directly and using the time-delay property of Fourier coefficients and the results you obtained in Problem 4 of Subsection 6.2.3. Use these Fourier coefficients to construct the Fourier series expansion of $x(t)$.

6. Verify that the convolution of the square-wave function (6.26) with itself is the triangular-wave function (6.32) from Problem 9 of Subsection 6.2.1. What is the relationship between the Fourier coefficients of the square-wave and triangular-wave functions?

6.2.5 T-Periodic Signals

Up until now, the focus in our exploration of frequency analysis has been on 1-periodic signals. What if, instead, the given signal is T-periodic in the sense that

$$x(t + T) = x(t)$$

for all $t \in \mathbb{R}$? In that case, a discussion similar to the one in Subsections 6.2.1 and 6.2.3 would lead to the definition of trigonometric Fourier coefficients by the formulas

$$a_n = \frac{1}{T}\int_0^T x(t)\cos(n \cdot 2\pi t/T)dt, \tag{6.66}$$

and

$$b_n = \frac{1}{T}\int_0^T x(t)\sin(n \cdot 2\pi t/T)dt, \tag{6.67}$$

and of the complex Fourier coefficients by

$$c_n = \frac{1}{T}\int_0^T x(t)e^{-n \cdot 2\pi i t/T}dt, \tag{6.68}$$

provided that the integrals exist. The Fourier series expansion of $x(t)$ can then be written as

$$x(t) = a_0 + 2\sum_{n=1}^{\infty} a_n \cos(n \cdot 2\pi t/T) + 2\sum_{n=1}^{\infty} b_n \sin(n \cdot 2\pi t/T), \tag{6.69}$$

or, equivalently, as

$$x(t) = \sum_{n=-\infty}^{\infty} c_n e^{n \cdot 2\pi i t/T}, \tag{6.70}$$

if complex Fourier coefficients are used. The following example is intended to clarify the details of working with these new formulas (6.66) – (6.70) .

Example 6.7 *We will calculate the Fourier series expansion of the periodic signal $x(t)$ given by*

$$x(t) = t - 2n, \quad n - 1 \le t < n + 1 \tag{6.71}$$

for all $n \in \mathbb{Z}$.

It is clear that $x(t)$ is 2-periodic. Since integrals of odd functions over intervals symmetric with respect to the origin are zero, the trigonometric Fourier coefficients a_n are

$$a_0 = \frac{1}{2}\int_{-1}^{1} t dt = 0$$

and

$$a_n = \frac{1}{2}\int_{-1}^{1} t \cos(n \cdot 2\pi t/2) dt = 0$$

for all $n \in \mathbb{N}$. To calculate the coefficients b_n, we use integration by parts and obtain

$$\begin{aligned}
b_n &= \frac{1}{2}\int_{-1}^{1} t \sin(n \cdot 2\pi t/2) dt \\
&= \frac{1}{2}\int_{-1}^{1} t \sin(n \cdot \pi t) dt \\
&= -\frac{1}{2\pi n}\int_{-1}^{1} t[\cos(n \cdot \pi t)]' dt \\
&= -\frac{t\cos(n \cdot \pi t)}{2\pi n}\bigg|_{-1}^{1} + \frac{1}{2\pi n}\int_{-1}^{1} \cos(n \cdot \pi t) dt \\
&= -\frac{2\cos(n \cdot \pi)}{2\pi n} + \frac{1}{2\pi^2 n^2}\sin(n \cdot \pi t)\bigg|_{-1}^{1} \\
&= \frac{(-1)^{n+1}}{\pi n},
\end{aligned}$$

since $\cos(n \cdot \pi) = (-1)^n$ and $\sin(n \cdot \pi) = 0$ for all $n \in \mathbb{N}$. The Fourier series expansion of $x(t)$ is, therefore,

$$\begin{aligned}
x(t) &= 2\sum_{n=1}^{\infty} \frac{(-1)^{n+1}}{\pi n} \sin(n \cdot \pi t) \qquad (6.72) \\
&= \frac{2}{\pi}\left(\sin(\pi t) - \frac{1}{2}\sin(2\pi t) + \frac{1}{3}\sin(3\pi t) - \frac{1}{4}\sin(4\pi t) + ... \right).
\end{aligned}$$

It might be worth noting that substituting $1/2$ for t in both sides of this Fourier series expansion yields the **Leibnitz summation formula**

$$1 - \frac{1}{3} + \frac{1}{5} - \frac{1}{7} + \frac{1}{9} - \frac{1}{11} + ... = \frac{\pi}{4},$$

which the reader may have already derived in Problem 7 at the end of Subsection 6.2.1.

For completeness, we also calculate the complex Fourier coefficients of $x(t)$, which are

$$c_0 = \frac{1}{2}\int_{-1}^{1} t dt = 0$$

and (using integration by parts)

$$\begin{aligned}
c_n &= \frac{1}{2}\int_{-1}^{1} te^{-n\cdot 2\pi it/2}dt = \frac{1}{2}\int_{-1}^{1} te^{-n\cdot\pi it}dt \\
&= -\frac{1}{2\pi in}\int_{-1}^{1} t(e^{-n\cdot\pi it})'dt \\
&= -\frac{t}{2\pi in}e^{-n\cdot\pi it}\Big|_{-1}^{1} + \frac{1}{2\pi in}\int_{-1}^{1} e^{-n\cdot\pi it}dt \\
&= \frac{-e^{n\cdot\pi i} - e^{-n\cdot\pi i}}{2\pi in} + \frac{1}{2\pi^2 n^2}e^{-n\cdot\pi it}\Big|_{-1}^{1} \\
&= \frac{i\cos(\pi n)}{n\pi} + 0 \\
&= \frac{(-1)^n i}{\pi n},
\end{aligned}$$

where we used the 2-periodicity of $e^{-n\cdot\pi it}$ and the facts that $\cos(\pi n) = (-1)^n$ for all $n \in \mathbb{Z}$ in the last two steps of the calculation. Since the pairwise sums of symmetric terms of the Fourier series expansion

$$\begin{aligned}
x(t) &= \sum_{n=-\infty}^{\infty} c_n e^{-n\cdot 2\pi it/2} \\
&= c_0 + (c_1 e^{\pi it} + c_{-1}e^{-\pi it}) + (c_2 e^{2\pi it} + c_{-2}e^{-2\pi it}) + \ldots
\end{aligned}$$

equal

$$\begin{aligned}
c_n e^{n\cdot\pi it} + c_{-n}e^{-n\cdot\pi it} &= \frac{(-1)^n i}{\pi n}e^{n\cdot\pi it} - \frac{(-1)^n i}{\pi n}e^{-n\cdot\pi it} \\
&= \frac{2(-1)^{n+1}}{\pi n}\sin(\pi nt), \qquad (6.73)
\end{aligned}$$

we get

$$x(t) = 2\sum_{n=1}^{\infty}\frac{(-1)^{n+1}}{\pi n}\sin(n\cdot\pi t),$$

which agrees with the expansion (6.72) that we obtained earlier using trigonometric Fourier coefficients a_n and b_n. Figure 6.7 illustrates the approximation

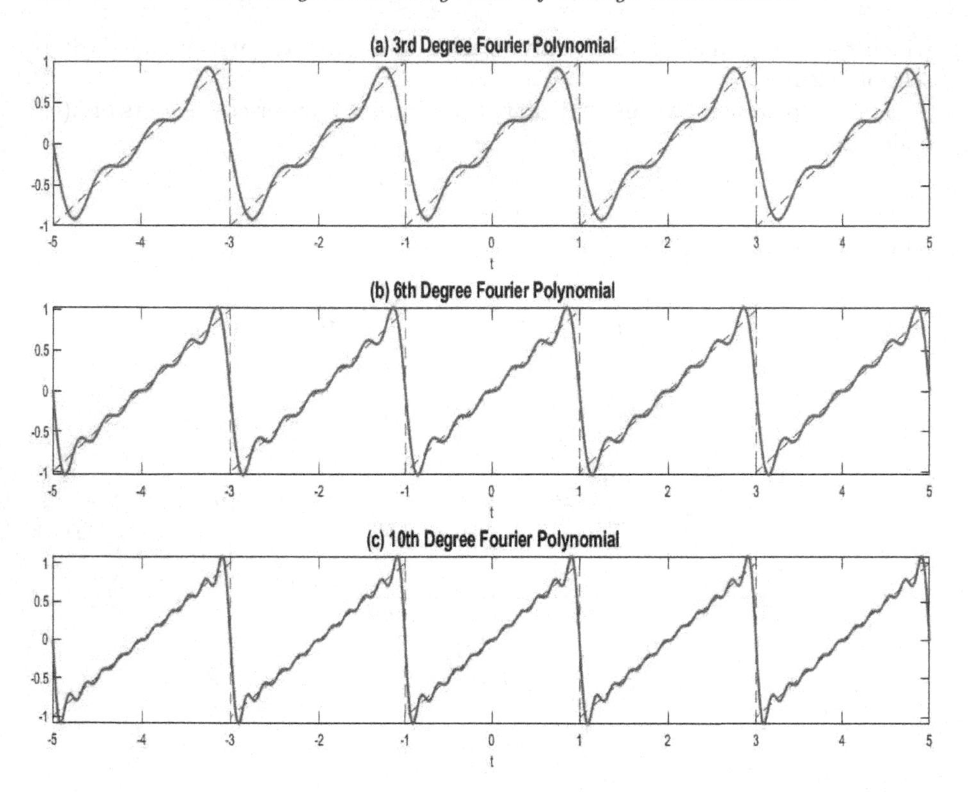

FIGURE 6.7: Approximations of the 2-periodic saw-tooth function by Fourier polynomials.

of several periods of the 2-periodic saw-tooth function (6.71) by means of the third-degree, sixth-degree, and tenth-degree Fourier polynomials.

In the exercises at the end of this section, the reader will be asked to calculate Fourier coefficients and to write Fourier series expansions for a variety of oft-encountered T-periodic functions.

It is also important to note that all the results derived in this subsection also apply to functions defined on the semi-open interval $[0, T)$, because any such function $x(t)$ is equivalent to its periodic extension $x_p(t)$ defined by

$$x_p(t) = x(t - nT)$$

whenever $nT \leq t < (n+1)T$. It is evident that the domain of $x_p(t)$ is the entire real line, that $x_p(t)$ is T-periodic, and that $x_p(t)$ agrees with $x(t)$ on the interval $[0, T)$ (and even on the closed interval $[0, T]$ if the function $x(t)$ is defined at the point T and has the additional property that $x(0) = x(T)$).

The exercises at the end of this subsection are very similar to Problems 8 – 15 of Subsections 6.2.1 and also to the related exercises of Subsection 6.2.3, but they involve functions/signals of periods other than $T = 1$. The reader is encouraged to compare the Fourier coefficients and Fourier series for similar functions with different periods.

Subsection 6.2.5 Exercises.

1. Calculate the Fourier coefficients a_n and b_n as well as the complex Fourier coefficients c_n for square-pulse functions of period $T = 10$ defined by

   (a)

$$x(t) = \begin{cases} 1, & 10n - 1 \leq t \leq 10n + 1 \\ 0, & \text{otherwise,} \end{cases} \tag{6.74}$$

   (b)

$$y(t) = \begin{cases} 1, & 10n \leq t \leq 10n + 1 \\ 0, & \text{otherwise,} \end{cases} \tag{6.75}$$

   for all $n \in \mathbb{Z}$. Use these Fourier coefficients to construct the Fourier series expansions of $x(t)$ and $y(t)$. Compare your answers to the formulas derived in Problem 8 at the end of Subsection 6.2.3.

2. Calculate the Fourier coefficients a_n and b_n as well as the complex Fourier coefficients c_n for a triangular-wave function of period $T = 2$ defined by

$$x(t) = \begin{cases} 1 + (t - 2n), & 2n - 1 \leq t \leq 0 \\ 1 - (t - 2n), & 0 \leq t \leq 2n + 1 \end{cases} \tag{6.76}$$

   for all $n \in \mathbb{Z}$. Use these Fourier coefficients to construct the Fourier series expansion of $x(t)$. Comment on the symmetric properties of c_n. Compare your answers to the formulas derived in Problem 9 at the end of Subsection 6.2.3.

3. Calculate the Fourier coefficients a_n and b_n as well as the complex Fourier coefficients c_n for the function of period $T = 2\pi$ defined by

$$y(t) = \begin{cases} -1, & 2\pi n - \pi \leq t \leq 0 \\ 1, & 0 \leq t < 2\pi n + \pi \\ 0, & \text{elsewhere} \end{cases} \tag{6.77}$$

   for all $n \in \mathbb{Z}$. Use these Fourier coefficients to construct the Fourier series expansion of $y(t)$. Comment on the symmetric properties of c_n. Compare your answers to the formulas derived in Problem 10 at the end of Subsection 6.2.3.

4. Calculate the Fourier coefficients a_n and b_n as well as the complex Fourier coefficients c_n for the ramp-shape functions of period $T = 2$ defined by

   (a)
$$r_1(t) = t - 2n, \qquad 2n - 1 \le t < 2n + 1 \tag{6.78}$$

   (b)
$$r_2(t) = \begin{cases} t - 2n, & 2n \le t < 2n + 1 \\ 0, & \text{otherwise} \end{cases} \tag{6.79}$$

   for all $n \in \mathbb{Z}$. Use these Fourier coefficients to construct the Fourier series expansion of $r_1(t)$ and $r_2(t)$. Compare your answers to the formulas derived in Problem 11 at the end of Subsection 6.2.3.

5. Calculate the Fourier coefficients a_n and b_n as well as the complex Fourier coefficients c_n for the 2π-periodic function defined by
$$p(t) = (t - 2\pi n)^2, \qquad n - \pi \le t \le n + \pi \tag{6.80}$$
   for all $n \in \mathbb{Z}$. Use these Fourier coefficients to construct the Fourier series expansion of $p(t)$. Comment on the symmetric properties of c_n. Compare your answers to the formulas derived in Problem 12 at the end of Subsection 6.2.3.

6. Calculate the Fourier coefficients a_n and b_n as well as the complex Fourier coefficients c_n for the 2π-periodic function defined by
$$s(t) = \begin{cases} (t - 2\pi n)^2, & 2\pi n \le t < 2\pi n + \pi \\ 0, & \text{otherwise} \end{cases} \tag{6.81}$$
   for all $n \in \mathbb{Z}$. Use these Fourier coefficients to construct the Fourier series expansion of $s(t)$. Compare your answers to the formulas derived in Problem 13 at the end of Subsection 6.2.3.

7. Calculate the Fourier coefficients a_n and b_n as well as the complex Fourier coefficients c_n for the 2π-periodic functions defined by

   (a)
$$q_1(t) = (t - 2\pi n)^3, \qquad 2\pi n - \pi \le t < 2\pi n + \pi \tag{6.82}$$

   (b)
$$q_2(t) = \begin{cases} (t - 2\pi n)^3, & 2\pi n \le t < 2\pi n + \pi \\ 0, & \text{otherwise} \end{cases} \tag{6.83}$$

   for all $n \in \mathbb{Z}$. Use these Fourier coefficients to construct the Fourier series expansions of $q_1(t)$ and $q_2(t)$.

After you have completed some (or all) of the problems 1 – 7, try to write equations of functions whose graphs have similar shapes but arbitrary periods T and amplitudes (or maxima) A. Derive expressions for the Fourier coefficients and Fourier series expansions for those functions. Your formulas should depend on the parameters T and A.

MATLAB Exercises.

1. Write MATLAB functions to create graphs of the Fourier polynomial approximations of the specified degree n for the following T-periodic signals:

   (a) The pulse-train function defined by (6.74).
   (b) The triangular-wave function defined by (6.76).
   (c) The step function defined by (6.77).
   (d) The ramp-shape functions defined by (6.78) and (6.79).
   (e) The two-branch periodic parabola defined by (6.80).
   (f) The one-branch periodic parabola defined by (6.81).
   (g) The periodic third-power function defined by (6.82).

   Your MATLAB functions should accept the value of n as a parameter and plot both the function to be approximated and the approximating nth degree Fourier polynomial on the same set of axes. Test your program using several different values of n. Comment on the quality of approximation.

2. Repeat Problem 2, but this time, generalize it to construct Fourier approximations for T-periodic functions for the specified period T. Your MATLAB functions should accept both the value of T and the value of n as parameters.

3. Write a MATLAB program that will do the following:

   (a) Ask the user which type of signal (from the list in the previous problems) needs to be approximated by a Fourier polynomial.
   (b) Ask the user to input the value T of the period.
   (c) Ask the user to input the value A of the amplitude/max.
   (d) Ask the user for the degree n of the Fourier polynomial to be used in approximation.
   (e) Plot both the function to be approximated and the approximating nth degree Fourier polynomial on the same set of axes.

   Test your program using several different types of signals and several different values of T, A, and n. Comment on the quality of approximation.

6.3 Inner Products, Orthogonal Bases, and Fourier Coefficients

In Section 6.2, we approached the tasks of formulating the notions of Fourier coefficients and Fourier series expansion in a completely intuitive manner. As a consequence, it took us a lot of time and effort to come up with the relevant definitions and to derive the relevant formulas. Now, with the fundamental notions and techniques of frequency analysis firmly established and the formulas in place, it may be worth looking back and inquiring whether a more abstract approach could have yielded the same results in a more concise and efficient manner. The purpose of this section is, therefore, to show the reader how the fairly abstract language and techniques of linear algebra can indeed greatly reduce the amount of work required to develop frequency analysis. The reader will also get valuable theoretical insights into what seems to be a purely applied topic.

We can recall from an introductory course in linear algebra that a vector space V over a scalar field F is a set of objects (called vectors) that is closed under two operations (which we usually call addition and multiplication by a scalar). It is assumed that the operations have the familiar and expected identity, inverse, commutative, associative, and distributive properties.

Common examples of vector spaces include the Euclidean spaces $\mathbb{R}^n$ (such as the two-dimensional plane and the three-dimensional physical space) and the spaces $\mathbb{P}_n$ of polynomials. Of primary importance in this text are the vector spaces of physically realizable signals over the real (or complex) numbers and the spaces of matrices that can be used to model digital images.

Other than the addition and multiplication by a scalar (which are there by default), other operations may be defined on a vector space. Of those the single most desirable one to have is the inner product, which the reader has already encountered as the "dot product" in calculus, analytic geometry, and introductory linear algebra. As a reminder, the dot product of two vectors $\mathbf{x}$ and $\mathbf{y}$ in $\mathbb{R}^n$ is defined by

$$\mathbf{x} \cdot \mathbf{y} = \begin{bmatrix} x_1 \\ x_2 \\ \dots \\ x_n \end{bmatrix} \cdot \begin{bmatrix} y_1 \\ y_2 \\ \dots \\ y_n \end{bmatrix} = \sum_{k=1}^{n} x_k y_k = x_1 y_1 + x_2 y_2 + \dots + x_n y_n, \tag{6.84}$$

and the dot product of $\mathbf{x}, \mathbf{y} \in \mathbb{C}^n$ is similarly defined by

$$\mathbf{x} \cdot \mathbf{y} = \begin{bmatrix} x_1 \\ x_2 \\ \dots \\ x_n \end{bmatrix} \cdot \begin{bmatrix} y_1 \\ y_2 \\ \dots \\ y_n \end{bmatrix} = \sum_{k=1}^{n} x_k \bar{y}_k = x_1 \bar{y}_1 + x_2 \bar{y}_2 + \dots + x_n \bar{y}_n, \tag{6.85}$$

where the bar denotes complex conjugation.

The dot product is an amazingly useful operation. Among other things, it allows us to define the notions of magnitude of a vector, of distance between vectors, and of the angle between vectors. Thanks to the dot product, we are able to extend the definition of orthogonality of vectors to abstract settings. We also rely on the dot product for calculations of orthogonal projections and of coordinates with respect to orthogonal bases.

Specifically, as a matter of reviewing those concepts, we recall that the magnitude of a vector $\mathbf{x} \in \mathbb{R}^n$ is defined by means of

$$||\mathbf{x}||^2 = \mathbf{x} \cdot \mathbf{x} = \sum_{k=1}^{n} |x_k|^2 = x_1^2 + x_2^2 + ... + x_n^2, \tag{6.86}$$

and this defining formula can be regarded as an extension of the Pythagorean Theorem. The distance between two vectors $\mathbf{x}$ and $\mathbf{y}$ in $\mathbb{R}^n$ is defined by

$$\begin{aligned} dist(\mathbf{x}, \mathbf{y}) &= ||\mathbf{y} - \mathbf{x}|| = \sqrt{\sum_{k=1}^{n} |y_k - x_k|^2} \qquad (6.87) \\ &= \sqrt{(y_1 - x_1)^2 + (y_2 - x_2)^2 + ... + (y_n - x_n)^2}. \end{aligned}$$

Taking advantage of the geometric definition of the dot product in the familiar three-dimensional Euclidean space $\mathbb{R}^3$ (motivated by such concepts from physics as work done by a force displacing an object), namely, that

$$\mathbf{x} \cdot \mathbf{y} = ||\mathbf{x}|| \cdot ||\mathbf{y}|| \cdot \cos\phi,$$

where ϕ is the angle between the vectors $\mathbf{x}$ and $\mathbf{y}$, we can use the dot product to *define* the angle between two vectors $\mathbf{x}, \mathbf{y} \in \mathbb{R}^n$ by means of

$$\cos\phi = \frac{\mathbf{x} \cdot \mathbf{y}}{||\mathbf{x}|| \cdot ||\mathbf{y}||}, \tag{6.88}$$

and this definition works for spaces of any dimension n. In particular, since the cosine of the right angle is zero, we can use the dot product to define orthogonality as follows:

We call vectors $\mathbf{x}$ and $\mathbf{y}$ *orthogonal* if and only if their dot product $\mathbf{x} \cdot \mathbf{y} = 0$.

In this way, the concept of orthogonality of vectors can be extended to higher dimensions where its intuitive and visual meanings are not as transparent as in the familiar three-dimensional space.

In this section, we are primarily interested in the vector spaces of continuous periodic signals. Therefore, we are naturally curious whether the familiar notion of the dot product in $\mathbb{R}^n$ can be generalized for more abstract vector spaces. The answer is yes for a certain class of spaces (called inner-product spaces), and such an operation, a generalized dot product, is called the *inner product*. It is defined as follows:

Definition 6.5 *An inner product on a (real) vector space V is a binary operation that assigns a real number $\langle \boldsymbol{x}, \boldsymbol{y} \rangle$ to each pair of vectors $\boldsymbol{x}$ and $\boldsymbol{y}$ in V and satisfies the following properties:*

1. *For all vectors $\boldsymbol{x}, \boldsymbol{y} \in V$, $\langle \boldsymbol{x}, \boldsymbol{y} \rangle = \langle \boldsymbol{y}, \boldsymbol{x} \rangle$.*
2. *For all vectors $\boldsymbol{x}, \boldsymbol{y}, \boldsymbol{w} \in V$, $\langle (\boldsymbol{x} + \boldsymbol{y}), \boldsymbol{w} \rangle = \langle \boldsymbol{x}, \boldsymbol{w} \rangle + \langle \boldsymbol{y}, \boldsymbol{w} \rangle$.*
3. *For all vectors $\boldsymbol{x}, \boldsymbol{y} \in V$, and all scalars $c \in \mathbb{R}$, $\langle c\boldsymbol{x}, \boldsymbol{y} \rangle = c\langle \boldsymbol{x}, \boldsymbol{y} \rangle$.*
4. *For all vectors $\boldsymbol{x} \in V$, $\langle \boldsymbol{x}, \boldsymbol{x} \rangle \geq 0$. Moreover, $\langle \boldsymbol{x}, \boldsymbol{x} \rangle = 0$ if and only if $\boldsymbol{x} = \boldsymbol{0}$.*

It is easy to see that the dot product on $\mathbb{R}^n$ satisfies all the properties of the inner product, and that the inner product, therefore, is indeed a generalization of the notion of the dot product. The main purpose of introducing this new concept of inner product is to extend the notions of magnitude, distance, and orthogonality to abstract mathematical objects for which they do not have immediate intuitive meaning.

For instance, if the space V is equipped with an inner product, we can define the magnitude of any element $\mathbf{x}$ of V by means of

$$||\mathbf{x}||^2 = \langle \mathbf{x}, \mathbf{x} \rangle \tag{6.89}$$

and the distance between any two elements $\mathbf{x}$ and $\mathbf{y}$ in V by

$$dist(\mathbf{x}, \mathbf{y}) = ||\mathbf{y} - \mathbf{x}||. \tag{6.90}$$

We can also define orthogonality of vectors in terms of their inner product in a way completely analogous to how it is done in $\mathbb{R}^n$:

Definition 6.6 *Two elements $\boldsymbol{x}$ and $\boldsymbol{y}$ in V are called* ***orthogonal*** *if and only if $\langle \boldsymbol{x}, \boldsymbol{y} \rangle = 0$.*

The next concept from linear algebra that we need to review is that of a basis – a linearly independent set $\mathcal{B} = \{\mathbf{b}_k\}$ of vectors in a vector space V that spans all of V. A typical problem studied in introductory linear algebra courses is that of finding coordinates of the given vector $\mathbf{x}$ with respect to the given basis $\mathcal{B}$, that is, finding the coefficients c_k in the equation

$$\sum_k c_k \mathbf{b}_k = \mathbf{x}. \tag{6.91}$$

In general, this is equivalent to the problem of solving a linear system and requires the use of a row-reduction algorithm. However, the procedure for finding the coordinates turns out to be much simpler if $\mathcal{B}$ happens to be an orthogonal basis. In that case, taking the inner product of both sides of (6.91) with any element $\mathbf{b}_m \in \mathcal{B}$ yields

$$\langle \mathbf{x}, \mathbf{b}_m \rangle = \langle \sum_k c_k \mathbf{b}_k, \mathbf{b}_m \rangle$$

$$= \sum_{k \neq m} c_k \langle \mathbf{b}_k, \mathbf{b}_m \rangle + c_m \langle \mathbf{b}_m, \mathbf{b}_m \rangle$$
$$= c_m ||\mathbf{b}_m||^2,$$

because the orthogonality of the basis $\mathcal{B}$ implies that $\langle \mathbf{b}_k, \mathbf{b}_m \rangle = 0$ whenever $m \neq k$. Consequently,

> The coordinates c_m of a vector $\mathbf{x}$ with respect to an **orthogonal** basis $\mathcal{B}$ are
> $$c_m = \frac{\langle \mathbf{x}, \mathbf{b}_m \rangle}{||\mathbf{b}_m||^2} \tag{6.92}$$
> and, therefore, $\mathbf{x}$ can be written as
> $$\mathbf{x} = \sum_m c_m \mathbf{b}_m = \sum_m \frac{\langle \mathbf{x}, \mathbf{b}_m \rangle}{||\mathbf{b}_m||^2} \mathbf{b}_m \tag{6.93}$$
> with all the terms of the expansion orthogonal to each other.

So, we see that orthogonal bases are very convenient! It is easy (conceptually and computationally) to calculate the coordinates of a vector with respect to an orthogonal basis. But is every orthogonal set of vectors a basis for its span? We recall that in order for a given set of vectors to form a basis, it must first and foremost be linearly independent. Here too, orthogonality proves to be a very useful property. Whereas the general problem of determining whether a given set of vectors is linearly independent is equivalent to solving a certain linear system (by means of row-reduction) and establishing that it only has the trivial solution, any *orthogonal* set of vectors is *automatically* linearly independent! Indeed, suppose that the nonzero vectors $\{\mathbf{v}_k\}$ are pairwise-orthogonal and that

$$\sum_k c_k \mathbf{v}_k = \mathbf{0}$$

for some scalars c_k. Taking the inner product of both sides with an arbitrary $\mathbf{v}_m$, we obtain

$$0 = \langle \sum_k c_k \mathbf{v}_k, \mathbf{v}_m \rangle$$
$$= \sum_{k \neq m} c_k \langle \mathbf{v}_k, \mathbf{v}_m \rangle + c_m \langle \mathbf{v}_m, \mathbf{v}_m \rangle$$
$$= c_m ||\mathbf{v}_m||^2,$$

which implies that $c_m = 0$ for all m and, consequently, that the set of vectors $\{\mathbf{v}_k\}$ is linearly independent and thus forms a basis for its span.

Having reviewed the preliminaries from linear algebra, we can return to our vector space V of physically realizable 1-periodic signals. We define the inner product of two such signals $x(t)$ and $y(t)$ by

$$\langle x(t), y(t)\rangle = \int_0^1 x(t)y(t)dt, \tag{6.94}$$

and it is evident that this operation satisfies all the properties listed in the Definition 6.5 of an inner product. Once we have the definition (6.94) of the inner product in place, we can define the magnitude of any physically realizable signal $x(t)$ by means of

$$||x(t)||^2 = \langle x(t), x(t)\rangle = \int_0^1 |x(t)|^2 dt \tag{6.95}$$

and the distance between signals $x(t)$ and $y(t)$ by

$$dist(x, y) = ||x(t) - y(t)|| = \sqrt{\int_0^1 |y(t) - x(t)|^2 dt}. \tag{6.96}$$

In view of Definition 6.6, two signals $x(t)$ and $y(t)$ are orthogonal if their inner product

$$\langle x(t), y(t)\rangle = \int_0^1 x(t)y(t)dt = 0.$$

For example, the functions $x(t) = \cos(2\pi t)$ and $y(t) = \sin(2\pi t)$ are orthogonal because

$$\int_0^1 \cos(2\pi t)\sin(2\pi t)dt = 0$$

as the reader can easily verify. Furthermore, from the observations

$$\int_0^1 \sin(k \cdot 2\pi t)\sin(n \cdot 2\pi t)dt = 0$$

and

$$\int_0^1 \cos(k \cdot 2\pi t)\cos(n \cdot 2\pi t)dt = 0$$

for $k \neq n$, and also from the observation

$$\int_0^1 \sin(k \cdot 2\pi t)\cos(n \cdot 2\pi t)dt = 0$$

for all k and n, it follows that all 1-periodic sine and cosine functions are orthogonal to each other. Therefore, the set of functions

$$\mathcal{B} = \{\cos(k \cdot 2\pi t), \sin(n \cdot 2\pi t) : \quad k, n \in \mathbb{Z}^+\} \tag{6.97}$$

forms an orthogonal basis for its span, which, according to the Bold Assumption (6.1), is the entire space of physically-realizable 1-periodic signals. Since the inner products of a given signal $x(t)$ with the elements of $\mathcal{B}$ are

$$\langle x(t), \cos(n \cdot 2\pi t)\rangle = \int_0^1 x(t)\cos(n \cdot 2\pi t)dt$$

and

$$\langle x(t), \sin(n \cdot 2\pi t)\rangle = \int_0^1 x(t)\sin(n \cdot 2\pi t)dt,$$

the formula (6.92) tells us that the Fourier coefficients a_n and b_n are (up to a constant factor) the coordinates of $x(t)$ with respect to the orthogonal basis $\mathcal{B}$ defined by (6.97). The Fourier series expansion

$$x(t) = a_0 + 2\sum_{n=1}^{\infty} a_n \cos(n \cdot 2\pi t) + 2\sum_{n=1}^{\infty} b_n \sin(n \cdot 2\pi t)$$

of $x(t)$ then follows immediately from (6.93).

Unfortunately, the delicate convergence issues that arise when studying infinite sums of this form are outside the scope of this book, but we hope that the reader is motivated to undertake further study in applied mathematics and frequency analysis.

We hope that the reader appreciated how the abstract language of orthogonal bases helped reduce the derivation of the formulas for the Fourier coefficients and of the Fourier series to a two-sentence argument. If that is the case, there is probably an anticipation of similar magic to happen for complex Fourier coefficients and series expansions.

In that expectation, the reader will not be disappointed. A little bit of preliminary work is needed though. To begin with, we note that the notion of multiplication by a scalar in vector spaces is not limited to real scalars - complex scalars are perfectly usable. However, once complex values are allowed, the definition of the inner product must undergo a slight modification. Specifically, the commutativity property of the inner product in Definition 6.5 becomes

1. For all vectors $\mathbf{x}, \mathbf{y} \in V$, $\langle \mathbf{x}, \mathbf{y}\rangle = \overline{\langle \mathbf{y}, \mathbf{x}\rangle}$.

We can then define the inner product of the complex-valued functions $x(t)$ and $y(t)$ by

$$\langle x(t), y(t)\rangle = \int_0^1 x(t)\overline{y(t)}dt, \tag{6.98}$$

and we encourage the reader to verify that all the conditions in the definition of the inner product are satisfied. The magnitude of $x(t)$ and the distance between $x(t)$ and $y(t)$ are calculated the same way it is done for real-valued functions. The observations that

$$||e^{k \cdot 2\pi i t}||^2 = \int_0^1 |e^{k \cdot 2\pi i t}|^2 dt = 1$$

for all $k \in \mathbb{Z}$ and that

$$\int_0^1 e^{k \cdot 2\pi it} e^{n \cdot 2\pi it} = 0$$

for $k \neq n$ prove that the family

$$\mathcal{C} = \{e^{k \cdot 2\pi it} : \quad k \in \mathbb{Z}\} \tag{6.99}$$

of the complex exponentials forms an orthonormal basis for its span, which, according to the Bold Assumption (6.1), is the entire space of physically realizable signals. Since the inner products of any given signal $x(t)$ with the elements of $\mathcal{C}$ are

$$\langle x(t), e^{k \cdot 2\pi it} \rangle = \int_0^1 x(t) e^{-k \cdot 2\pi it} dt,$$

the formula (6.92) tells us that the Fourier coefficients c_n are precisely the coordinates of $x(t)$ with respect to the orthonormal basis $\mathcal{C}$. The Fourier series expansion

$$x(t) = \sum_{n=-\infty}^{\infty} c_n e^{n \cdot 2\pi it}$$

of $x(t)$ then follows immediately from (6.93). As we may have expected based on our experience in Subsection 6.2.3, the derivation of all the formulas proved to be easier and shorter in the complex setting.

Just as in the case of trigonometric Fourier series, the convergence issues that arise when studying complex Fourier series are unfortunately outside the scope of this book. Nevertheless, a look at the Fourier series from a slightly different point of view brings us to the very threshold of fully grasping the meaning of the Fourier series convergence.

We may recall that the **orthogonal projection** $\hat{\mathbf{x}}$ of a vector $\mathbf{x}$ onto the line L determined by a vector $\mathbf{v}$ can be calculated using the formula

$$\hat{\mathbf{x}} = \frac{\langle \mathbf{x}, \mathbf{v} \rangle}{||\mathbf{v}||^2} \mathbf{v} \tag{6.100}$$

and is the closest point of L to $\mathbf{x}$. The distance from $\mathbf{x}$ to L is then given by the magnitude $||\mathbf{x} - \hat{\mathbf{x}}||$. By extension, the orthogonal projection $\hat{\mathbf{x}}$ of the vector $\mathbf{x}$ onto the subspace $W = Span\{\mathbf{v}_1, \mathbf{v}_2, ..., \mathbf{v}_n\}$ can be calculated using the formula

$$\hat{\mathbf{x}} = \frac{\langle \mathbf{x}, \mathbf{v}_1 \rangle}{||\mathbf{v}_1||^2} \mathbf{v}_1 + \frac{\langle \mathbf{x}, \mathbf{v}_2 \rangle}{||\mathbf{v}_2||^2} \mathbf{v}_2 + ... + \frac{\langle \mathbf{x}, \mathbf{v}_n \rangle}{||\mathbf{v}_n||^2} \mathbf{v}_n \tag{6.101}$$

and is the closest point of W to $\mathbf{x}$, with the distance from $\mathbf{x}$ to W equal to the magnitude $||\mathbf{x} - \hat{\mathbf{x}}||$.

What does all of that have to do with Fourier series? We can define the subspace W_n of the space of physically realizable signals to be the span of the family of trigonometric polynomials of degree at most n, that is

$$W_n = Span\{\cos(k \cdot 2\pi t)\}, \sin(l \cdot 2\pi t) : \quad k = 0, \ldots, n \text{ and } l = 1, \ldots n\}$$

or, equivalently and more concisely,

$$W_n = Span\{e^{k\cdot 2\pi it}: \quad k = -n, \ldots, 0, \ldots, n\}$$

if the complex setting is used. The orthogonal projection $\hat{x}(t) = Proj_{W_n} x(t)$ of a function $x(t)$ onto W_n is then the best approximation of $x(t)$ by an nth-degree trigonometric polynomial from the point of view of minimizing the magnitude

$$||\hat{x}(t) - x(t)|| = \sqrt{\int_0^1 |\hat{x}(t) - x(t)|^2 dt} \tag{6.102}$$

of the approximation error and, according to (6.101), it coincides precisely with the nth-degree Fourier polynomial of $x(t)$. The Fourier series of $x(t)$ is said to converge whenever the approximation errors (6.102) converge to zero or, less formally, when the approximations of $x(t)$ by Fourier polynomials of degree n get arbitrarily close as $n \to \infty$.

Thus, all we have to do to prove the Bold Conjecture (6.1) and to complete our understanding of the meaning of the Fourier series convergence is to establish the conditions under which

$$\lim_{n\to\infty} \sqrt{\int_0^1 |\hat{x}_n(t) - x(t)|^2 dt} = 0,$$

where, once again, $\hat{x}_n(t)$ denotes the approximation of $x(t)$ by the nth-degree Fourier polynomial. This topics lies just outside the scope of this book, and we hope that the reader is motivated to make the next step in the study of mathematical analysis.

What about the application of techniques of linear algebra to T-periodic functions? We can define the inner product of two such functions $x(t)$ and $y(t)$ by

$$\langle x(t), y(t)\rangle = \int_0^T x(t)y(t)dt \tag{6.103}$$

if they are real-valued and by

$$\langle x(t), y(t)\rangle = \int_0^T x(t)\overline{y(t)}dt, \tag{6.104}$$

if they are complex-valued.

We leave it to the reader to develop the frequency analysis of T-periodic signals based on the concepts of linear algebra and to verify that the formulas (6.66), (6.67) and (6.68) for the Fourier coefficients and the formulas (6.69) and (6.70) for the Fourier series expansions for T-periodic functions follow directly from Equations (6.92) and (6.93).

In the exercises at the end of this section, the reader will also be asked to review the use of the dot product in $\mathbb{R}^n$ and to perform calculations of angles and distances between vectors and to determine coordinates with respect to orthogonal bases. The reader will then be asked to extend those skills to perform similar operations with inner products in more abstract spaces of periodic functions.

Section 6.3 Exercises.

1. Calculate the dot product of the given vectors $\mathbf{x}$ and $\mathbf{y}$, the magnitude of $\mathbf{x}$ and $\mathbf{y}$, and the angle between $\mathbf{x}$ and $\mathbf{y}$:

   (a) $\mathbf{x} = [2, -2]^T$ and $\mathbf{y} = [3, 3]^T$.

   (b) $\mathbf{x} = [2, -2]^T$ and $\mathbf{y} = [-3, 0]^T$.

   (c) $\mathbf{x} = [2, -2, 6]^T$ and $\mathbf{y} = [5, 2, -5]^T$.

   (d) $\mathbf{x} = [3, 4, -3]^T$ and $\mathbf{y} = [-2, 2, 7]^T$.

   (e) $\mathbf{x} = [3, -3, 5]^T$ and $\mathbf{y} = [2, -3, 7]^T$.

   (f) $\mathbf{x} = [1, -3, 7]^T$ and $\mathbf{y} = [5, 2, -5]^T$.

   (g) $\mathbf{x} = [3, -1, 4]^T$ and $\mathbf{y} = [1, 7, -2]^T$.

   (h) $\mathbf{x} = [9, 4, 1]^T$ and $\mathbf{y} = [-1, 4, -9]^T$.

   (i) $\mathbf{x} = [1, 4, 2]^T$ and $\mathbf{y} = [5, 3, -1]^T$.

   Use MATLAB to check your answers.

2. For the following complex vectors $\mathbf{x}$ and $\mathbf{y}$, calculate their magnitudes $||\mathbf{x}||$ and $||\mathbf{y}||$ and their dot product $\mathbf{x} \cdot \mathbf{y}$:

   (a) $\mathbf{x} = [2 - i, -2 + i]^T$ and $\mathbf{y} = [5 - 3i, 2 - i]^T$.

   (b) $\mathbf{x} = [3 - 5i, 7 + 2i, 6]^T$ and $\mathbf{y} = [1 - 3i, 3 - i, -5]^T$.

   Use MATLAB to check your answers.

3. Check that the given basis $\mathcal{B}$ is orthogonal and calculate the coordinates of the given vector $\mathbf{x}$ with respect to $\mathcal{B}$. Use those coordinates to construct the orthogonal expansion of $\mathbf{x}$ expressing it as a linear combination of the elements of $\mathcal{B}$.

   (a) $\mathcal{B} = \left\{ \begin{bmatrix} 3 \\ 2 \end{bmatrix}, \begin{bmatrix} 4 \\ -6 \end{bmatrix} \right\}$ and $\mathbf{x} = \begin{bmatrix} 24 \\ -10 \end{bmatrix}$.

   (b) $\mathcal{B} = \left\{ \begin{bmatrix} 4 \\ 1 \end{bmatrix}, \begin{bmatrix} 1 \\ -4 \end{bmatrix} \right\}$ and $\mathbf{x} = \begin{bmatrix} 10 \\ 11 \end{bmatrix}$.

(c) $\mathcal{B} = \left\{ \begin{bmatrix} 4 \\ 1 \\ 1 \end{bmatrix}, \begin{bmatrix} 1 \\ -2 \\ -2 \end{bmatrix}, \begin{bmatrix} 0 \\ -1 \\ 1 \end{bmatrix} \right\}$ and $\mathbf{x} = \begin{bmatrix} 1 \\ -3 \\ 5 \end{bmatrix}$.

(d) $\mathcal{B} = \left\{ \begin{bmatrix} 2 \\ -2 \\ 0 \end{bmatrix}, \begin{bmatrix} 3 \\ 3 \\ -1 \end{bmatrix}, \begin{bmatrix} 1 \\ 1 \\ 6 \end{bmatrix} \right\}$ and $\mathbf{x} = \begin{bmatrix} 5 \\ -2 \\ 1 \end{bmatrix}$.

(e) $\mathcal{B} = \left\{ \begin{bmatrix} 1 \\ -1 \\ 0 \end{bmatrix}, \begin{bmatrix} 2 \\ 2 \\ -1 \end{bmatrix}, \begin{bmatrix} 1 \\ 1 \\ 4 \end{bmatrix} \right\}$ and $\mathbf{x} = \begin{bmatrix} 8 \\ -6 \\ 2 \end{bmatrix}$.

(f) $\mathcal{B} = \left\{ \begin{bmatrix} 2 \\ -2 \\ 1 \\ 2 \end{bmatrix}, \begin{bmatrix} 1 \\ -4 \\ 4 \\ -7 \end{bmatrix}, \begin{bmatrix} 4 \\ 7 \\ 6 \\ 0 \end{bmatrix} \right\}$ and $\mathbf{x} = \begin{bmatrix} 13 \\ 6 \\ 18 \\ -3 \end{bmatrix}$.

(g) $\mathcal{B} = \left\{ \begin{bmatrix} 1 \\ -2 \\ 1 \\ 1 \end{bmatrix}, \begin{bmatrix} 1 \\ -2 \\ 1 \\ -1 \end{bmatrix}, \begin{bmatrix} 2 \\ 1 \\ 0 \\ 0 \end{bmatrix} \right\}$ and $\mathbf{x} = \begin{bmatrix} 8 \\ -1 \\ 2 \\ 0 \end{bmatrix}$.

4. Determine the point of the line L passing through the origin and the given vector $\mathbf{v}$ that is closest to the given $\mathbf{x}$. Also, calculate the distance from $\mathbf{x}$ to L. How would you go about determining the reflection of $\mathbf{x}$ across L?

   (a) $\mathbf{v} = \begin{bmatrix} 4 \\ -6 \end{bmatrix}$ and $\mathbf{x} = \begin{bmatrix} 12 \\ -5 \end{bmatrix}$.

   (b) $\mathbf{v} = \begin{bmatrix} 3 \\ 2 \end{bmatrix}$ and $\mathbf{x} = \begin{bmatrix} 12 \\ -5 \end{bmatrix}$.

   (c) $\mathbf{v} = \begin{bmatrix} 1 \\ -4 \end{bmatrix}$ and $\mathbf{x} = \begin{bmatrix} 10 \\ 11 \end{bmatrix}$.

5. Determine the point of the plane W passing through the origin and the given vectors $\mathbf{v}_1$ and $\mathbf{v}_2$ that is closest to the given $\mathbf{x}$. Also, calculate the distance from $\mathbf{x}$ to W. How would you go about determining the reflection of $\mathbf{x}$ across W?

   (a) $\mathbf{v}_1 = \begin{bmatrix} 4 \\ 1 \\ 1 \end{bmatrix}, \mathbf{v}_2 = \begin{bmatrix} 1 \\ -2 \\ -2 \end{bmatrix}$, and $\mathbf{x} = \begin{bmatrix} 1 \\ -3 \\ 5 \end{bmatrix}$.

   (b) $\mathbf{v}_1 = \begin{bmatrix} 2 \\ -2 \\ 0 \end{bmatrix}, \mathbf{v}_2 = \begin{bmatrix} 1 \\ 1 \\ 6 \end{bmatrix}$, and $\mathbf{x} = \begin{bmatrix} 15 \\ -6 \\ 3 \end{bmatrix}$.

(c) $\mathbf{v}_1 = \begin{bmatrix} 2 \\ 2 \\ -1 \end{bmatrix}, \mathbf{v}_2 = \begin{bmatrix} 1 \\ 1 \\ 4 \end{bmatrix}$, and $\mathbf{x} = \begin{bmatrix} 4 \\ -3 \\ 1 \end{bmatrix}$.

6. Calculate the orthogonal projection of the given vector $\mathbf{x}$ onto the subspace of $\mathbb{R}^4$ spanned by the vectors $\mathbf{v}_1$ and $\mathbf{v}_2$. How would you go about determining the reflection of $\mathbf{x}$ across W and what would be its geometric meaning?

   (a) $\mathbf{v}_1 = \begin{bmatrix} 2 \\ -2 \\ 1 \\ 2 \end{bmatrix}, \mathbf{v}_2 = \begin{bmatrix} 1 \\ -4 \\ 4 \\ -7 \end{bmatrix}$, and $\mathbf{x} = \begin{bmatrix} 13 \\ 6 \\ 18 \\ -3 \end{bmatrix}$.

   (b) $\mathbf{v}_1 = \begin{bmatrix} 1 \\ -2 \\ 1 \\ -7 \end{bmatrix}, \mathbf{v}_2 = \begin{bmatrix} 2 \\ 1 \\ 0 \\ 0 \end{bmatrix}$, and $\mathbf{x} = \begin{bmatrix} 8 \\ -6 \\ 2 \\ -8 \end{bmatrix}$.

7. Verify that the inner products of two 1-periodic signals $x(t)$ and $y(t)$ defined by (6.94) and by (6.98) satisfy all the properties listed in the Definition 6.5.

8. Verify that the inner products of two T-periodic signals $x(t)$ and $y(t)$ defined by (6.103) and by (6.104) satisfy all the properties listed in the Definition 6.5.

9. Calculate the magnitudes and the pairwise inner products of the members of the following set of 1-periodic functions:

$$\mathcal{C} = \{1, \cos(2\pi t), \cos^2(2\pi t), \cos^3(2\pi t)\}.$$

   Are any of these functions orthogonal to each other?

10. For the following pairs of functions $x(t)$ and $y(t)$ calculate their magnitudes $||x(t)||$ and $||y(t)||$ and their inner product $\langle x(t), y(t)\rangle$. Determine whether any of these pairs of functions are orthogonal.

   (a) The 1-periodic square-wave function defined by

$$x(t) = \begin{cases} 1, & n - 1/4 \leq t \leq n + 1/4 \\ 0, & \text{otherwise} \end{cases}$$

   for all $n \in \mathbb{Z}$ and the 1-periodic triangular-wave function defined by

$$y(t) = \begin{cases} 1 + 2(t - n), & n - 1/2 \leq t \leq n \\ 1 - 2(t - n), & n \leq t \leq n + 1/2 \end{cases}$$

   for all $n \in \mathbb{Z}$.

(b) The 1-periodic step-function defined by

$$x(t) = \begin{cases} -1, & n - 1/4 \le t \le n \\ 1, & n \le t \le n + 1/4 \\ 0, & \text{elsewhere,} \end{cases}$$

for all $n \in \mathbb{Z}$ and the 1-periodic parabola defined by

$$y(t) = 4(t - n)^2, \qquad n - 1/2 \le t \le n + 1/2$$

for all $n \in \mathbb{Z}$.

(c) The 1-periodic square-wave function defined by

$$x(t) = \begin{cases} 1, & n - 1/4 \le t \le n + 1/4 \\ 0, & \text{otherwise,} \end{cases}$$

for all $n \in \mathbb{Z}$ and the 1-periodic third-power function defined by

$$y(t) = 8(t - n)^3, \qquad n - 1/2 \le t < n + 1/2$$

for all $n \in \mathbb{Z}$.

(d) The 1-periodic ramp function defined by

$$x(t) = \begin{cases} 2(t - n), & n \le t \le n + 1/2 \\ 0, & \text{otherwise} \end{cases}$$

for all $n \in \mathbb{Z}$ and the 1-periodic saw-tooth function defined by

$$y(t) = 2(t - n), \quad n - 1/2 < t \le n + 1/2$$

for all $n \in \mathbb{Z}$.

11. Calculate the magnitudes of the following set of 2-periodic functions whose values on the interval $[-1, 1]$ are given by

$$\mathcal{C} = \{1, t, 3t^2 - 1, 5t^3 - 3t\}.$$

Verify that $\mathcal{C}$ is an orthogonal basis by calculating the pairwise inner products. Calculate the $\mathcal{C}$-coordinates of the following functions:

(a) $x(t) = 25x^3 - 6x^2 - 11x + 5$.

(b) $x(t) = x^3 - 5x^2 + 3x - 1$.

12. Verify that the formulas (6.66), (6.67), and (6.68) for the Fourier coefficients and the formulas (6.69) and (6.70) for the Fourier series expansions of T-periodic functions follow directly from Equations (6.92) and (6.93).

6.4 Discrete Fourier Transform

In the previous sections, we developed methods of frequency analysis for continuous periodic signals. However, it is quite possible that at this point, the reader still does not see any obvious connection between continuous signals and digital images, which are, after all, the central focus of this text. So why did we invest so much time and effort in studying continuous signals and in developing techniques that, after all, seem to have but a remote connection to our main topic?

This is a completely legitimate question, and the answer is three-fold. First of all, due to the way mathematical curricula are structured, the reader probably has more experience with continuous functions than with discrete mathematical structures. It, therefore, seems more appropriate to approach a new topic from a more familiar starting point. Second, although as mentioned in the introduction to this chapter, frequency methods play a very important role in image processing, the development of those methods is more intuitive in the setting of continuous periodic signals. And, finally, all the formulas we are going to need for frequency analysis of digital images follow effortlessly from the results we have obtained for continuous functions.

6.4.1 Discrete Periodic Sequences

Our digital cameras, computers, tablets, smartphones, and other electronic devices cannot work with continuous (analog) signals and images directly. Those signals and images first need to undergo an analog-to-digital conversion in order to be transformed into their digital versions by means of sampling (and also quantization, which is, unfortunately, not discussed in this text). The result of this conversion is a finite sequence or an $M \times N$ matrix of discrete values.

Discrete sequences have certain peculiar features and properties, which some readers might be unfamiliar with. We devote this subsection to a brief discussion of those peculiarities because they can appear unexpected and even surprising to a reader whose mathematical background consists exclusively of topics related to continuous functions.

Throughout this section, we will be working with finite sequences $x(k)$ where the integer variable k ranges from 0 to $N-1$ for some integer N, which determines the length of the sequence. In Section 6.2, we saw that a continuous function defined on a finite interval is equivalent to its periodic extension. In a similar manner, for any finite sequences $x(k)$, we can construct its periodic extension $x_p(k)$ – a doubly-infinite sequence defined by

$$x_p(k) = x(k \mod N)$$

for all $k \in \mathbb{Z}$. It is clear from the definition that

$$x_p(k) = x_p(k + N)$$

for all $k \in \mathbb{Z}$ and, therefore, $x_p(k)$ is N-periodic. As an illustration of this concept of periodic extension, Figure 6.8 shows a graph of the 20-point unit ramp sequence defined by

$$r(k) = k, \quad 0 \leq k < 20 \tag{6.105}$$

alongside the graph of three periods of its periodic extension $r_p(k)$.

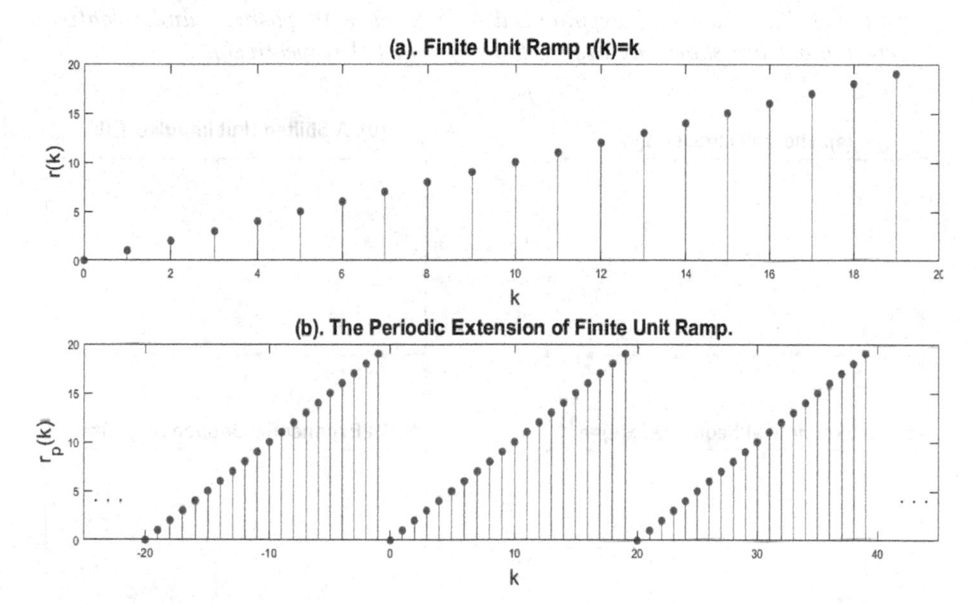

FIGURE 6.8: An illustration of the concept of the periodic extension of a finite sequence.

In view of this equivalence of finite and periodic sequences, in the subsequent discussion, we will be referencing both types of sequences interchangeably, and any results obtained for the former will also apply to the latter (and vice versa). Before proceeding to the development of frequency analysis of finite or periodic sequences, we mention a few more standard examples.

Example 6.8 *In Chapter 5, we have already met the unit impulse sequence* $\boldsymbol{\delta}$*. Its* N*-periodic version is defined by*

$$\delta(k) = \begin{cases} 1, & k = 0 \mod N \\ 0, & k \neq 0 \mod N, \end{cases} \tag{6.106}$$

and its graph is shown in Figure 6.9 (a). A shifted periodic unit impulse is similarly denoted by $\boldsymbol{\delta}_a$ *and defined by*

$$\delta_a(n) = \begin{cases} 1, & k = a \mod N \\ 0, & k \neq a \mod N \end{cases} \tag{6.107}$$

for a fixed integer a*, and its graph is shown in Figure 6.9 (b). The* N*-periodic (or, equivalently, finite) unit ramp sequence defined by* (6.105) *has already been illustrated in Figure 6.8. The* N*-periodic (or, equivalently, finite) exponential sequence is defined by*

$$x(k) = e^{ak} \mod N \tag{6.108}$$

for all $k \in \mathbb{Z}$*. Graphs of exponential sequences with positive and negative growth factors are shown in Figure 6.9 (c) and (d) respectively.*

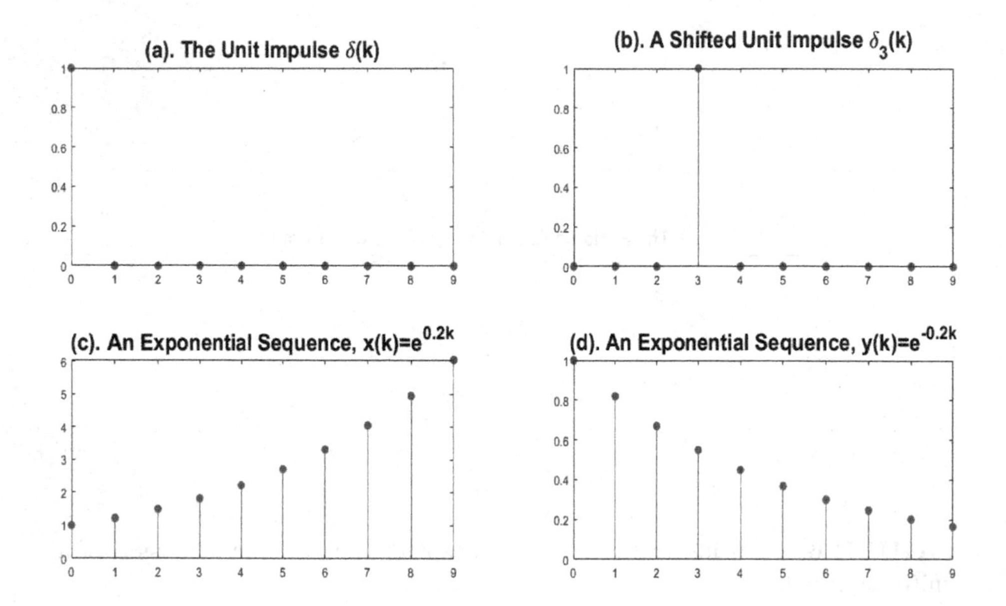

FIGURE 6.9: Graphs of one period of 10-periodic sequences in Example 6.8.

Since the theme of this chapter is frequency analysis, of greatest interest to us are the cosine sequences defined by

$$c(k) = A\cos(2\pi f k), \tag{6.109}$$

the sine sequences defined by

$$s(k) = A\sin(2\pi f k), \tag{6.110}$$

and the generalized cosine sequence defined by

$$g(k) \quad = \quad A\cos(2\pi f k - \phi)$$

$$= A\cos\phi\cos(2\pi fk) + A\sin\phi\sin(2\pi fk), \tag{6.111}$$

where A is the amplitude and f is the frequency (measured in revolutions per *sample*). In order to avoid having to work separately with sine and cosine sequences of the same frequency, we follow the method that worked for us in Section 6.2 and combine them both together into a complex exponential sequence

$$w(k) = Ae^{2\pi ifk} = A\big[\cos(2\pi fk) + i\sin(2\pi fk)\big] \tag{6.112}$$

with the familiar consequences that

$$A\cos(2\pi fk) = \frac{A}{2}\big[e^{2\pi ifk} + e^{-2\pi ifk}\big]$$

and

$$A\sin(2\pi fk) = \frac{A}{2i}\big[e^{2\pi ifk} - e^{-2\pi ifk}\big].$$

It is important to note that the trigonometric sequences $c(k)$, $s(k)$, and $g(k)$, as well as the complex exponential $w(k)$, have certain properties that set them apart from their continuous counterparts:

1. A continuous sine or cosine signal is guaranteed to be periodic regardless of the value of its frequency. This is not the case for their discrete counterparts. It is evident that the signals $c(k)$, $s(k)$, $g(k)$, and $w(k)$ are N-periodic if and only if the frequency f satisfies the condition

$$f = \frac{n}{N}$$

for some integer n. In other words, trigonometric and complex exponential sequences are periodic if and only if they have rational frequencies. Because of this, going forward, we will usually use the notation $f = \frac{n}{N}$ to denote the frequencies of sequences of those types.

2. We know that the cosine function is even, which implies that

$$c(-k) = A\cos(-2\pi nk/N) = A\cos(2\pi nk/N) = c(k).$$

In addition, any N-periodic cosine sequence is circular-even, which means that

$$c(N-k) = A\cos(2\pi n(N-k)/N) = A\cos(2\pi nk/N) = c(k),$$

any N-periodic sine sequence is circular-odd, which means that

$$s(N-k) = A\sin(2\pi n(N-k)/N) = -A\sin(2\pi nk/N) = -s(k),$$

whereas an N-periodic complex exponential sequence merely enjoys the periodicity property

$$w(N-k) = Ae^{2\pi in(N-k)/N} = Ae^{-2\pi ink/N} = w(-k),$$

since $e^{2\pi in} = 1$ for all $n \in \mathbb{Z}$.

3. We are used to the situation where two continuous sine, cosine, or complex exponential signals are different whenever they have different frequencies. On the contrary, their discrete counterparts are identical if their frequencies f_1 and f_2 differ by an integer (or, equivalently, n_1 and n_2 differ by an integer multiple of N), which follows from the observation that

$$w(k) = e^{2\pi ifk} = Ae^{2\pi i(f+m)k}$$

for any $m \in \mathbb{Z}$ with similar relations following for $c(k)$, $s(k)$, and $g(k)$. Moreover (and this fact might appear even more surprising when first encountered), the two N-periodic cosine sequences

$$c_1(k) = A\cos(2\pi n_1 k/N) \quad \text{and} \quad c_2(k) = A\cos(2\pi n_2 k/N)$$

are identical whenever their frequencies n_1 and n_2 satisfy the condition

$$n_1 + n_2 = N,$$

since under this condition

$$\begin{aligned} c_2(k) &= A\cos(2\pi n_2 k/N) \\ &= A\cos(2\pi(N - n_1)k/N) \\ &= A\cos(2\pi k - 2\pi n_1 k/N) = c_1(k) \end{aligned}$$

due to the periodicity and symmetry properties of the cosine function. Similarly, the two N-periodic sine sequences

$$s_1(k) = A\sin(2\pi n_1 k/N) \quad \text{and} \quad s_2(k) = A\sin(2\pi n_2 k/N)$$

enjoy the property that

$$s_1(k) = -s_2(k) \quad \text{whenever} \quad n_1 + n_2 = N$$

due to the periodicity and odd-symmetry of the sine function.

The three peculiarities of the trigonometric sequences just discussed are illustrated by Figure 6.10 and Figure 6.11, which show the graphs of sine and cosine 8-periodic sequences of all the frequencies n ranging from 0 to 7.

It follows from those three observations that given the period N, the number of distinct N-periodic complex exponenetial sequences of the form (6.112) (or, equivalently, the number of independent sine and cosine sequences of the form (6.109) and (6.110)) is precisely N. It, therefore, seems reasonable to use the set of the N complex exponential sequences of the form

$$w(k) = Ae^{2\pi ink/N}, \quad n = 0, ..., N-1$$

as the building blocks for developing techniques of frequency analysis of finite (or, equivalently, N-periodic) sequences along the lines of the paradigms established in the previous sections.

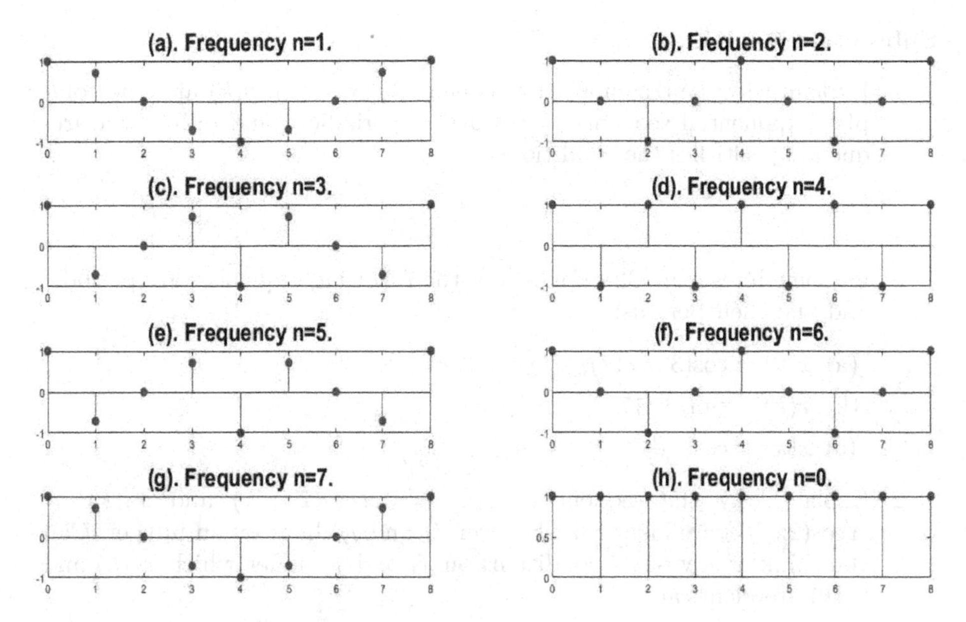

FIGURE 6.10: The graphs of discrete 8-periodic cosine sequences of frequencies ranging from 0 to 7.

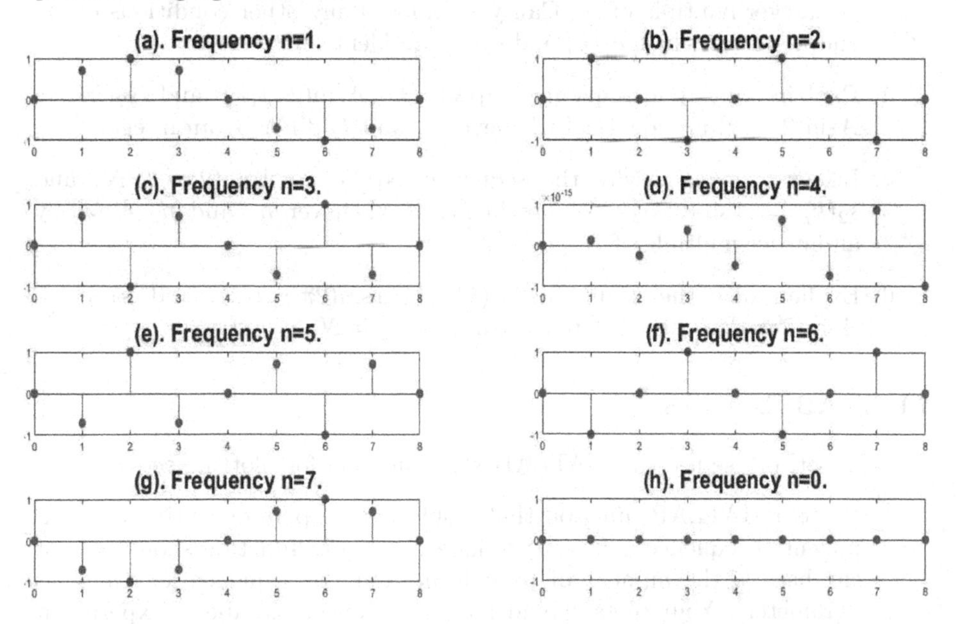

FIGURE 6.11: The graphs of discrete 8-periodic sine sequences of frequencies ranging from 0 to 7.

Subsection 6.4.1 Exercises.

1. Explain why the trigonometric sequences $c(k)$, $s(k)$, $g(k)$ and the complex exponential sequences $w(k)$ are N-periodic if and only if the frequency f satisfies the condition

$$f = \frac{n}{N}$$

for some integer n. Check which of the following sequences are periodic and find their periods:

   (a) $x(k) = \cos(3\pi k/17)$,

   (b) $x(k) = \sin(3k/17)$,

   (c) $x(k) = \cos\left(\frac{3k}{17\pi}\right)$.

2. Explain why the sequences $x_1(k) = A\cos(2\pi f_1 k)$ and $x_2(k) = A\cos(2\pi f_2 k)$ are identical whenever f_1 and f_2 differ by an integer. Can you think of any other conditions on f_1 and f_2 under which $x_1(k)$ and $x_2(k)$ are identical?

3. Likewise, explain why the sequences $c_1(k) = A\cos(2\pi n_1 k/N)$ and $c_2(k) = A\cos(2\pi n_2 k/N)$ are identical whenever n_1 and n_2 differ by an integer multiple of N. Can you think of any other conditions on n_1 and n_2 under which $c_1(k)$ and $c_2(k)$ are identical?

4. Explain why the sequences $s_1(k) = A\sin(2\pi f_1 k)$ and $s_2(k) = A\sin(2\pi f_2 k)$ are identical whenever f_1 and f_2 differ by an integer.

5. Likewise, explain why the sequences $s_1(k) = A\sin(2\pi n_1 k/N)$ and $s_2(k) = A\sin(2\pi n_2 k/N)$ are identical whenever n_1 and n_2 differ by an integer multiple of N.

6. Explain how the sequences $s_1(k) = A\sin(2\pi n_1 k/N)$ and $s_2(k) = A\sin(2\pi n_2 k/N)$ are related when $n_1 + n_2 = N$.

MATLAB Exercises.

For this set, please use the MATLAB *stem* function for plotting sequences.

1. Write a MATLAB function that would plot m periods of the finite exponential sequence $x(k) = b^k$ of length N. Your function should accept the base b, the number of periods m, and the sequence length N as parameters. Your plots should have titles and axis labels. Experiment with several different combinations of parameter values and compile a PowerPoint presentation.

2. Write MATLAB functions that would plot m periods of the following N-periodic sine, cosine, and generalized cosine sequences:

(a) $c(k) = A\cos(2\pi nk/N)$,

(b) $s(k) = A\sin(2\pi nk/N)$,

(c) $g(k) = A\cos(2\pi nk/N - \phi)$.

Your function should accept the period N, the amplitude A, the frequency n, the phase shift ϕ, and the number of periods m as parameters. Your plots should have titles and axis labels. Experiment with several different combinations of parameter values and compile a PowerPoint presentation.

3. Use the function you created and tested in Problem 2 to demonstrate the symmetry and circular symmetry properties of the sine and cosine sequences.

4. Use the function you created and tested in Problem 2 to demonstrate that sine and cosine sequences with different frequencies can be identical.

6.4.2 DFT: Definition, Examples, and Basic Properties

In order to analyze frequency content of continuous periodic functions, we devised Fourier coefficients and Fourier series expansions. What would be an analogue for a finite (or, equivalently, an N-periodic) **sequence** $x(k)$? It would seem reasonable to replace integration in the formula

$$X(n) = \frac{1}{T}\int_0^T x(t)e^{-2\pi int/T}dt$$

for the nth Fourier coefficient with summation. If we do so, the result will be

$$X(\omega) = \frac{1}{N}\sum_{k=0}^{N-1} x(k)e^{-2\pi ik\omega/N}. \tag{6.113}$$

What kind of function is $X(\omega)$ and what kind of variable is ω?

First of all, by analogy with Fourier coefficients, we would expect $X(\omega)$ to measure the contribution of frequency ω to the signal $x(k)$. When working with continuous periodic signals, we dealt exclusively with integer frequencies. What type of frequencies should we expect to encounter when working with N-periodic sequences?

1. It might come as a surprise to discover that $X(\omega)$ is also N-periodic, just like $x(k)$. Indeed, because the continuous variable t has been replaced with an integer-valued counter k, we have

$$X(\omega + N) = \frac{1}{N}\sum_{k=0}^{N-1} x(k)e^{-2\pi ik(\omega+N)/N}$$

$$= \frac{1}{N}\sum_{k=0}^{N-1} x(k)e^{-2\pi ik\omega/N}e^{-2\pi ik}$$
$$= X(\omega),$$

because $e^{-2\pi ik} = 1$ when k is an integer-valued (as opposed to a continuous) variable.

2. We recall that the Fourier coefficients $X(n)$ contain the complete information about the continuous signal $x(t)$, and, therefore, $x(t)$ can be reconstructed from its Fourier coefficients by means of its Fourier series expansion. We would likewise expect the function $X(\omega)$ to contain the same amount of information as the discrete signal $x(k)$. Since $x(k)$ is given by N values, it would seem reasonable to only evaluate $X(\omega)$ at N values of ω. The most straightforward choice seems to be the uniformly-spaced integer values of ω from 0 to $N-1$.

Based on this discussion, it would seem reasonable to propose that for a finite N-point sequence $x(k)$ with the indices $0 \le k < N$, the role of Fourier coefficients can be played by the quantities defined by

$$X(n) = \frac{1}{N}\sum_{k=0}^{N-1} x(k)e^{-2\pi ikn/N}. \tag{6.114}$$

Due to the similarity of this definition to that of Fourier coefficients, we would expect $X(n)$ to measure the contribution of the component $e^{2\pi in/N}$ to the signal $x(k)$ and, consequently, we would expect to be able to reconstruct $x(k)$ by means analogous to the Fourier series expansion, specifically, by applying the formula

$$x(k) = \sum_{n=0}^{N-1} X(n)e^{2\pi ikn/N}. \tag{6.115}$$

Our intuition will be proved correct in the next subsection, where we will use methods of linear algebra to place the formulas (6.114) and (6.115) on a solid mathematical foundation.

It turns out, however, that the presence of the factor $\frac{1}{N}$ in the defining formula (6.114) poses certain technical difficulties, particularly when one wishes to extend the theory to include infinite non-periodic sequences. For this reason, most texts omit the factor $\frac{1}{N}$ in the definition of $X(n)$. Most importantly for us, MATLAB also omits it. We, therefore, follow the trend and also omit this factor in the following definition:

Definition 6.7 *The* ***discrete Fourier transform (DFT)*** *of the N-point sequence $x(k)$ is given by*

$$\hat{x}(n) \equiv X(n) = \sum_{k=0}^{N-1} x(k)e^{-2\pi ikn/N}. \tag{6.116}$$

As has just been discussed, similar to the Fourier coefficients, $X(n)$ measures the contribution of the component $e^{2\pi ink/N}$ to the signal $x(k)$, but because of the absence of the factor $\frac{1}{N}$ in the formula (6.116), $X(n)$ gives us N times its amplitude. As mentioned a few paragraphs earlier, it would, therefore, seem plausible for $x(k)$ to have an expansion analogous to the Fourier series of continuous signals, namely

$$x(k) = \frac{1}{N}\sum_{n=0}^{N-1} X(n)e^{2\pi ink/N}, \tag{6.117}$$

which is known as the **Inversion Formula** for the discrete Fourier transform. The Inversion Formula enables us to reconstruct the original discrete signal from its frequency content.

If all of that sounds a bit abstract, we hope that the following examples will help the reader become more familiar and comfortable with the concepts of discrete Fourier transform and its inversion formula.

Example 6.9 *As the first and simplest example, we calculate the DFT of the periodic* ***unit impulse*** $\boldsymbol{\delta}$*, which, we recall, is defined by*

$$\delta(k) = \begin{cases} 1, & k = 0 \mod N \\ 0, & k \neq 0 \mod N. \end{cases} \tag{6.118}$$

It is evident that

$$\hat{\delta}(n) = \delta(0)e^{-2\pi in\cdot 0/N} = 1$$

for all n and, therefore, the unit impulse is comprised of all the frequencies in the spectrum (see Figure 6.12 (a,b)). As a variation of this example, we may also consider the shifted unit impulse $\boldsymbol{\delta}_a$ *defined by*

$$\delta_a(k) = \begin{cases} 1, & k = a \mod N \\ 0, & k \neq a \mod N \end{cases} \tag{6.119}$$

and observe that its DFT

$$\hat{\delta}_a(n) = \delta(a)e^{-2\pi ian/N} = e^{-2\pi in\cdot a/N}$$

for all n, which equals the DFT of the unit impulse times a complex exponential.

Example 6.9 makes us wonder what would happen in general if the original periodic sequence $x(k)$ were shifted (delayed) by a positions. Formally, for a fixed integer a, we define

$$y(k) = x(k-a)$$

for all $k \in \mathbb{Z}$. Then, using the substitution $m = k - a$, we get

$$Y(n) = \sum_{k=0}^{N-1} y(k)e^{-2\pi ikn/N}$$

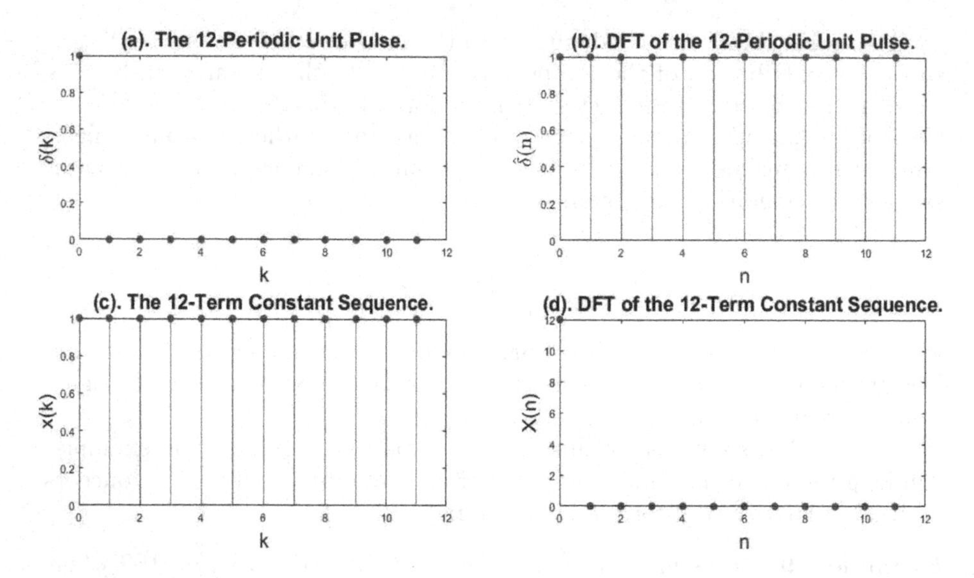

FIGURE 6.12: The discrete Fourier transforms of basic signals.

$$
\begin{aligned}
&= \sum_{k=0}^{N-1} x(k-a)e^{-2\pi ikn/N} \\
&= \sum_{m=0}^{N-1} x(m)e^{-2\pi i(m+a)n/N} \\
&= e^{-2\pi ian/N} \sum_{m=0}^{N-1} x(m)e^{-2\pi imn/N} \\
&= e^{-2\pi ian/N} X(n) \qquad (6.120)
\end{aligned}
$$

due to the N-periodicity of $x(k)$. We have thus established the following **time-delay rule** for the discrete Fourier transform:

> The delay/shift of a finite or an N-periodic discrete signal by a positions is equivalent to the multiplication of its DFT by the factor $e^{-2\pi ian/N}$. More formally, if
>
> $$y(k) = x(k-a)$$
>
> then
>
> $$Y(n) = e^{-2\pi ian/N} X(n). \qquad (6.121)$$

Example 6.10 *Because it does not vary and, hence, only consists of the zero-frequency component, it is intuitively apparent (see Figure 6.12(c,d)) that the*

DFT of the constant sequence

$$x(k) = 1$$

of length N is

$$X(n) = N\delta(n)$$

with the unit impulse $\delta(n)$ defined in (6.118). *This can be verified by means of the calculations*

$$X(0) = \sum_{k=0}^{N-1} 1 \cdot e^{-2\pi ik\cdot 0/N} = \sum_{k=0}^{N-1} 1 = N$$

for $n = 0$ and

$$\begin{aligned} X(n) &= \sum_{k=0}^{N-1} 1 \cdot e^{-2\pi ik\cdot n/N} \\ &= \frac{1 - e^{-2\pi in}}{1 - e^{-2\pi in/N}} = 0, \end{aligned}$$

for $n \neq 0$, where we used the formula for the sum of the first N terms of the geometric series with the initial term 1 *and the ratio $e^{-2\pi in/N}$.*

Since among the most important mathematical structures used in this chapter are trigonometric sequences and their relatives, the reader is probably eagerly awaiting the derivation of the discrete Fourier transform of sine, cosine, and complex exponential sequences.

Example 6.11 *It is intuitively apparent (from the meaning and definition of the discrete Fourier transform) that the DFT of the complex exponential sequence*

$$w_m(k) = e^{2\pi ik\cdot m/N}$$

of length N and frequency m is $N\delta_m(n)$, which can also be confirmed by the calculation

$$\begin{aligned} \hat{w}_m(n) &= \sum_{k=0}^{N-1} e^{2\pi ik\cdot m/N} \cdot e^{-2\pi ik\cdot n/N} \\ &= \sum_{k=0}^{N-1} e^{-2\pi ik\cdot (n-m)/N} \\ &= N\delta_m(n), \end{aligned}$$

as the reader can easily verify. It follows that the DFT of the discrete cosine wave of frequency m given by

$$c_m(k) = A\cos(m \cdot 2\pi k/N) = \frac{A}{2}\left(e^{m\cdot 2\pi ik/N} + e^{-m\cdot 2\pi ik/N}\right)$$

is

$$\hat{c}_m(n) = \frac{AN}{2}\left[\delta_m(n) + \delta_m(-n)\right]$$

(see Figure 6.13(a-d)), and that the DFT of the analogous discrete sine wave given by

$$s_m(k) = A\sin(m \cdot 2\pi k/N) = \frac{A}{2i}\left(e^{m\cdot 2\pi ik/N} - e^{-m\cdot 2\pi ik/N}\right)$$

is

$$\hat{s}_m(n) = \frac{AN}{2i}\left[\delta_m(n) - \delta_m(-n)\right]$$

as illustrated in Figure 6.13(e,f).

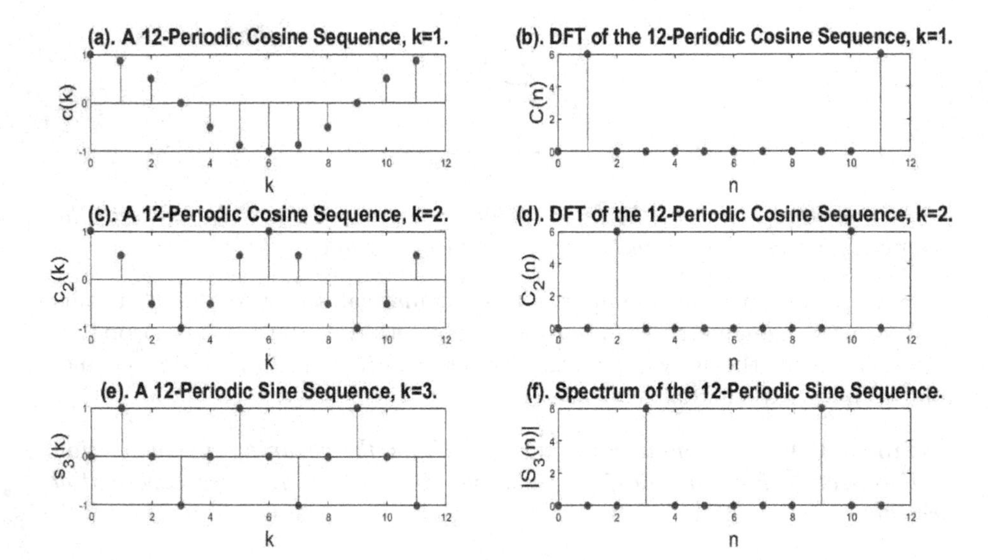

FIGURE 6.13: The discrete Fourier transforms of discrete sine and cosine waves of different frequencies.

What would happen in general if an N-periodic sequence $x(k)$ were multiplied by the complex exponential $e^{2\pi ik\cdot m/N}$? To answer this question, we define

$$y(k) = x(k)e^{2\pi ik\cdot m/N}$$

and calculate

$$\begin{aligned} Y(n) &= \sum_{k=0}^{N-1} x(k)e^{2\pi ik\cdot m/N} \cdot e^{-2\pi ik\cdot n/N} \\ &= \sum_{k=0}^{N-1} x(k)e^{-2\pi ik\cdot(n-m)/N} \end{aligned}$$

$$= \quad X(n-m), \tag{6.122}$$

which leads to the following frequency-shift rule:

A frequency modulation (that is, multiplication by $e^{2\pi ik\cdot m/N}$) of an N-periodic discrete signal results in the shift of its DFT by m positions. More formally, if

$$y(k) = x(k)e^{2\pi ik\cdot m/N}$$

then

$$Y(n) = X(n-m). \tag{6.123}$$

Since in real life, we are more likely to encounter a real-valued cosine wave than a complex exponential, we would also like to know what would happen if a periodic sequence $x(k)$ were to be multiplied by $c_m(k) = \cos(2\pi km/N)$. In this case, we define

$$y(k) = x(k)\cos(2\pi k\cdot m/N) = \frac{x(k)}{2}\left(e^{2\pi ikm/N} + e^{-2\pi ikm/N}\right)$$

and readily obtain

$$Y(n) = \frac{1}{2}\big[X(n-m) + X(n+m)\big],$$

which leads to the following general rule:

Multiplying an N-periodic sequence $x(k)$ by the discrete N-periodic cosine sequence of frequency m is equivalent to taking the average of the shifts of its DFT by m and $-m$ positions. More formally, if

$$y(k) = x(k)\cos(2\pi km/N)$$

then

$$Y(n) = \frac{1}{2}\big[X(n-m) + X(n+m)\big]. \tag{6.124}$$

As an illustration of this rule, we consider the 13-point sequences $x(k)$ whose discrete Fourier transform is

$$X(n) = (0,0,0,6,8,0,0,0,0,8,6,0,0)$$

and its modulated version

$$y(k) = x(k)\cos(4\pi k/13),$$

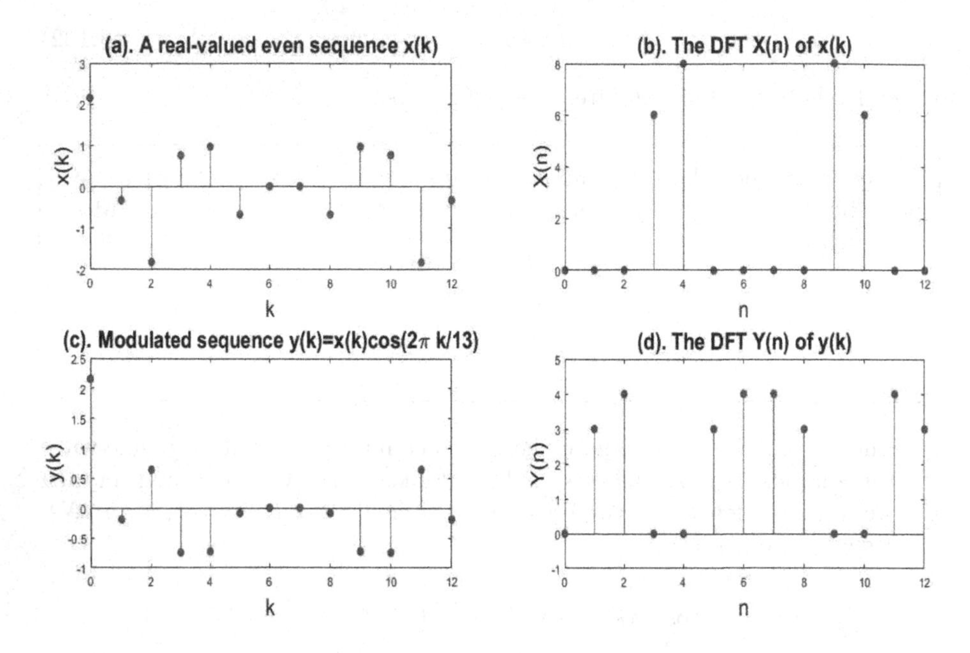

FIGURE 6.14: Illustration of the modulation rule described by (6.124).

formed by multiplying $x(k)$ by the cosine sequence of frequency $n = 2$. Figure 6.14 shows graphs of $x(k)$ and $y(k)$ alongside the graphs of their discrete Fourier transforms.

Because the quantities we encounter in real life are almost exclusively real-valued, it is certainly of interest to us to determine the conditions under which the discrete Fourier transform of a finite sequence is real-valued. Similar to what we have already observed in Subsection 6.2.4 in the context of Fourier coefficients, the real-valuedness of the DFT is strongly related to the symmetry properties of the signal and vice versa.

As we saw in Subsection 6.4.1, there are certain peculiarities about finite and N-periodic sequences. To help us recall the peculiar meaning of symmetry in this context, the following example might prove helpful:

Example 6.12 *The finite sequence $x(k) = (1, 2, 4, 4, 2)$ is of length $N = 5$, it is real-valued and even in the sense that its periodic extension is even or, equivalently, that $x(k) = x(N - k)$ for all k. The values of the discrete Fourier transform $X(n)$ of $x(k)$ are*

$$
\begin{aligned}
X(0) &= \sum_{k=0}^{4} x(k) = 1 + 2 + 4 + 4 + 2 + 1 = 13, \\
X(1) &= \sum_{k=0}^{4} x(k) e^{-2\pi i k/5}
\end{aligned}
$$

$$= \quad 1 + 2e^{-2\pi i/5} + 4e^{-4\pi i/5} + 4e^{-6\pi i/5} + 2e^{-8\pi i/5}$$

$$= \quad 1 + \cos(2\pi/5) + 2\cos(4\pi/5),$$

$$= \quad \sum_{k=0}^{4} x(k)e^{-2\pi ik\cdot 2/5}$$

$$= \quad 1 + 2e^{-4\pi i/5} + 4e^{-8\pi i/5} + 4e^{-12\pi i/5} + 2e^{-16\pi i/5}$$

$$= \quad 1 + \cos(4\pi/5) + 2\cos(8\pi/5),$$

$$= \quad \sum_{k=0}^{4} x(k)e^{-2\pi ik\cdot 3/5}$$

$$= \quad 1 + 2e^{-6\pi i/5} + 4e^{-12\pi i/5} + 4e^{-18\pi i/5} + 2e^{-24\pi i/5}$$

$$= \quad 1 + \cos(4\pi/5) + 2\cos(8\pi/5) = X(2), \quad \textit{and}$$

$$= \quad \sum_{k=0}^{4} x(k)e^{-2\pi ik\cdot 4/5}$$

$$= \quad 1 + 2e^{-8\pi i/5} + 4e^{-16\pi i/5} + 4e^{-24\pi i/5} + 2e^{-32\pi i/5}$$

$$= \quad 1 + \cos(2\pi/5) + 2\cos(4\pi/5) = X(1).$$

or this specific real-valued and even sequence $x(k)$, its discrete orm $X(n)$ is also real-valued and even.

y, suppose that an N-periodic discrete signal $x(k)$ is **real-**

$$N - n) = X(-n) \quad = \quad \sum_{k=0}^{N-1} x(k)e^{2\pi ink/N}$$

$$= \quad \overline{\sum_{k=0}^{N-1} x(k)e^{-2\pi ink/N}} = \overline{X(n)},$$

$x(k)$ is *conjugate-symmetric.* If in addition to being real-valued, $x(k)$ is also **even**, substituting $l = N - k$ yields

$$\overline{X(n)} \quad = \quad \overline{\sum_{k=0}^{N-1} x(k)e^{-2\pi ink/N}}$$

$$= \quad \sum_{l=0}^{N-1} x(l)e^{-2\pi inl/N}$$

$$= \quad X(n),$$

ablishes that

T of an even real-valued sequence is itself real-valued and even.

contrary, the sequence $x(k)$ is **real-valued and odd**, substituting k yields

$$\begin{aligned}\overline{X(n)} &= \overline{\sum_{k=0}^{N-1} x(k)e^{-2\pi ink/N}} \\ &= -\sum_{l=0}^{N-1} x(l)e^{-2\pi inl/N} \\ &= -X(n)\end{aligned}$$

conclusion that

T of a real-valued and odd sequence is purely imaginary and odd.

g studied Fourier series of continuous periodic functions, we can ex-
e sort of duality between multiplication in the time domain and con-
in the frequency domain (and vice versa) to also exist in the context
e signals. To investigate this matter, we suppose that $x(k)$ and $y(k)$
or N-periodic sequences and consider their component-wise product
y

$$g(k) = x(k) \cdot y(k).$$

te Fourier transform is

$$\begin{aligned}G(n) &= \sum_{k=0}^{N-1} g(k)e^{-2\pi ink/N} \\ &= \sum_{k=0}^{N-1} x(k) \cdot y(k)e^{-2\pi ink/N} \\ &= \sum_{k=0}^{N-1} \Big(\frac{1}{N}\sum_{m=0}^{N-1} X(m)e^{2\pi ikm/N}\Big) y(k)e^{-2\pi ink/N} \\ &= \frac{1}{N}\sum_{m=0}^{N-1} X(m) \sum_{k=0}^{N-1} y(k)e^{-2\pi i(n-m)k/N} \\ &= \frac{1}{N}\sum_{m=0}^{N-1} X(m) \cdot Y(n-m) = \frac{1}{N}(X * Y)(n),\end{aligned}$$

shes that

> wise multiplication of finite or N-periodic sequences is equiv-
> the constant factor $\frac{1}{N}$) to the circular convolution of their
> rier transforms. Formally, if
>
> $$g(k) = x(k) \cdot y(k)$$
>
> $$G(n) = \frac{1}{N}(X * Y)(n). \tag{6.125}$$

nind the reader that the circular convolution of two finite se-
e same lengths is equivalent to the linear convolution or their
sions with the summation performed over one period (and, for
is also called "periodic" convolution). In order to calculate the
lution of sequences of different lengths, the shorter one must be
o extend it to the length of the longer one.

ne single most important property of all variants of Fourier trans-
g the DFT, is that they provide a way to replace convolution
ation. If we calculate the discrete Fourier transform of the cir-
ion

$$g(k) = (x * y)(k) = \sum_{m=0}^{N-1} x(m)y(k-m)$$

r N-periodic sequences $x(k)$ and $y(k)$, we obtain

$$\begin{aligned}
&= \sum_{k=0}^{N-1} g(k)e^{-2\pi ink/N} \\
&= \sum_{k=0}^{N-1} \Big(\sum_{m=0}^{N-1} x(m)y(k-m) \Big) e^{-2\pi inm/N} e^{-2\pi in(k-m)/N} \\
&= \sum_{m=0}^{N-1} x(m)e^{-2\pi inm/N} \Big(\sum_{k=0}^{N-1} y(k-m)e^{-2\pi in(k-m)/N} \Big) \\
&= X(n) \cdot Y(n)
\end{aligned}$$

which proves that

r convolution of finite or N-periodic sequences is equivalent to the
se product of their discrete Fourier transforms. Formally, if

$$g(k) = (x * y)(k)$$

$$G(n) = X(n) \cdot Y(n). \quad (6.126)$$

mportance of this property is difficult to overstate. Not only does
us to replace the computationally expensive operation of convolu-
a faster operation of pointwise multiplication, but it also provides
reverse the effect of convolution (when necessary and under cer-
litions). We will have an opportunity to appreciate its significance in
on 6.4.4 and in Section 6.5.
ection on the definition and basic properties of discrete Fourier trans-
ld not be complete without at least a brief set of instructions on how
ATLAB for DFT calculations. The MATLAB function of choice is
ich uses a very efficient algorithm to calculate the discrete Fourier
n of the vector **x**. The following MATLAB code creates a superposi-
of three 60-periodic cosine sequences with three different amplitudes
$A_2 = 7$, and $A_3 = 5$) and three different frequencies ($n_1 = 2$, $n_2 = 6$,
15) and calculates its DFT $X(n)$. The program also plots the graphs
(k) and $X(n)$ side by side.

```
           % the signal length
 k=k';       % the array of sequence indices from 0 to 60
2=7; A3=5;        % the amplitudes of signal components
=6; f3=15;       % the frequencies of signal components
s(2*pi*f1*k/N)+A2*cos(2*pi*f2*k/N)+A3*cos(2*pi*f3*k/N);
,1,1);
'filled');       %plotting x
,'FontSize',18);
(k)','FontSize',18);
 Superposition of three cosine sequences.','FontSize',18);
x);       % calculation of the DFT of x
,1,2);
hat,'filled');       %plotting DFT of x
,'FontSize',18);
(n)','FontSize',18);
 DFT X(n) of x(k)','FontSize',18);
```

ram output is shown in Figure 6.15. We hope that this example will

l reference for completing the relevant parts of the exercises at
e remaining sections of this chapter.

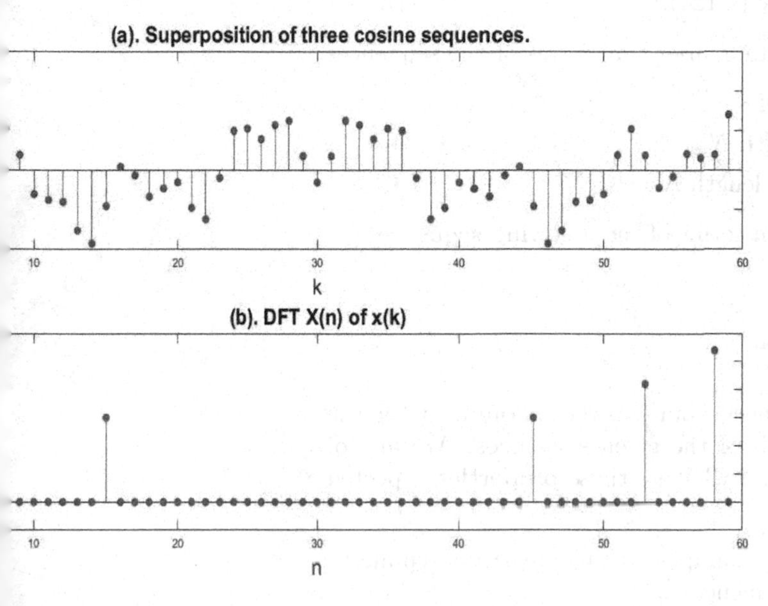

15: Example of using MATLAB for calculating and plotting
er transforms.

rcises below, the reader will be asked to verify some of the
ound in this subsection and to deepen their understanding of
ourier transform by proving several additional properties.

.4.2 Exercises.

at the discrete Fourier transform is a linear transformation of
N-periodic sequences by establishing the following properties:

pose $x(k)$ and $y(k)$ are finite or N-periodic sequences. Define
$) = x(k) + y(k)$. Prove that $G(n) = X(n) + Y(n)$.

pose $x(k)$ is a finite or N-periodic sequence and α is a constant
l or complex). Define $g(k) = \alpha \cdot x(k)$. Prove that $G(n) = \alpha X(n)$.

e calculation of the discrete Fourier transform of the unit im-
d the shifted unit impulse in Example 6.9.

at the DFT of the discrete complex exponential function

$$e_m(k) = e^{2\pi i k \cdot m/N}$$

N and frequency m is $N\delta_m(n)$ as stated in Example 6.11.

fine $y(k) = x(k)\sin(2\pi km/N)$ where $x(k)$ is an N-periodic sequence. lculate the discrete Fourier transform $Y(n)$ of $y(n)$ and state a rule alogous to the one expressed by (6.124).

lculate and compare the discrete Fourier transforms of the sequences

) $x(k) = (1, 2, 3, 4)$ of length $N = 4$,

) $x(k) = (1, 2, 3, 4, 0, 0)$ of length $N = 6$,

) $x(k) = (1, 2, 3, 4, 0, 0, 0, 0)$ of length $N = 8$.

lculate the discrete Fourier transforms of the following sequences:

) $x(k) = (0, 5, 7, 0, 0, 7, 5)$,

) $x(k) = (1, 1, 3, 2, 2, 3, 1)$,

) $x(k) = (0, 1, 2, 3, -3, -2, -1)$.

e MATLAB function *fft* to check your answers. Comment on the nmetry properties of the DFTs of the given sequences. Are any of m real-valued or purely imaginary? Were those properties expected l why?

nstruct an example similar to Example 6.12 to illustrate symmetry perties of purely imaginary sequences.

neralize the example you constructed in the previous problem and de- e symmetry properties of discrete Fourier transforms of purely imag- ry sequences analogous to the properties of real-valued sequences.

sume that a finite or N-periodic sequence $x(k)$ and its discrete Fourier nsform $X(n)$ are both complex-valued. Then they can be expressed

$$x(k) = x_R(k) + x_I(k) \quad \text{and} \quad X(n) = X_R(n) + X_I(n)$$

ng their real and imaginary parts. Show that

$$X_R(n) = \sum_{k=0}^{N-1} \left[x_R(k)\cos(2\pi kn/N) + x_I(k)\sin(2\pi kn/N)\right],$$

$$X_I(n) = \sum_{k=0}^{N-1} \left[x_I(k)\cos(2\pi kn/N) - x_R(k)\sin(2\pi kn/N)\right]$$

$$x_R(k) = \frac{1}{N}\sum_{n=0}^{N-1} \left[X_R(n)\cos(2\pi kn/N) - X_I(n)\sin(2\pi kn/N)\right],$$

$$x_I(k) = \frac{1}{N}\sum_{n=0}^{N-1} \left[X_I(n)\cos(2\pi kn/N) + X_R(n)\sin(2\pi kn/N)\right].$$

e these expressions to provide an alternative derivation of the sym- ry properties of the discrete Fourier transform.

10. Suppose that $x(k)$ is a finite or an N-periodic sequence and $y(k)$ is defined by
$$y(k) = x(N-k).$$
Show that the discrete Fourier transforms of the two sequences satisfy
$$Y(n) = X(N-n).$$
In other words, reversing a sequence is equivalent to reversing its DFT. Construct several examples to illustrate this rule using real-valued sequences.

11. Suppose that $x(k)$ is a finite or an N-periodic sequence and $y(k)$ is defined by
$$y(k) = \overline{x(N-k)}.$$
Show that the discrete Fourier transforms of the two sequences satisfy
$$Y(n) = \overline{X(N-n)}.$$

12. Suppose N is even and that $x(k)$ is a finite or N-periodic real-valued sequence with the property that the "right" half of the sequence is the negative of the "left" half or, more formally,
$$x(k+N/2) = -x(k).$$
Show that its discrete Fourier transform
$$X(n) = 0$$
whenever n is even.

13. Construct examples illustrating the rule (6.121) by calculating and comparing the discrete Fourier transforms of the following sequences:
    (a) $x_1(n) = (0,1,2,3)$ and $x_2(n) = (1,2,3,0)$,
    (b) $x_1(n) = (0,1,2,3)$ and $x_2(n) = (2,3,0,1)$,
    (c) $x_1(n) = (1,1,1,1,0,0,0,0)$ and $x_2(n) = (1,1,0,0,0,0,1,1)$.

14. Calculate the discrete Fourier transforms of the following finite sequences:
    (a) $x(k) = (1/2)^k, \quad 0 \le k \le 7$,
    (b) $x(k) = \cos(3\pi k/8), \quad 0 \le k \le 7$,
    (c) $x(k) = \sin(3\pi k/8), \quad 0 \le k \le 7$.

    Recover the original sequence using the DFT Inversion Formula and check all your calculations using the MATLAB functions *fft* and *ifft*.

15. Construct an example illustrating the rule (6.124) by calculating and plotting the discrete Fourier transforms of the following sequences:

    (a) $x(k) = (0, 0, 0, 4, 8, 0, 0, 0, 0, 8, 4, 0, 0)$,

    (b) $y(k) = x(k)\cos(2\pi k/13)$.

16. Consider the following two sequences of length 5:

$$x_1(k) = (0, 2, 1, 1, 2) \quad \text{and} \quad x_2(k) = (0, 1, 3, 3, 1).$$

    Construct an example illustrating the rule (6.125) by performing the following:

    (a) Calculate the discrete Fourier transforms $X_1(n)$ and $X_2(n)$ of the sequences $x_1(k)$ and $x_2(k)$ respectively.

    (b) Calculate the component-wise product of the sequences $x_1(k)$ and $x_2(k)$ and its discrete Fourier transform.

    (c) Calculate the circular convolutions $(X_1 * X_2)(n)$.

    (d) Check whether your results in parts (b) and (c) agree.

17. Provide several more examples of the rule (6.125) similar to the one in the previous problem by using pairs of sequences in Exercise 14. If the sequences are of different lengths, use zero-padding to calculate their circular convolutions and pairwise products. Check all your calculations using the MATLAB functions *fft* and *ifft*

18. Consider again the following two sequences of length 5:

$$x_1(k) = (0, 2, 1, 1, 2) \quad \text{and} \quad x_2(k) = (0, 1, 3, 3, 1).$$

    Construct an example of the rule (6.126) by performing the following:

    (a) Calculate the discrete Fourier transforms $X_1(n)$ and $X_2(n)$ of the sequences $x_1(k)$ and $x_2(k)$ respectively.

    (b) Calculate the circular convolution of the sequences $x_1(n)$ and $x_2(n)$ and its discrete Fourier transform.

    (c) Calculate the component-wise product of $X_1(n)$ and $X_2(n)$.

    (d) Check whether your results in parts (b) and (c) agree.

19. Provide several more examples of the rule (6.126) similar to the one in the previous problem by using pairs of sequences from Exercise 14. Check all your calculations using the MATLAB functions *fft* and *ifft*.

MATLAB Exercises.

1. Write a MATLAB function that would calculate and plot a graph of the discrete N-periodic cosine wave of amplitude A and frequency n alongside a graph of its discrete Fourier transform. Your function should accept the values of N, A, and n as parameters. Test your function with several different values of the parameters and compile a PowerPoint presentation.

2. Write and test a similar MATLAB function for plotting a graph of the discrete N-periodic sine wave alongside a graph of its discrete Fourier transform.

3. Write a MATLAB function to calculate the discrete Fourier transform of the sequence $x(k)$ of length N using the defining formula (6.116). Your function should accept $x(k)$ as a parameter and should calculate its length N internally.

4. Similarly, write a MATLAB function to calculate the inverse discrete Fourier transform of the specified sequence $\hat{x}(n)$ of length N using the Inversion Formula (6.117). Your function should accept $\hat{x}(n)$ as a parameter and should calculate its length N internally.

5. Use the MATLAB functions you created in the two previous exercises to calculate the DFT and the Inverse DFT of a variety of sequences of your choice. Check your results with the help of the MATLAB functions *fft* and *ifft*.

6.4.3 Placing the DFT on a Firm Foundation

Definition 6.7 of the discrete Fourier transform and the DFT Inversion Formula (6.117) that followed were motivated by the experience of Section 6.2. Even though the parallels with Section 6.2 have been convincing, we still have to admit that the Inversion Formula was accepted uncritically and without any rigorous proof. In this subsection, we use techniques of linear algebra to remedy that deficiency.

As the first step in this endeavor, we introduce the vector notation

$$\mathbf{x} = \begin{bmatrix} x(0) \\ x(1) \\ \vdots \\ x(N-1) \end{bmatrix} \quad \text{and} \quad \hat{\mathbf{x}} = \begin{bmatrix} \hat{x}(0) \\ \hat{x}(1) \\ \vdots \\ \hat{x}(N-1) \end{bmatrix}$$

for the sequence $\mathbf{x}$ and its discrete Fourier transform $\hat{\mathbf{x}}$ and express the DFT Inversion Formula

$$x(k) = \frac{1}{N} \sum_{n=0}^{N-1} \hat{x}(n) e^{2\pi i n k/N},$$

in matrix form by writing

$$\mathbf{x} = A \cdot \hat{\mathbf{x}},$$

where the inverse discrete Fourier transform matrix A is evidently

$$A = \begin{bmatrix} 1 & 1 & 1 & \dots & 1 \\ 1 & e^{\frac{2\pi i}{N}} & \left(e^{\frac{2\pi i}{N}}\right)^2 & \dots & \left(e^{\frac{2\pi i}{N}}\right)^{N-1} \\ 1 & \left(e^{\frac{2\pi i}{N}}\right)^2 & \left(e^{\frac{2\pi i}{N}}\right)^{2\cdot 2} & \dots & \left(e^{\frac{2\pi i}{N}}\right)^{2(N-1)} \\ 1 & \left(e^{\frac{2\pi i}{N}}\right)^3 & \left(e^{\frac{2\pi i}{N}}\right)^{3\cdot 2} & \dots & \left(e^{\frac{2\pi i}{N}}\right)^{3(N-1)} \\ \vdots & \vdots & \vdots & \ddots & \vdots \\ 1 & \left(e^{\frac{2\pi i}{N}}\right)^{N-1} & \left(e^{\frac{2\pi i}{N}}\right)^{(N-1)\cdot 2} & \dots & \left(e^{\frac{2\pi i}{N}}\right)^{(N-1)\cdot(N-1)} \end{bmatrix}, \tag{6.127}$$

which (after we alleviate notation by denoting $w_N = e^{\frac{2\pi i}{N}}$) becomes

$$A = \begin{bmatrix} 1 & 1 & 1 & \dots & 1 \\ 1 & w_N & w_N^2 & \dots & w_N^{N-1} \\ 1 & w_N^2 & w_N^{2\cdot 2} & \dots & w_N^{2(N-1)} \\ 1 & w_N^3 & w_N^{3\cdot 2} & \dots & w_N^{3(N-1)} \\ \vdots & \vdots & \vdots & \ddots & \vdots \\ 1 & w_N^{N-1} & w_N^{(N-1)\cdot 2} & \dots & w_N^{(N-1)\cdot(N-1)} \end{bmatrix} \tag{6.128}$$

or, even more compactly,

$$A = \begin{bmatrix} \mathbf{a}_0 & \mathbf{a}_1 & \mathbf{a}_2 & \dots & \mathbf{a}_{N-1} \end{bmatrix}.$$

The matrix form of the DFT Inversion formula can then be written as

$$\begin{bmatrix} x(0) \\ x(1) \\ x(2) \\ x(3) \\ \vdots \\ x(N-1) \end{bmatrix} = \begin{bmatrix} 1 & 1 & 1 & \dots & 1 \\ 1 & w_N & w_N^2 & \dots & w_N^{N-1} \\ 1 & w_N^2 & w_N^{2\cdot 2} & \dots & w_N^{2(N-1)} \\ 1 & w_N^3 & w_N^{3\cdot 2} & \dots & w_N^{3(N-1)} \\ \vdots & \vdots & \vdots & \ddots & \vdots \\ 1 & w_N^{N-1} & w_N^{(N-1)\cdot 2} & \dots & w_N^{(N-1)\cdot(N-1)} \end{bmatrix} \times \begin{bmatrix} \hat{x}(0) \\ \hat{x}(1) \\ \hat{x}(2) \\ \hat{x}(3) \\ \vdots \\ \hat{x}(N-1) \end{bmatrix}, \tag{6.129}$$

and the equivalent vector form as

$$\mathbf{x} = \hat{x}(0)\mathbf{a}_0 + \hat{x}(1)\mathbf{a}_1 + \dots + \hat{x}(N-1)\mathbf{a}_{N-1}, \tag{6.130}$$

where $\mathbf{a}_n$ is the nth column of the Inverse DFT matrix A. A straightforward calculation shows that the columns $\mathbf{a}_n$ of A are mutually orthogonal. For example, the inner product of the second and the third columns is

$$\begin{aligned}
\mathbf{a}_2 \cdot \mathbf{a}_3 &= \begin{bmatrix} 1 \\ w_N \\ w_N^2 \\ w_N^3 \\ \vdots \\ w_N^{N-1} \end{bmatrix} \cdot \begin{bmatrix} 1 \\ w_N^2 \\ w_N^4 \\ w_N^6 \\ \vdots \\ w_N^{(N-1)\cdot 2} \end{bmatrix} \\
&= 1 + w_n^3 + w_N^6 + w_N^9 + ... + w_N^{(N-1)\cdot 3} \\
&= \frac{1 - \left(w_N^3\right)^N}{1 - w_N^3} = 0, \qquad (6.131)
\end{aligned}$$

where we used the formula for the sum of the first N terms of the geometric sequence with the ratio $r = w_N^3$ together with the fact that

$$\left(w_N^3\right)^N = \left(e^{\frac{2\pi i}{N}}\right)^{3N} = e^{6\pi i} = 1.$$

It follows that the Inversion Formula is just a representation of the given discrete periodic signal $\mathbf{x}$ as a linear combination of the vectors from the orthogonal basis

$$\mathcal{B} = \{\mathbf{a}_0, ..., \mathbf{a}_{N-1}\},$$

which consists of the columns of the Inverse DFT matrix A. Therefore, in order to verify the Inversion Formula (and, in doing so, to confirm that we have the appropriate definition of the discrete Fourier transform), all we need to do is calculate the coordinates of $\mathbf{x}$ with respect to $\mathcal{B}$, check that they coincide with the values of the DFT as we defined it, and see to it that the Inversion Formula uses those precise coordinates as the coefficients at the discrete complex exponentials.

So, let us check all of that. The magnitudes $||\mathbf{a}_n||$ of the columns of A are

$$||\mathbf{a}_n||^2 = \sum_{k=0}^{N-1} |w_N^{nk}|^2 = N,$$

and the inner products of $\mathbf{x}$ with the columns of A equal

$$\begin{aligned}
\langle \mathbf{x}, \mathbf{a}_n \rangle &= \sum_{k=0}^{N-1} x(k)\overline{w_N^{nk}} \\
&= \sum_{k=0}^{N-1} x(k)e^{-2\pi ink/N}.
\end{aligned}$$

Therefore, according to the formula (6.92) of Section 6.3, the coefficients c_n

of $\mathbf{x}$ with respect to the basis $\mathcal{B}$ are

$$c_n = \frac{\langle \mathbf{x}, \mathbf{a}_n \rangle}{||\mathbf{a}_n||^2} = \frac{1}{N} \sum_{k=0}^{N-1} x(k) e^{-2\pi i n k/N},$$

which coincides precisely with the definitions of the discrete Fourier transform in the form (6.114). The Inversion Formula in the form (6.115) then follows directly from (6.93). As we discussed in Subsection 6.4.2, the choice of placement of the factor $\frac{1}{N}$ in the definition of the DFT or in the Inversion Formula is largely a matter of convenience.

Subsection 6.4.3 Exercises.

1. Verify formulas (6.127) and (6.128) for the inverse discrete Fourier transform matrix A.

2. Verify formulas (6.129) and (6.130) for the matrix and vector forms of the inverse discrete Fourier transform.

3. Express discrete Fourier transform in matrix and vector forms similar to equations (6.129) and (6.130). Also, derive formulas similar to (6.127) and (6.128) for the DFT matrix.

MATLAB Exercises.

1. Write a MATLAB function to calculate the discrete Fourier transform of sequence $x(k)$ of length N using the transform matrix you derived in Exercise 3 above. Your function should accept $x(k)$ in vector form as a parameter and should calculate its length N internally.

2. Similarly, write a MATLAB function to calculate the inverse discrete Fourier transform of sequence $\hat{x}(n)$ of length N using the transform matrix (6.129) derived in this subsection. Your function should accept $\hat{x}(n)$ in vector form as a parameter and should calculate its length N internally.

3. Use the MATLAB functions you created in the two previous exercises to calculate the DFT and the inverse DFT of a variety of sequences of your choice. Compare the results with those you obtained when using the functions created in the exercises following Subsection 6.4.2 and with the results obtained using the MATLAB functions *fft* and *ifft*.

6.4.4 Linear Time-Invariant Transformations and the DFT

Having established some of the foundations of signal analysis, we are now ready to explore the basics of signal and image processing. After all, our ultimate goal is to apply the concepts and techniques we are learning here to the task of enhancing digital images, and, therefore, it is the methods of signal processing (extended to two dimensions) that, we hope, might prove useful to us.

What types of signal (and, ultimately, image) transformations should we focus on? Our experience in linear algebra suggests that easiest to handle are linear transformations (usually denoted by the letter T), which are characterized by the property that

$$T(a\mathbf{x} + b\mathbf{y}) = aT(\mathbf{x}) + bT(\mathbf{y})$$

for all vectors $\mathbf{x}$ and $\mathbf{y}$ in the given vector space (such as, for example, the space of N-periodic sequences) and all real (or complex) scalars a and b.

In the context of signal or image processing, one usually makes an additional fundamental simplifying assumption proceeding from expectations that seem reasonable in a wide class of situations. The basis of this assumption is that one would expect the transformation to work the same way both in the morning and in the evening; one would expect it to work the same way tomorrow as it did yesterday. This property is called *time-invariance*. It means that delaying the input by a specified amount of time (or number of positions) causes the output to be delayed by the same amount with no other alterations to the output.

More formally, we denote the sequence $x(k)$ by $\mathbf{x}$ and its image $T(\mathbf{x})$ under the transformation T by $\mathbf{y}$. We also denote by $\mathbf{x}_\tau$ and $\mathbf{y}_\tau$ the sequences defined by

$$x_\tau(k) = x(k - \tau) \quad \text{and} \quad y_\tau(k) = y(k - \tau)$$

respectively. Then the transformation T is called **time-invariant** if

$$T(\mathbf{x}_\tau) = \mathbf{y}_\tau \tag{6.132}$$

for all values of τ. It is worth noting that the definition of time-invariance applies both to operators that act on periodic (and, equivalently, finite) sequences and to those that act on non-periodic infinite sequences.

The assumption of time-invariance may seem eminently reasonable and even obvious, but it has profound and far-reaching implications. First of all, it is important to note that any transformation defined by convolution, that is, by

$$T(\mathbf{x}) = \mathbf{x} * \mathbf{h}$$

is automatically linear and time-invariant, which the reader will prove in the exercises at the end of this subsection. But the converse is also true. Suppose that we are going to transform the input signals $\mathbf{x}$ by means of a linear time-invariant (LTI) transformation T. We recall from Chapter 5 that the unit impulse $\boldsymbol{\delta}$ is the convolutional identity. Therefore, the sequence $\mathbf{x}$ can be expressed as the convolution $\mathbf{x} * \boldsymbol{\delta}$, specifically, as

$$x(k) = \sum_m x(m)\delta(k - m),$$

and, consequently, its image $T(\mathbf{x})$ can be expanded and reformulated as

$$\begin{aligned}
T(\mathbf{x})(k) &= T\Big(\sum_m x(m)\delta(k-m)\Big) \\
&\quad \text{[by the linearity of } T\text{]} \\
&= \sum_m x(m)T(\boldsymbol{\delta}_m)(k) \\
&\quad \text{[by the time-invariance of } T\text{]} \\
&= \sum_m x(m)T(\boldsymbol{\delta})(k-m) \\
&= (\mathbf{x} * T(\boldsymbol{\delta}))(k) = (\mathbf{x} * \mathbf{h})(k),
\end{aligned} \tag{6.133}$$

where we introduced the notation

$$\mathbf{h} = T(\boldsymbol{\delta})$$

in the last step of the calculation. The image $\mathbf{h}$ of the unit impulse sequence $\boldsymbol{\delta}$ under the transformation T is called **the impulse response** of T. The result we just derived is so important that we reiterate and frame it for easy reference:

The Fundamental Property of the LTI Transformations

The output $\mathbf{y}$ of the input $\mathbf{x}$ under the linear time-invariant transformation T with the impulse response $\mathbf{h}$ is the convolution

$$\mathbf{y} = \mathbf{x} * \mathbf{h} \tag{6.134}$$

of the input with the impulse response of the transformation.

Remark. In the derivation (6.133), we were somewhat ambiguous in our use of the indices in the summation formulas. That ambiguity was intentional, as the derivation and the resulting Fundamental Property of the LTI Transformations are equally valid for both finite and infinite sequences, and for both linear and circular convolution.

In this section, we are working with finite (or, equivalently, N-periodic) sequences, and impulse responses are no exception. The discrete Fourier transform $\mathbf{H}$ of the impulse response $\mathbf{h}$ is called **the transfer function** of the transformation T, and it equals

$$H(n) = \sum_{k=0}^{N-1} h(k)e^{-2\pi ink/N}. \tag{6.135}$$

Since the circular convolution of two N-periodic sequences is equivalent to the pointwise product of their discrete Fourier transforms, the rule (6.134) can be restated for this type of sequences in the following equivalent form:

> The DFT $\mathbf{Y}$ of the output $\mathbf{y}$ generated by the input $\mathbf{x}$ under the linear time-invariant transformation T with the impulse response $\mathbf{h}$ is the product
>
> $$\mathbf{Y} = \mathbf{X} \cdot \mathbf{H} \tag{6.136}$$
>
> of the DFT of the input and the transfer function of the transformation.

In practice, since products are easier to compute than convolutions, the rule (6.136) is much more convenient to apply than its convolution-based equivalent (6.134). Moreover, while it is not at all clear how to solve equation (6.134) for $\mathbf{x}$, equation (6.136) can be solved for $\mathbf{Y}$ quite readily (provided, of course, that $\mathbf{H}$ is everywhere nonzero).

Equation (6.134) suggests a way for an effective strategy for transforming signals by means of manipulating their frequency content. The key is to design an appropriate sequence $\mathbf{h}$ whose discrete Fourier transform $\mathbf{H}$ will emphasize the desirable frequencies and suppress (or eliminate) the undesirable ones. Afterwards, the linear time-invariant transformation with the impulse responses $\mathbf{h}$ can be applied to the input signal. The process of enhancing a given signal $\mathbf{x}$ using methods of frequency analysis works as follows:

1. The discrete Fourier transform $\mathbf{X}$ of the given signal $\mathbf{x}$ is calculated; $\mathbf{X}$ represents the frequency spectrum of the signal we are trying to enhance.

2. The pointwise product $\mathbf{Y} = \mathbf{X} \cdot \mathbf{H}$ is computed; $\mathbf{Y}$ represents the frequency spectrum of the enhanced version $\mathbf{y}$ of $\mathbf{x}$.

3. The DFT Inversion Formula is used to calculate $\mathbf{y}$, which, it is hoped, is free from most of the deficiencies and undesirable artifacts of $\mathbf{x}$.

There often develops an alternative scenario, when the original signal $\mathbf{x}$ is distorted by a system (or a transmission channel) with an impulse response $\mathbf{h}$ that is either known or can be approximated. As a result, instead of receiving the "clean" signal $\mathbf{x}$, we receive its distorted version $\mathbf{y}$, which equals the convolution of $\mathbf{x}$ with the system impulse response $\mathbf{h}$. Under those circumstances, one executes the following procedure:

1. The discrete Fourier transform $\mathbf{Y}$ of the received signal $\mathbf{y}$ is calculated; the goal is to "cleanse" it of the undesirable distortions.

2. To do so, $\mathbf{Y}$ is *divided* component-wise by the system transfer function $\mathbf{H}$, provided that the latter one is nonzero, resulting in the DFT $\mathbf{X}$ of the original signal. If $H(n) = 0$ for any n, then, unfortunately, the original signal cannot be fully reconstructed and can only be approximated.

3. Finally, the DFT Inversion formula is used to reconstruct $\mathbf{x}$.

The following example is intended to illustrate both scenarios. Consider the sequence $\mathbf{x}$ of length $N = 59$ defined by

$$x(k) = 10\cos\left(\frac{4\pi k}{59}\right) + 3\cos\left(\frac{54\pi k}{59}\right)$$

whose graph is shown in Figure 6.16(a). Evidently, the given sequence $\mathbf{x}$ consists of one high-frequency component and one low-frequency component. Suppose that we decided that the high-frequency component was undesirable and we would like to suppress it. How can this task be accomplished?

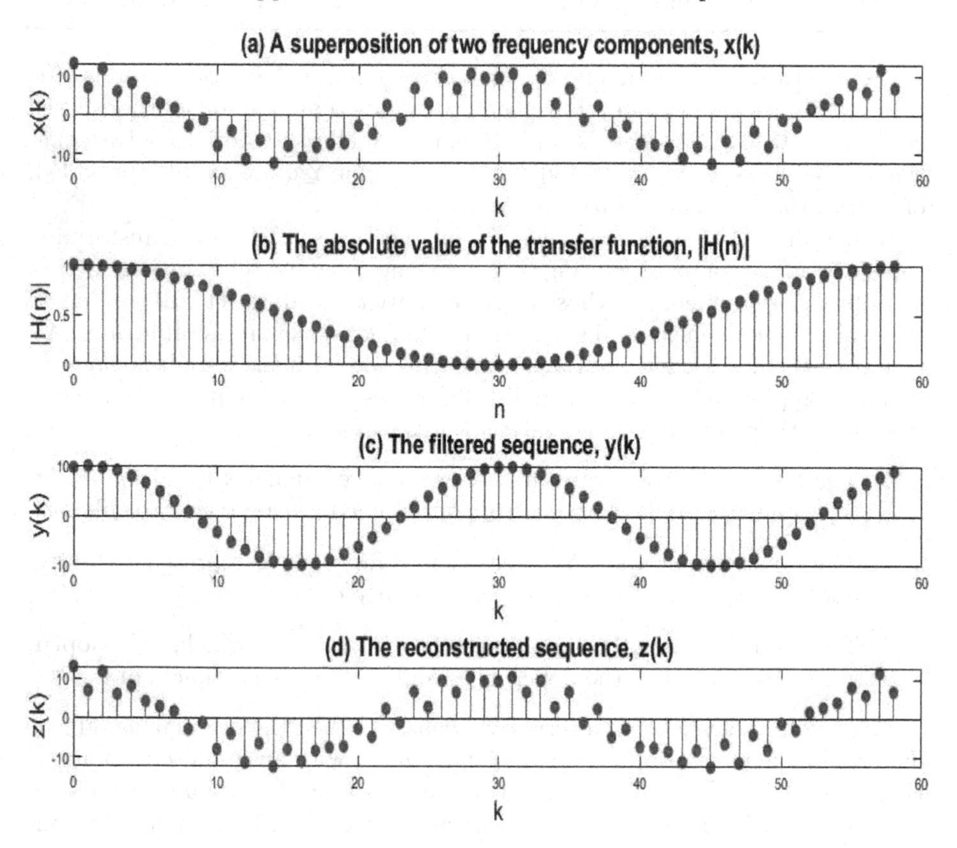

FIGURE 6.16: Filtering and restoration of a sequence with two frequency components.

It follows from the previous discussion, that we need to design a transformation with the right type of impulse response. We recall from Chapter 5 that taking averages (or weighted averages) of successive values has a smoothing effect. Therefore, we immediately think of using a linear time-invariant operator $T = \mathbf{x} * \mathbf{h}$ defined by taking the convolution of $\mathbf{x}$ with an appropriate averaging sequence, say

$$\mathbf{h} = (1/4, 1/2, 1/4).$$

We can recall from Chapter 5 that this process is called "filtering" and, in this context, the sequence $\mathbf{h}$ that we are convolving our signal $\mathbf{x}$ with is called a "filter". In order to be able to perform circular convolution of $\mathbf{x}$ with $\mathbf{h}$, we must zero-pad the shorter of the two to extend its length to match that of the longer of the two.

What characteristic properties does the averaging filter $\mathbf{h}$ possess? Figure 6.16(a) shows the graph of the absolute value of the discrete Fourier transform $H(n)$ of its zero-padded version. What immediately stands out is that the values of $H(n)$ at low frequencies (near $n = 0$ or $n = N$) are close to 1, whereas the values of $H(n)$ at high frequencies (near $N/2$) are close to zero. Such filters are called "low-pass filters", and they are commonly used to suppress high-frequency components, such as small-scale details and noise. The graph of $|H(n)|$ looks promising, and we can expect this filter to be effective in suppressing the high-frequency component of $\mathbf{x}$.

Next, we calculate the filtered sequence $\mathbf{y} = \mathbf{x} * \mathbf{h}$ (either by directly computing the circular convolution or by means of taking the inverse discrete Fourier transform of the pointwise product of $X(n)$ with the transfer function $H(n)$). The result is displayed in Figure 6.16(c). All that is visible now is the low-frequency component, with the high-frequency component having been filtered out almost without a trace.

But what if we come to regret the action just taken? What if we decided that getting rid of the high-frequency component was a mistake, and that it would be better to restore the original signal? Is it possible to reverse the convolution with the filter $\mathbf{h}$?

Fortunately, in this example, the transfer function $H(n)$ is never zero, and, consequently, the answer is yes. All we have to do is divide the discrete Fourier transform $Y(n)$ of the filtered signal by the transfer function, and we will obtain the DFT of the reconstructed signal $\mathbf{z}$, which, hopefully, is indistinguishable from the original signal $\mathbf{x}$. The graph of $\mathbf{z}$ is shown in Figure 6.16(d).

We hope that the reader has been impressed by the combined opportunities offered by the interplay of filtering and frequency analysis. Unfortunately, we have to limit our exploration of the rich and fascinating area of digital signal processing to the most fundamental definitions and facts, but we hope that the reader is motivated to undertake further study. There are many excellent texts on the subject (such as, for example, [14], [21], and [29]), and we hope that the readers will pursue further study of this field.

In the next section, we will apply all those ideas to digital images.

Subsection 6.4.4 Exercises.

1. Prove that the operator T defined by

$$T(\mathbf{x}) = \mathbf{x} * \mathbf{h}$$

on the space of physically realizable sequences (finite or infinite) is linear and time-invariant.

2. Decide whether the transformations defined by the following formulas are linear and time-invariant:

   (a) $y(k) = T(\mathbf{x}) = 2x(k-1) + x(k)$,
   (b) $y(k) = T(\mathbf{x}) = nx(k-1) + x(k)$,
   (c) $y(k) = T(\mathbf{x}) = ny(k-1) + x(k)$,
   (d) $y(k) = T(\mathbf{x}) = a^k x(k)$ for $0 < a < 1$.

   If the transformation is linear and time-invariant, determine its impulse response $\mathbf{h}$.

3. Determine the transfer function of the operator whose impulse response is given by

   (a) $\mathbf{x} = (1\ 2, 1\ 2)$,
   (b) $\mathbf{x} = (1, -1)$,
   (c) $h(k) = (1/2)^k, \quad 0 \leq k \leq 7$,
   (d) $h(k) = \cos(3\pi k/8), \quad 0 \leq k \leq 7$,
   (e) $h(k) = \sin(3\pi k/8), \quad 0 \leq k \leq 7$.

4. Determine the images of

   (a) the constant sequence $x(k) = 1$,
   (b) the unit ramp sequence $r(k) = k$,
   (c) the exponential sequence $y(k) = (1/4)^k$

   under the operators given in Exercise 3.

5. Use the Gaussian function

$$g(x) = \frac{1}{\sigma\sqrt{2\pi}} e^{-\frac{x^2}{2\sigma^2}}$$

   with judiciously chosen values of σ to construct several rational approximations of low-pass filters of lengths $N = 3$ and $N = 5$.

6. Calculate discrete Fourier transforms of the low-pass filters you constructed in Exercise 5. Feel free to shift the filter as necessary to ensure that their DFTs are real-valued.

7. A *high-pass filter* $\mathbf{h}$ is characterized by the property that its discrete Fourier transform $H(n)$ is zero for low frequencies (near $n = 0$ or $n = N$) and $H(n) = 1$ for high frequencies (near $n = N/2$). Use the same considerations as the ones used in Section 5.3 to design a few high-pass filters of lengths $N = 2$, $N = 3$, and $N = 5$. Feel free to shift your filter as needed to ensure odd-symmetry. Zero-pad your filters as needed to help determine the properties of $H(n)$.

MATLAB Exercises.

1. Write a MATLAB function that would create a superposition of two cosine sequences of length N with specified amplitudes and specified frequencies. Your function should accept the sequence length, the amplitudes, and the frequencies as inputs.

2. Write a MATLAB function that would create a rational approximation of the specified Gaussian low-pass filter based on your result in Exercise 5. Your function should have the following inputs: the value of σ, the length of the filter, and the length to which it is zero-padded.

3. Similarly, write a MATLAB function that would create a high-pass filter based on your result in Exercise 7. Your function should have the following inputs: the length of the filter and the length to which it is zero-padded.

4. Write a MATLAB program that will do the following:

   (a) Ask the user to input the amplitudes and frequencies of the components and the length of the sequence to be generated.

   (b) Use the function you created in MATLAB Exercise 1 to create the superposition of the two cosine functions with parameters specified in Part (a). Plot the generated sequence.

   (c) Ask the user whether it is the low-pass or the high-pass filtering that is desired. Depending on the answer, ask the user to choose the filter from among the ones you created in MATLAB Exercises 2 and 3.

   (d) Use either your own circular convolution function (created in MATLAB exercises in Chapter 5) or the MATLAB function *cconv* to filter the generated sequence using the chosen filter. Plot the filtered sequence.

   (e) Attempt the restoration (or an approximate restoration) of the original sequence using the transfer function of your chosen filter in the spirit of the example at the end of this subsection. Plot the restored sequence.

5. Run the program you just created with several different input sequences and several different low-pass and high-pass filters. Comment on the relative quality of the filters.

6. Compile the results of the previous MATLAB exercises into a PowerPoint presentation.

6.5 Discrete Fourier Transform in 2D

Having established all the relevant formulas, results, and techniques in the previous sections, we are ready to apply them to digital images, which are, we recall, just $M \times N$ matrices for all the practical purposes of this text. Our approach to the discrete Fourier transform of matrices and digital images is completely analogous to what we did for finite and periodic sequences.

6.5.1 Definition, Examples, and Properties

In order to extend the notion of discrete Fourier transform to $M \times N$ matrices, we can apply the DFT first to the columns and then to the rows of the given matrix. The result is the so-called *two-dimensional discrete Fourier transform* formally defined as follows:

Definition 6.8 *The 2-D discrete Fourier transform of an $M \times N$ matrix A is given by*

$$\hat{A}(m,n) = \sum_{k=0}^{M-1} \sum_{l=0}^{N-1} A(k,l) e^{-2\pi imk/M - 2\pi inl/N}. \tag{6.137}$$

Thanks to the hard work we have done over the several previous sections, we can intuitively foresee a method for reconstructing a matrix from its 2-D discrete Fourier transform using the **2-D Inversion Formula**

$$A(k,l) = \frac{1}{MN} \sum_{m=0}^{M-1} \sum_{n=0}^{N-1} \hat{A}(m,n) e^{2\pi imk/M + 2\pi inl/N}, \tag{6.138}$$

and this formula can indeed be verified by reconstructing the matrix A separately in each dimension(along the rows and along the columns).

Most of the properties of the 2-D discrete Fourier transform are also completely analogous to those of its 1-D counterpart, and we refer the reader to the exercises at the end of this section. For example, the 2-D DFT of the 2-D unit impulse $\boldsymbol{\delta}$ defined by

$$\delta(k,l) = \begin{cases} 1, & \text{if } (k,l) = (0,0) \\ 0, & \text{if } (k,l) \neq (0,0) \end{cases} \tag{6.139}$$

is the 2-D unit constant sequence

$$\hat{\delta}(m,n) = 1 \quad \text{for all} \quad 0 \leq k < M, \quad 0 \leq l < N, \tag{6.140}$$

and, vice versa, the 2-D DFT of the 2-D unit constant sequence (6.140) is the 2-D unit impulse $\boldsymbol{\delta}$ (times $M \times N$), as the reader will be asked to verify in the exercises at the end of this section.

The 2-D analogy of a time shift of a sequence is a spatial shift of a matrix, and it is equivalent to the multiplication of its 2-D discrete Fourier transform by a complex exponential. Conversely, a frequency shift (multiplication by a complex exponential) of a matrix is equivalent to the corresponding shift in its 2-D DFT. Readers are encouraged to work out the details both in the exercises and on their own.

Most important for our purposes are the rules that establish the equivalence of convolution in the spatial domain and multiplication in the frequency domain (and vice versa). Based on the previous section, we can conjecture the following two rules, whose verification is also left as an exercise.

> Component-wise multiplication of two $M \times N$ matrices is equivalent (up to a constant factor) to the cyclic convolution of their 2-D discrete Fourier transforms. Formally, if
>
> $$C(k,l) = A(k,l) \cdot B(k,l)$$
>
> then
>
> $$\hat{C}(m,n) = \frac{1}{MN}(\hat{A} * \hat{B})(m,n). \tag{6.141}$$

> Cyclic convolution of two $M \times N$ matrices is equivalent to the component-wise multiplication of their 2-D discrete Fourier transforms. Formally, if
>
> $$C(k,l) = \big(A * B\big)(k,l).$$
>
> then
>
> $$\hat{C}(m,n) = \hat{A}(m,n) \cdot \hat{B}(m,n). \tag{6.142}$$

As an analogue to time-invariant transformations of sequences, we can define *space-invariant transformations* operating on matrices. Let T be a linear transformation acting on $M \times N$ matrices and let $B = T(A)$. We also denote by $A_{(u,v)}$ and $B_{(u,v)}$ the matrices defined by

$$A_{(u,v)}(k,l) = A(k-u, l-v) \quad \text{and} \quad B_{(u,v)}(k) = B(k-u, l-v)$$

with subtraction modulo M and N respectively (or combined with zero-padding depending on the circumstances). A space-invariant transformation T is characterized by the property that

$$T(A_{(u,v)}) = B_{(u,v)} \tag{6.143}$$

for all values of u and v.

Just as any finite (or infinite) sequence can be written as a convolution of itself with the unit impulse $\delta(k)$, any matrix A can be written as the convolution of itself with the 2-D unit impulse $\delta(k,l)$. Therefore, the image of the matrix A under the linear space-invariant transformation T is

$$B = T(A) = A * H,$$

where $H = T(\boldsymbol{\delta})$ is the impulse response of the transformation T, which in the two-dimensional setting, particularly in the context of digital images, is called **the point-spread function(PSF)** of the transformation T. It follows from (6.142) that

$$\hat{B}(m,n) = \hat{A}(m,n) \cdot \hat{H}(m,n),$$

with the discrete Fourier transform $\hat{H}$ of the PSF H called **the optical transfer function(OTF)**, especially in the context of digital images.

Section 6.5.1 Exercises.

1. Verify that the defining formula (6.137) in the definition of the 2-D discrete Fourier transform can indeed be obtained by first applying the one-dimensional DFT to the rows of A and then to its columns (or vice versa).

2. Verify that the 2-D DFT Inversion Formula (6.138) can indeed be obtained by first applying the one-dimensional Inversion Formula to the rows of $\hat{A}$ and then to its columns (or vice versa).

3. Verify that the 2-D discrete Fourier transform of the 2-D unit impulse $\boldsymbol{\delta}$ defined by (6.139) is indeed $\hat{\delta}(m,n) = 1$.

4. Verify that the 2-D discrete Fourier transform of the 2-D constant

$$x(k,l) = 1 \quad \text{for all} \quad 0 \le k < M, \quad 0 \le l < N$$

is indeed $M \times N \times \boldsymbol{\delta}$ (that is, the 2-D unit impulse up to a constant factor).

5. Formulate and verify the spatial shift rule for matrices similar to Rule (6.121) for finite sequences.

6. Formulate and verify symmetry properties for the discrete Fourier transform of real-valued matrices similar to those for finite real-valued sequences.

7. Verify the rules (6.141) and (6.142) by using the defining formulas of 2-D discrete Fourier transform and 2-D convolution.

6.5.2 Frequency Domain Processing of Digital Images

Before we get fully immersed in the application of frequency analysis to digital image processing, we must accustom ourselves with certain peculiar techniques of displaying the 2-D discrete Fourier transforms of digital images. There are three main points that we need to pay special attention to, and they will be discussed further in the text.

1. The most common way to display the discrete Fourier transform of a digital image is by interpreting its values as brightness levels of grayscale pixels and by using image-displaying functions (like the MATLAB *imshow* function). However, the values of the DFT are not always non-negative (and not necessarily real either). The common way out of this difficulty is to display the **absolute values** of the discrete Fourier transforms.

2. The absolute value of the so-called DC-coefficient, which is defined as the value

$$\hat{A}(0,0) = \sum_{k=0}^{M-1}\sum_{l=0}^{N-1} A(k,l)$$

of the discrete Fourier transform at the origin, is often much larger than the rest of the DFT values. As a result, a raw image of the DFT is akin to that of a night sky - a few bright specks of light with very few other visible features. The common solution is to apply a logarithmic transform to the image of the DFT in the spirit of Section 2.4. A good starting point is to use the transformation

$$x \to \log(1 + |x|)$$

followed by rescaling to fit the range from 0 to 255.

3. Another peculiarity has to do with the position of the origin. All our experience with mathematics has conditioned us to expect the origin to be *in the middle of the graph/image.* Therefore, that is where we are conditioned to expect to see the DC coefficient. However, digital images are equivalent to matrices, and the initial values of matrix indices that specify the row and column position of the pixels correspond to *the upper left corner* of the image. In order to reconcile our mathematical expectations with the image displaying conventions, it is customary to use the MATLAB function

```
>> shiftfft
```

when displaying 2-D discrete Fourier transforms.

Our game plan for the rest of the chapter is as follows: we begin this subsection by calculating and displaying discrete Fourier transforms of a few very basic images and, in doing so, familiarize ourselves with the conventions of displaying images and their 2-D DFTs that will be used once we move on to the task of genuine image restoration.

As our first example, we generate a 15×15 image A consisting of a bright source of light at the origin (which, and we have to get used to it, is located in the upper left corner) against a dark background (Figure 6.17 (a)) and a similar image B where the source of light was shifted to the center (Figure 6.17 (b)). Mathematically the image A is represented by the matrix

$$A = 255 \cdot \boldsymbol{\delta} = \begin{bmatrix} 255 & 0 & \cdots & 0 \\ 0 & 0 & \cdots & 0 \\ \vdots & \vdots & \ddots & \vdots \\ 0 & 0 & \cdots & 0 \end{bmatrix},$$

where $\boldsymbol{\delta}$ is the 2-D unit impulse. Therefore, as the reader has verified in Problem 3 at the end of the previous subsection, its discrete Fourier transform is just a scalar multiple of the 15×15 unit constant matrix, specifically

$$\hat{A} = \begin{bmatrix} 255 & \cdots & 255 \\ \vdots & \ddots & \vdots \\ 255 & \cdots & 255 \end{bmatrix},$$

which represents a white square shown in Figure 6.17 (c). Now, the image B is just a cyclic shift of the image A by seven positions to the right and seven positions down. Therefore, according to Problem 5 at the end of the previous section, its discrete Fourier transform is

$$\hat{B}(m,n) = e^{14\pi i(m+n)/15}\hat{A}(m,n),$$

which is complex-valued but has the same absolute value as $\hat{A}$ at every point. It follows that the images corresponding to the DFTs of the light in the upper-left corner and to the light in the center are identical (Figure 6.17 (c)).

As our second introductory example, we create a 15×15 gray square, which is given by the matrix

$$B = c \cdot \begin{bmatrix} 1 & \cdots & 1 \\ \vdots & \ddots & \vdots \\ 1 & \cdots & 1 \end{bmatrix},$$

where c is a chosen gray level, and display it alongside its discrete Fourier transform

$$\hat{B} = c \cdot M \cdot N \cdot \boldsymbol{\delta}$$

in Figure 6.17 (d,e). We may find the image of the DFT somewhat unsatisfactory because we are used to the origin being in the middle of the screen as opposed to the upper-left corner. Therefore, purely for display purposes, we shift the discrete Fourier transform in such a way that the value $\hat{A}(0,0)$ moves to the middle of the image. This is typically done with the help of the MATLAB command *shiftfft*. Figure 6.17(e) shows the shifted version of the DFT of the gray square.

Below is the MATLAB code used for generating Figure 6.17.

```
M=15;N=15;
A=zeros(M,N);B=A;A(1,1)=250;B(8,8)=250;
FA=fft2(A);FB=fft2(B);
subplot(2,3,1);imshow(A);
title('(a). 2-D Unit Impulse','FontSize',18);
subplot(2,3,2);imshow(B);
title('(b). Shifted 2-D Unit Impulse','FontSize',18);
subplot(2,3,3);imshow(abs(FB),[0,255]);
title('(c). DFT of the 2-D Unit Impulse','FontSize',18);
gray=120;C=gray*ones(M,N);FC=fft2(C);
subplot(2,3,4);imshow(C,[0,255]);
title('(d). A gray square','FontSize',18);
subplot(2,3,5);imshow(abs(FC),[]);
title('(e). The DFT of the gray square','FontSize',18);
FCshift=fftshift(FC);
subplot(2,3,6);imshow(abs(FCshift),[]);
title('(f). Shifted DFT of the gray square','FontSize',18);
```

We are ready to move on to genuine frequency-domain image processing. Out notation and mathematical model are exactly as in Subsection 6.5.1: we denote the "clean" image by A, the blurry image by B, and the point-spread function of the blurring filter by H. Then, in absence of any additional noise,

$$B = A * H \quad \text{and} \quad \hat{B} = \hat{A} \cdot \hat{H},$$

and, consequently, the "clean" image can be restored by means of

$$\hat{A} = \hat{B} \cdot \hat{G},$$

where the filter G, defined by its 2-D discrete Fourier transform

$$\hat{G} = 1/\hat{H}, \tag{6.144}$$

is called *the inverse* of the filter H. Clearly, the inverse filter G only exists if $\hat{G}(m,n) \neq 0$ for all frequency pairs (m,n); otherwise, the best we can do is find an "approximate inverse" of H.

Let us take a look at the photo of Emanuel Lasker in Figure 6.18(a) that we have already seen in the introduction to this chapter. We may consider

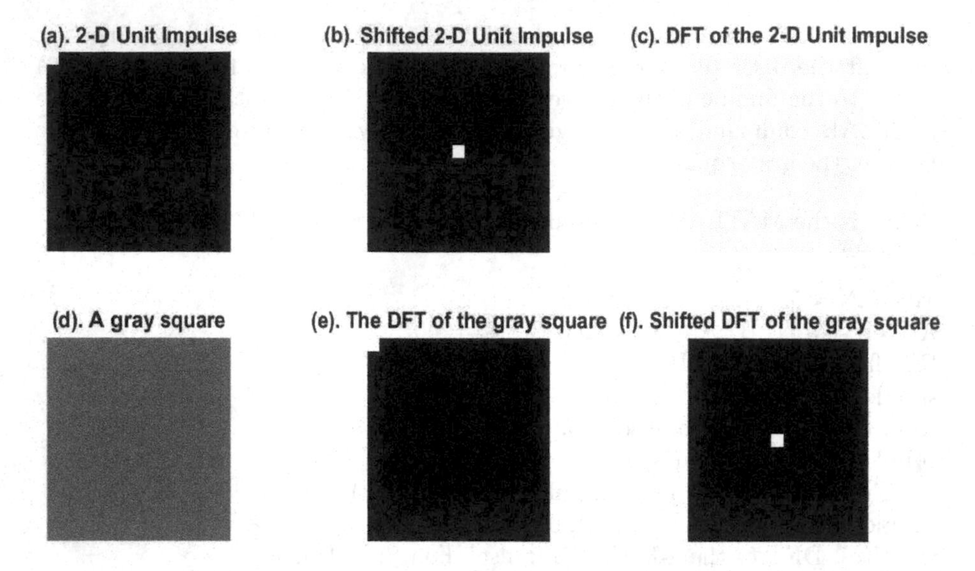

FIGURE 6.17: Discrete Fourier Transforms of elementary images.

it too sharp for a hundred-year old photo, and would like to make it look a little more hazy for reasons of "authenticity". Taking a look at the 2-D discrete Fourier transform of the photo, we observe a significant presence of light in the high-frequency ranges (away from the origin), and, therefore, conclude that a little bit of low-pass filtering might do the job. We recall from Chapter 5 that a good way to produce a bit of a static blur is to take the convolution of the image with the (approximated) Gaussian filter

$$H = \frac{1}{25}\begin{bmatrix} 2 & 3 & 2 \\ 3 & 5 & 3 \\ 2 & 3 & 2 \end{bmatrix}, \tag{6.145}$$

zero-padded to the size of the image. In the parlance of this chapter, we would be talking about applying the point-spread function H to the image A, and the formal way of expressing this transformation would be

$$A \to A * H.$$

However, working in the frequency domain is often more computationally efficient than taking convolutions in the spatial domain, and that is the approach we will take here. To that effect, we calculate the 2-D discrete Fourier transform of the image and multiply it pixel-by-pixel by the optical transfer function of the Gaussian filter. The filtered image B will then be obtained by taking the inverse 2-D DFT of the product. Denoting the operation of taking the 2-D discrete Fourier transform by $\mathcal{F}$ and its inverse by $\mathcal{F}^{-1}$, we can formally express this process by

$$B = \mathcal{F}^{-1}(\hat{A} \cdot \hat{H}),$$

where $\hat{A}$ and $\hat{H}$ are the 2-D DFT of the image and the optical transfer function respectively. The final outcome is shown in Figure 6.18(c).

FIGURE 6.18: An example of frequency-domain image processing.

Upon seeing the filtered image, however, we realize that we might have made a mistake. We like the original image much better. Can we restore it from its blurred version? Well, we can certainly try. One of the nicest features of Gaussian filters is that their Fourier transforms are everywhere nonzero, and, hence, we can divide the DFT of the filtered image by the optical transfer function at every frequency pair (m, n). The result of the restoration is shown in Figure 6.18(d).

Below is the MATLAB code used to generate Figure 6.18, which, we hope, the reader will find useful while working on the exercises at the end of this section.

```
A=imread('Lasker.jpg');
FA=fft2(A); FAshifted=fftshift(FA);
subplot(2,2,1);
imshow(A);
title('(a) Original Image','FontSize',12);
subplot(2,2,2);
imshow(log(1+abs(FAshifted)),[]);
title('(b) DFT of the Original Image','FontSize',12)
H=zeros(size(A)); H(1:3,1:3)=[2,3,2;3,5,3;2,3,2];
FH=fft2(H);
FHshifted=fftshift(FH);
subplot(2,2,3);
FB=FA.*FH; FBshifted=fftshift(FB);
B=ifft2(FB);
imshow(B,[]);
title('(c) Filtered Image','FontSize',12)
FC=FB./FH;
C=ifft2(FC);
subplot(2,2,4);
imshow(C,[0,255]);
title('(d) Restored Image','FontSize',12);
```

We can certainly claim that the situation considered in the last example was a little unrealistic and that in real life, we usually do not know the optical transfer function that was responsible for the blur in the first place. This is certainly true, but that is because the method described here is just a gateway to a vast and exciting field of image restoration based on frequency-domain analysis and processing.

Having said that, there is a lot we can do with the instruments we have already developed in this course. Even if we do not know what caused the blur, we can be fairly certain it must have been a low-pass filter. Why not try to reverse its effect, at least partially, by an application of one of the high-pass filters we developed in Exercise 3 at the end of Subsection 6.4.4?

For the sake of a quick example, we use the Laplacian edge detector, which happens to be a high-pass filter. We apply it to the blurry image to produce the edges and add the edges (scaled by a suitable scaling factor) to the blurry image in an attempt to partially offset the effects of the blur. The result is shown in Figure 6.19, and the MATLAB code (the continuation of the code for the previous example) used to generate Figure 6.19 is given below.

```
L=zeros(size(A));
L(1:3,1:3)=[-1,-1,-1;-1,8,-1;-1,-1,-1];
FL=fft2(L);
FD=FB.*FL;
D=ifft2(FD)/8+B;
subplot(1,2,1);
imshow(B,[]);
title('A Blurry Image','FontSize',12);
subplot(1,2,2);
imshow(D,[]);
title('A Sharpened Image','FontSize',12);
```

FIGURE 6.19: An attempt at image sharpening.

Another example of an image badly in need of restoration is shown in Figure 6.20(a). This time, the cause of the blur is obvious – it is motion, most probably resulting from taking the picture while riding in a moving vehicle. A careful examination of the outlines of the structure reveals that the motion was in the horizontal direction over approximately eight pixels. We can, therefore, guess that the point-spread function that caused the motion blur is probably

$$H = \begin{bmatrix} 1 & 1 & 1 & 1 & 1 & 1 & 1 & 1 \end{bmatrix}.$$

If our guess is correct, then the 2-D discrete Fourier transform $\hat{A}$ of the "clean" image A can be obtained from the 2-D DFT $\hat{B}$ of the blurry image by means of

$$\hat{A} = \frac{\hat{B}}{\hat{H}} = \hat{B} \cdot \hat{G},$$

where G is the inverse of the filter H (provided that $\hat{H}(m,n) \neq 0$ at every frequency point).

Unfortunately, the motion-blur filter H is no Gaussian filter, and its 2-D DFT $\hat{H}$ does equal zero for a lot of frequencies. We, therefore, have two options: we can either approximate the value of $\hat{G}$ at all those frequency points where $\hat{H}$ is zero, or we can modify the formula (6.144) to prevent division by zero, for example, by means of adding a small positive constant τ to the denominator as in

$$\hat{G} = \frac{1}{\tau + \hat{H}} \tag{6.146}$$

to produce a *modified inverse filter*. Figure 6.20(b) shows the result of a restoration by means of a modified inverse filter of this type with the value of $\tau = 0.1$. We can still perceive some periodic noise in the form of vertical bars, but there has obviously been a dramatic improvement in the visual quality of the photo.

FIGURE 6.20: Guessing the point-spread function of the motion blur and motion blur removal.

There exist several ways to further improve the image by removing the periodic noise, but they are outside the scope of this book, and we will not discuss them here. Besides, we have already made very significant progress and have amply illustrated the usefulness of frequency-domain processing in image restoration.

Even though we achieved good results with the photo in Figure 6.20, another look at the way we modified the filter G in (6.146) makes us grow uneasy. The problem is that it is not just the absolute value of $\hat{G}$ that gets modified (to prevent division by zero), but also its angle. And that might lead to unforeseen and undesirable consequences. Is it possible to only adjust the absolute value but preserve the angle of $\hat{G}$?

The answer is yes, and there is a standard way to do that. We recall that $1/z = \bar{z}/|z|^2$ for any complex number z. Consequently,

$$1/\hat{H} = \bar{\hat{H}}/|\hat{H}|^2,$$

and we can define the approximate inverse filter G by

$$\hat{G} = \frac{\bar{\hat{H}}}{\tau + |\hat{H}|^2}, \tag{6.147}$$

thus moderating its absolute value while preserving its angle.

Unfortunately, the method just described does not work very well in the presence of even a small amount of noise. Consider, once again, the photo of Emanuel Lasker, blurred by the same Gaussian filter (6.145), but this time, it also got corrupted by very slight additive white Gaussian noise with the standard deviation $\sigma = 10$ (Figure 6.21(a)). If we attempt to restore the image by using the inverse of the Gaussian filter, that is, by calculating

$$\hat{A} = \frac{\hat{B}}{\hat{H}} = \hat{B} \cdot \hat{G},$$

the results will be catastrophic (Figure 6.21) (b)), albeit, unfortunately, completely predictable.

To understand what is going on, we have to adjust our basic image restoration model. When noise is present, the blurry and noisy image B we are working with can be modeled in the spatial domain by

$$B = A * H + N,$$

where N denotes the noise, and in the frequency domain by

$$\hat{B} = \hat{A} \cdot \hat{H} + \hat{N}.$$

Our attempt to use the inverse filter G is equivalent to solving the last equation for $\hat{A}$, which yields

$$\hat{A}(m, n) = \frac{\hat{B}(m, n)}{\hat{H}(m, n)} + \frac{\hat{N}(m, n)}{\hat{H}(m, n)} \tag{6.148}$$

for every frequency pair (m, n). Let us examine both terms on the right-hand side of the Equation (6.148):

- For **low frequencies** (m, n), the values of both $\hat{B}(m, n)$ and $\hat{H}(m, n)$ are comparable, so the first term is of moderate size. We recall that since

(a) Blurry Image with Noise

(b) Inverse Filtering

(c) Adjusted Inverse Filtering

(d) W-H Filtering

FIGURE 6.21: Attempts to restore a noisy blurred photo.

noise values change wildly from pixel to pixel, noise is an inherently high-frequency phenomenon. Therefore, the values $\hat{N}(m,n)$ are vanishingly small and, consequently, the contribution of the second term is negligible. Thus, for low frequencies, the right-hand side of the Equation (6.148) is dominated by the valuable information contained in $\hat{B}$.

- For **high frequencies** (m,n), the value of $\hat{H}(m,n)$ is quite small (we recall here that the DC component is of the largerst value in the vast majority of images) and is comparable to the value of $\hat{B}(m,n)$. On the contrary, $\hat{N}(m,n)$ is quite large (noise being an inherently high-frequency phenomenon). Therefore, for the high frequencies, the right-hand side of the equation (6.148) is dominated by noise.

It is also evident that the junk values of the second term $\hat{N}(m,n)/\hat{H}(m,n)$, which tend to dominate the high frequencies, are going to be much larger than the values of the first term $\hat{B}(m,n)/\hat{H}(m,n)$, which tends to dominate the lower frequencies. Therefore, we can expect the "restored" image to be completely dominated by noise, which is exactly what we observed in Figure 6.21.

An attempt to use the modified inverse filter of the form (6.146) produces a better result (Figure 6.21(c)), but it can hardly be considered an improvement over the original image. In order to achieve a genuine solution, we must come up with a new type of a restoration filter that can mitigate the noise adaptively for all different frequency pairs (m,n).

How should we proceed in our search for the ideal restoration filter? At the very least, we would like such a filter G to have the following properties:

1. At the frequency pairs (m,n), where the (unknown) value of $\hat{N}(m,n)$ is much smaller than the known value of $\hat{H}(m,n)$, we want

$$\hat{G}(m,n) \approx \frac{1}{\hat{H}(m,n)},$$

that is, we want G to resemble the inverse of the filter H responsible for the blur.

2. At the frequency pairs (m,n), where the (unknown) value of $\hat{N}(m,n)$ is much larger than the known value of $\hat{H}(m,n)$, we want

$$\hat{G}(m,n) \approx 0.$$

3. Finally, at the frequency pairs (m,n), where the (unknown) value of $\hat{N}(m,n)$ is comparable to the known value of $\hat{H}(m,n)$, we want to find a compromise value for $\hat{G}(m,n)$.

It is evident that the key to constructing such a filter G is to find a suitable expression measuring the strength of the noise relative to the useful information at any given frequency. Such a measure exists, and it is called noise-to-signal ratio (NSR). Motivated by (6.147), a reasonable way to define the desired filter G would then be

$$G(m,n) = \frac{\overline{\hat{H}(m,n)}}{|\hat{H}(m,n)|^2 + NSR(m,n)}, \tag{6.149}$$

where the bar over the expression in the numerator denotes complex conjugation. Such a restoration filter is called the *Wiener-Helstrom filter*, and it can be shown to be optimal according to the criterion of minimizing the mean-square error of approximation. The reader is encouraged to check that the filter G defined by (6.149) meets all three criteria we have imposed on an ideal restoration filter. The result of Wiener-Helstrom filter-based restoration of our photo is shown in Figure 6.21(d).

Although optimal in a certain mathematical sense, Wiener-Helstrom filter does not always produce satisfactory results, as demonstrated by Figure 6.21(d). For that reason, numerous other filters have been developed, and they tend to perform better for specific types of images. Unfortunately, development and implementation of the Wiener-Helstrom filter or any of its successors is beyond the scope of this book. We have mentioned them in order to provide readers with a preview of the next topic in image restoration in case they are inspired to undertake a further study of digital image processes.

Subsection 6.5.2 MATLAB Exercises.

1. Create a few images showing white geometric shapes (such as squares, rectangles, circles, ovals, etc.) against a dark background. Display them alongside their 2-D discrete Fourier transforms (shifted so that the origin is positioned in the middle of the picture). Comment on the possible connection between the shape of an object and the features of its 2-D DFT.

2. Prepare a selection of images that were blurred using several different Gaussian filters you created in the exercises in Chapter 5. You should experiment with different types of blur, such as average, Gaussian, motion, etc.

3. Restore the blurry images you prepared using the method of inverse filtering discussed in this section. If you resort to using a modified inverse filter as in (6.146), experiment with different values of τ and try to determine the optimal value τ_0 for each given image.

4. Attempt to sharpen those images affected by average or Gaussian blur by means of applying a high-pass filter. Use any of the high-pass filters created in Chapter 5 or design new ones.

5. Next, try to find a selection of blurred images with features enabling you to guess the blurring filter (such as stars, point sources of light, and structures that can help identify the direction and strength of the motion blur). Use the method of inverse filtering to restore those images.

6. Apply additive white Gaussian noise to the blurred images and attempt restoration by means of inverse filtering and modified inverse filtering.

7. Compile the results of the previous MATLAB exercises into a PowerPoint presentation.

6.6 Chapter Summary

In this chapter, we have covered a tremendous amount of material. Having started with frequency analysis of continuous periodic functions, where we could rely on our experience with techniques of integral calculus, we expanded our study to include frequency analysis of sequences and matrices. In the process, we reviewed complex numbers and complex-valued functions, established connections with earlier chapters, and developed powerful methods of image restoration through manipulating image frequency content.

Thanks to our in-depth exploration of frequency analysis, we feel much more competent and confident applying frequency-domain filtering to the purpose of image restoration. Hopefully, the reader feels inspired to continue their studies in this field.

But the benefit derived from learning about Fourier coefficients, discrete Fourier transforms, image filtering, and other topics covered in this chapter extends way beyond digital image processing. Methods of Fourier analysis play a prominent role in numerous areas of pure and applied mathematics, science, technology, and even social sciences. No matter what the reader is planning to do after graduating from college – whether it is working in the industry or going to graduate school – the additional experience gained in this chapter will likely prove to be of lasting practical value.

Chapter 7

Wavelet-Based Methods in Image Compression

In this chapter, motivated by the goal of achieving better image compression, we leverage our experience with linear algebra and Fourier analysis to develop a variety of discrete wavelet transforms of ever-increasing sophistication, culminating in the CDF97 transform used in JPEG2000. We also briefly discuss other areas of applications of wavelet transforms, such as edge detection and image denoising.

7.1 Introduction

Suppose that we are looking to buy a new smartphone. All we need is a simple basic device which has internal memory of 16 gigabytes. After installing all the apps that will be necessary in the immediate future, we have 4 gigabytes of device memory left for our photos and videos. How many pictures are we going to be able to fit into our new phone?

We check the device specifications and discover that it comes with a very nice 16.1 megapixel camera. We recall that a color image has three channels (red, green, and blue) and also contains some additional data. All of that presumably means that each image will take more than

$$16.1\text{MP} * 3 \text{ channels } + \text{ extra } = \text{ almost } 50 \text{ Mb} \tag{7.1}$$

of storage per color image. Dividing the available 4 gigabytes of device memory by almost 50 megabytes per image, we come to a disappointing conclusion that it will only be possible to fit just over 80 images into our new phone.

Based on this calculation, should we immediately purchase the largest Micro-SD card available in the store? Possibly, but not just for the sake of storing photos. We all know that thousands and even tens of thousands of images can be stored in the phone memory. And when we check the file properties of our photos, we discover that images do not take nearly as much memory as 50 megabytes. So what did we miss when we were doing our back-of-the-envelope calculation (7.1)?

The answer, of course, is image compression, which is the main topic of this chapter. Over the past decades, many efficient methods of image compression have been developed. The most commonly used ones are based on various Fourier transforms (JPEG, for example, is based on the Discrete Cosine Transform, a close relative of the Discrete Fourier Transform that we studied in Chapter 6, and on various discrete wavelet transforms. It is the latter ones that this chapter is devoted to.

Keeping in mind the second principal objective of this book - a review of topics from college mathematics - we are confident that in this chapter, the reader will find plenty of opportunities to enjoy a beautiful interplay of elements of differential calculus, linear algebra, and frequency analysis. Even though this chapter refers to fairly recent mathematical developments, no further prerequisites are necessary beyond basic familiarity with derivatives, matrices, and complex numbers.

7.2 Naive Compression in One Dimension

We begin by trying to use general considerations and our intuition to come up with a working method of compressing a string of numbers, such as, for example,

$$\mathbf{v} = (7, 9, 1, 3)$$

with the ultimate goal of saving storage space. How about just replacing each successive pair of numbers with its average? In other words, we observe that $(7+9)/2 = 8$ and $(1+3)/2 = 2$ and perform the transformation

$$(7, 9, 1, 3) \longrightarrow (8, 2). \tag{7.2}$$

Unfortunately, this transform is not invertible: it is not possible to faithfully recover the original sequence of numbers from the sequence of averages of successive pairs. The same sequence of averages, $(8, 2)$, could have been obtained from the sequences $(5, 11, 0, 4)$, $(4, 12, 5, -1)$ or, indeed, from an infinite multitude of other sequences. Thus, in order to reconstruct the original data, we need more information than just the averages of successive pairs. How about calculating half-differences of the successive pairs of numbers in addition to calculating the averages? That is, how about noting that $(9-7)/2 = 1$ and $(3-1)/2 = 1$ and performing the transform

$$(7, 9, 1, 3) \longrightarrow (8, 2 \mid 1, 1)? \tag{7.3}$$

We can now reconstruct the original sequence from the available information by observing that $7 = 8 - 1$, $9 = 8 + 1$, $1 = 2 - 1$, and $3 = 2 + 1$. We have come across a fact of fundamental importance, which is worth reiterating:

A Fundamental Observation:

A sequence can be reconstructed faithfully from the averages and half-differences of the successive pairs of sequence elements!

But have we accomplished anything by transforming the original sequence? Has any compression been achieved? After all, we started with four numbers and ended up with four numbers!

It might be tempting to answer in the negative and abandon the entire approach. Before we do that, however, let us recall that numbers are stored (and transmitted) in their binary code. So let us try to assign binary codes to the numerical symbols in our sequence and in its transform as shown in Table 7.1. The table also reports the relative frequency of each symbol's occurrence in the sequence.

TABLE 7.1: Symbol Encoding

Symbol	Code	Frequency	Symbol	Code	Frequency
7	00	0.25	8	10	0.25
9	01	0.25	2	11	0.25
1	10	0.25	1	0	0.5
3	11	0.25	-	Average Length	1.5 bits

The original sequence $(7, 9, 1, 3)$ will then be encoded as 00011011 (eight bits), whereas its transform $(8, 2 \mid 1, 1)$ – as 101100 (six bits). What a surprise! It appears that we have come across a simple way to save 25 percent of storage space at no cost at all and without any loss of information! In other words, we have discovered a way to achieve *lossless data compression.*

As simple and naive as this idea of calculating averages and differences for successive pairs of sequence elements might appear, it should not be underestimated. Indeed, this approach does provide the basis for all the methods and techniques that will be developed further in this chapter.

Section 7.1 Exercises.

1. Follow the example of this section to calculate the averages and half-differences of the following strings of numbers:

   (a) (1, 3, 5, 7, 9, 11),
   (b) (1, 3, 5, 7, 9, 11, 13, 15),
   (c) (1, 3, 5, 7,. . . ,29, 31),
   (d) (-15, 23, -13, 21, -11, 19),
   (e) (1, 3, 0, 4, -1, 5, -2, 6),
   (f) (0, 1, 2, 4, 8, 16, 32, 64).

2. Use the averages and half-differences you constructed in Exercise 1 to reconstruct the original string of numbers.

3. For every part of Problem 1, try to construct a fixed-length binary code for the original string of numbers and a variable-length code for the transform. Compile a table similar to Table 7.1 and comment on the achieved compression ratio.

4. In this section, we saw that it is impossible to reconstruct the original even-length sequence $(x_1, x_2, ..., x_n)$ from the averages of *non-overlapping* successive pairs of its elements

$$\left(\frac{x_1 + x_2}{2}, \frac{x_3 + x_4}{2}, \dots, \frac{x_{n-1} + x_n}{2}\right).$$

   Moreover (and that might come as a big surprise), it turns out that it is also impossible to reconstruct the original sequence from the averages of *all* the pairs of successive elements, namely

$$\left(\frac{x_1 + x_2}{2}, \frac{x_2 + x_3}{2}, \dots, \frac{x_{n-1} + x_n}{2}\right).$$

   To convince yourself of this surprising fact, construct two *different* sequences, $(x_1, x_2, \dots, x_n)$ and $(y_1, y_2, \dots, y_n)$ that yield the *same* sequence of averages of pairs of successive entries.

5. In a similar way, demonstrate that it is impossible to faithfully reconstruct the original sequence from all the half-differences

$$\left(\frac{x_2 - x_1}{2}, \frac{x_3 - x_2}{2}, \dots, \frac{x_n - x_{n-1}}{2}\right)$$

   of its successive elements.

6. Explain how Problems 4 and 5 help us appreciate the Fundamental Observation of this section.

7.3 Entropy and Entropy Encoding

Reading over the lossless-compression example in the previous section, we might be wondering whether the binary encoding used there is optimal. In other words, would it be possible to achieve an average code length of less than 1.5 bits per symbol when encoding the sequence $(8, 2 \mid 1, 1)$? Or is the binary code given in Table 7.1 best possible and cannot possibly be improved upon?

In order to answer this question, we need to introduce the following measure of information content, which is fundamental in computer science and information theory:

Definition 7.1 *Suppose that the string* $\boldsymbol{v}$ *consists of symbols from the set* $S = \{s_1, ..., s_n\}$*, which occur with relative frequencies* $\{p_1, ..., p_n\}$ *respectively. Then the entropy of* $\boldsymbol{v}$ *is defined by*

$$Ent(\boldsymbol{v}) = -\sum_{k=1}^{n} p_k \log(p_k) = \sum_{k=1}^{n} p_k \log\left(\frac{1}{p_k}\right), \tag{7.4}$$

where the logarithm is base 2.

For example, let us consider the string

$$\mathbf{v} = (1, 1, 1, 1, 2, 2, 3, 4),$$

which uses the symbols

$$s_1 = 1, \quad s_2 = 2, \quad s_3 = 3, \quad \text{and} \quad s_4 = 4$$

with the corresponding frequencies

$$p_1 = \frac{1}{2}, \quad p_2 = \frac{1}{4}, \quad p_3 = \frac{1}{8}, \quad \text{and} \quad p_4 = \frac{1}{8}$$

respectively. According to Definition 7.1,

$$Ent(\mathbf{v}) = -\left[\frac{1}{2}\log\left(\frac{1}{2}\right) + \frac{1}{4}\log\left(\frac{1}{4}\right) + 2 * \frac{1}{8}\log\left(\frac{1}{8}\right)\right] = \frac{7}{4}.$$

The definition of entropy is not limited to strings of numerical symbols. For example, the symbols encountered in the word

$$\mathbf{v} = \text{"ABRACADABRA"}$$

are

$$s_1 = \text{"A"}, \quad s_2 = \text{"B"}, \quad s_3 = \text{"C"}, \quad s_4 = \text{"D"}, \text{ and } s_5 = \text{"R"}$$

with the corresponding frequencies

$$p_1 = \frac{5}{11}, \quad p_2 = \frac{2}{11}, \quad p_3 = \frac{1}{11}, \quad p_4 = \frac{1}{11}, \text{ and } p_5 = \frac{2}{11}$$

respectively. The entropy of $\mathbf{v}$ is

$$\begin{aligned} Ent(\mathbf{v}) &= -\left[\frac{5}{11}\log\left(\frac{5}{11}\right) + 2 * \frac{2}{11}\log\left(\frac{2}{11}\right) + 2 * \frac{1}{11}\log\left(\frac{1}{11}\right)\right] \\ &= 2.04. \end{aligned}$$

The entropy of a string $\mathbf{v}$ of symbols provides the lower limit for the average number of bits per symbol we can hope to achieve when using a binary encoding for $\mathbf{v}$. For example, referring to Table 7.1, a short calculation yields that the entropy of the string $(7, 9, 1, 3)$ is

$$H_1 = -4 \cdot [0.25 \log(0.25)] = 2,$$

whereas the entropy of the string $(8, 2 \mid 1, 1)$ is

$$H_2 = -\left[0.25 \log(0.25) + 0.25 \log(0.25) + 0.5 \log(0.5)\right] = 1.5,$$

which confirms that the symbol encoding shown in Table 7.1 is, indeed, optimal and cannot be improved.

The symbol sets that we worked with in our first example were rather small, and it was not difficult for us to come up with an optimal variable-length binary code - optimal in the sense that its average length (in bits per symbol) equaled the entropy of the string being encoded. But what should we do if we encounter a much larger set of symbols? Fortunately, many different coding methods have been developed to achieve this goal, for example, Huffman coding and context-adaptive binary arithmetic coding. Since 2014, asymmetric numerical systems (ANS) have been widely introduced due to their superior performance characteristics.

A detailed discussion of modern entropy encoding is beyond the scope of this book, but we will briefly describe the classical Huffman coding, which was developed by David A. Huffman while he was a student at MIT and published in his 1952 paper "A Method for the Construction of Minimum-Redundancy Codes"[16]. The basic algorithm has the following steps:

1. The symbols $\{s_1, ..., s_n\}$ are listed in the order of decreasing frequency. The two symbols of lowest frequency are assigned the binary digits 0 and 1.

2. The two symbols of lowest frequency are combined together into a "new symbol", and their frequencies are added together. This "new symbol" replaces both of the original symbols.

3. The procedure is repeated until we are left with just two symbols, which are assigned the binary digits 0 and 1.

The algorithm is best illustrated with the help of a binary decision tree. The original symbols $\{s_1, ..., s_n\}$ denote the leaves of the tree; a branch to the left corresponds to an assignment of a 0, whereas a branch to the right corresponds to an assignment of a 1. For example, the binary tree below illustrates the Huffman encoding of the string $(8, 2 \mid 1, 1)$.

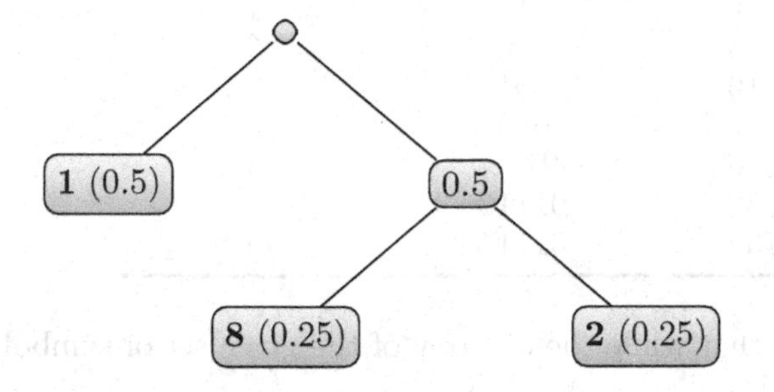

Following the principle that "left" means the binary digit 0 and "right" means the binary digit 1, we can read the variable-length binary encoding $\{\mathbf{1} \to 0, \mathbf{2} \to 11, \mathbf{8} \to 10, \}$ off the tree. As the reader has already seen, the average length of this particular Huffman encoding is 1.5 bits per symbol, which equals the entropy of the given symbol set with the given frequencies.

For a more complex example, we construct the Huffman tree for the set of symbols "A", "B", "C", "D", "E", "F", and "G", which occur with frequencies 3/8, 3/16, 3/16, 1/8, 1/16, 1/32, and 1/32 respectively.

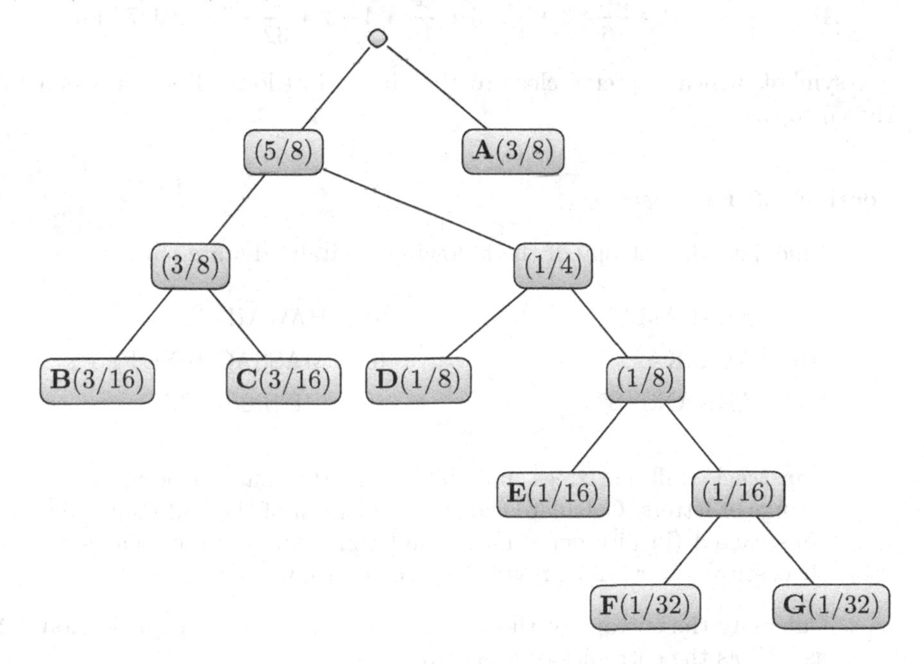

Again, following the principle that "left" means the binary digit 0 and "right" means the binary digit 1, we can read the variable-length binary encoding as summarized in Table 7.2 below.

TABLE 7.2: Huffman Encoding

Symbol	Frequency	Binary Code	Code Length
A	3/8	1	1
B	3/16	000	3
C	3/16	001	3
D	1/8	010	3
E	1/16	0110	4
F	1/32	01110	5
G	1/32	01111	5

It is interesting to note that while the entropy of the given set of symbols with the given frequencies is

$$\begin{aligned} Ent = & - \left[\frac{3}{8}\log\left(\frac{3}{8}\right) + 2 * \frac{3}{16}\log\left(\frac{3}{16}\right) + \frac{1}{8}\log\left(\frac{1}{8}\right)\right. \\ & + \left.\frac{1}{16}\log\left(\frac{1}{16}\right) + 2 * \frac{1}{32}\log\left(\frac{1}{32}\right)\right] = 2.3738, \end{aligned}$$

the average code length in the Huffman encoding constructed above is

$$Avg = \frac{3}{8} * 1 + 2 * \frac{3}{16} * 3 + \frac{1}{8} * 3 + \frac{1}{16} * 4 + 2 * \frac{1}{32} * 5 = 2.4375 \text{ bits}$$

per symbol, which is pretty close to the theoretical lower limit provided by the entropy.

Section 7.3 Exercises.

1. Calculate the entropy of the following capitalized state names:

   (a) "ALABAMA",
   (b) "ALASKA",
   (c) "ARKANSAS",
   (d) "HAWAII",
   (e) "MASSACHUSETTS",
   (f) "MISSISSIPPI".

   Construct Huffman trees and determine Huffman encoding for these strings of letters. Calculate the average length of the Huffman code you constructed (in bits per symbol) and comment on its closeness to the theoretical lower limit provided by the entropy.

2. Calculate the entropy of the vectors given in Exercise 1 for Section 7.2 as well as the entropies of their transforms.

3. For every part of Problem 1 at the end of Section 7.2, construct a Huffman encoding for the transforms of the given data, compile a table similar to Table 7.1, and comment on the achieved compression ratio: is the Huffman code you constructed optimal or close to optimal?

7.4 The Discrete Haar Wavelet Transform

Since the method introduced in Section 7.1 appears to hold promise for applications related to data compression, it seems worthwhile to devote time and effort to its further development. To start the process, we formalize the leading example of that section by introducing the following definition:

Definition 7.2 *Given the real-valued even-length sequence $\boldsymbol{x} = (x_1, ..., x_N)$, we define its (non-normalized)* ***discrete Haar wavelet transform (HWT)*** $(\boldsymbol{y} \mid \boldsymbol{z})$ *by*

$$y_n = \frac{1}{2}(x_{2n} + x_{2n-1}) \text{ and } z_n = \frac{1}{2}(x_{2n} - x_{2n-1}) \tag{7.5}$$

for $1 \leq n \leq N/2$.

For reasons that will become clearer towards the end of this section, we call the sequence $\mathbf{y}$ the *low-pass* part of the transform and the sequence $\mathbf{z}$ the *high-pass* part of the transform.

It is easy to see that the Haar wavelet transform is invertible by means of the formulas

$$x_{2n-1} = y_n - z_n \quad \text{and} \quad x_{2n} = (y_n + z_n), \tag{7.6}$$

as the reader will be asked to verify in the exercises at the end of this section.

The formulas (7.5) and (7.6) are fairly easy to implement in MATLAB or in any other programming language (and the reader will be asked to do just that in the exercises at the end of this section). However, in the interest of developing our methodology of signal and image compression, we will proceed to discuss alternative representations of the Haar wavelet transform.

Let us revisit the transformation of the sequence $\mathbf{v} = (7, 9, 1, 3)$ into its Haar wavelet transform $(8, 2 \mid 1, 1)$, but this time, we will recast $\mathbf{v}$ as a column vector and will rewrite the transformation in the form of matrix multiplication as

$$\begin{bmatrix} 8 \\ 2 \\ 1 \\ 1 \end{bmatrix} = \begin{bmatrix} 1/2 & 1/2 & 0 & 0 \\ 0 & 0 & 1/2 & 1/2 \\ -1/2 & 1/2 & 0 & 0 \\ 0 & 0 & -1/2 & 1/2 \end{bmatrix} \begin{bmatrix} 7 \\ 9 \\ 1 \\ 3 \end{bmatrix}. \tag{7.7}$$

If, instead, we consider $\mathbf{v}$ a *row* vector, then the same process can be expressed by the product

$$\begin{bmatrix}8 & 2 & 1 & 1\end{bmatrix} = \begin{bmatrix}7 & 9 & 1 & 3\end{bmatrix} \begin{bmatrix}1/2 & 0 & -1/2 & 0 \\ 1/2 & 0 & 1/2 & 0 \\ 0 & 1/2 & 0 & -1/2 \\ 0 & 1/2 & 0 & 1/2\end{bmatrix}, \tag{7.8}$$

and we can observe that the two transform matrices in (7.7) and (7.8) are transposes of each other.

For a more general column vector $\mathbf{v}$ of an *even* dimension N, we can formally describe its Haar wavelet transform in the matrix form by means of

$$HWT(\mathbf{v}) = W_N \times \mathbf{v} = \left[\frac{H_N}{G_N}\right] \mathbf{v}, \tag{7.9}$$

where the Haar wavelet transform matrix,

$$W_N = \left[\begin{array}{ccccccc} 1/2 & 1/2 & 0 & 0 & \cdots & 0 & 0 \\ 0 & 0 & 1/2 & 1/2 & \cdots & 0 & 0 \\ \vdots & \vdots & \vdots & \vdots & \ddots & \vdots & \vdots \\ 0 & 0 & 0 & 0 & \cdots & 1/2 & 1/2 \\ \hline -1/2 & 1/2 & 0 & 0 & \cdots & 0 & 0 \\ 0 & 0 & -1/2 & 1/2 & \cdots & 0 & 0 \\ \vdots & \vdots & \vdots & \vdots & \ddots & \vdots & \vdots \\ 0 & 0 & 0 & 0 & \cdots & -1/2 & 1/2 \end{array}\right], \tag{7.10}$$

consists of the "averaging" and "differencing" $N/2 \times N$ parts H_N and G_N given by

$$H_N = \begin{bmatrix} 1/2 & 1/2 & 0 & 0 & \cdots & 0 & 0 \\ 0 & 0 & 1/2 & 1/2 & \cdots & 0 & 0 \\ \vdots & \vdots & \vdots & \vdots & \ddots & \vdots & \vdots \\ 0 & 0 & 0 & 0 & \cdots & 1/2 & 1/2 \end{bmatrix} \tag{7.11}$$

and by

$$G_N = \begin{bmatrix} -1/2 & 1/2 & 0 & 0 & \cdots & 0 & 0 \\ 0 & 0 & -1/2 & 1/2 & \cdots & 0 & 0 \\ \vdots & \vdots & \vdots & \vdots & \ddots & \vdots & \vdots \\ 0 & 0 & 0 & 0 & \cdots & -1/2 & 1/2 \end{bmatrix} \tag{7.12}$$

respectively.

We can also recast the inversion formulas (7.6) in the form of matrix multiplication. For a given transformed vector $(\mathbf{y}|\mathbf{z})$, the original even-length vector $\mathbf{x}$ can be calculated by means of

$$\mathbf{x} = W_N^{-1} \left[\frac{\mathbf{y}}{\mathbf{z}}\right]. \tag{7.13}$$

In the exercises, we ask the reader to construct the inverse matrix, W_N^{-1} using the inversion formulas (7.6).

Ultimately, we plan to make use of the Haar wavelet transform in image processing applications and would like to relate this new instrument to the ideas and techniques discussed in earlier chapters. We recall, for example, that slowly-changing features of the images correspond to low frequencies and that abrupt changes and small details correspond to high frequencies. Therefore, it seems reasonable to wonder how the Haar wavelet transform affects different frequency components.

We can visualize the effect of the HWT on digital signals by examining Figure 7.1. The graph on the left represents a superposition $\mathbf{x}$ of a high-frequency, low-amplitude discrete cosine wave with a low-frequency, high-amplitude discrete cosine wave; the graph on the right represents the Haar wavelet transform of $\mathbf{x}$.

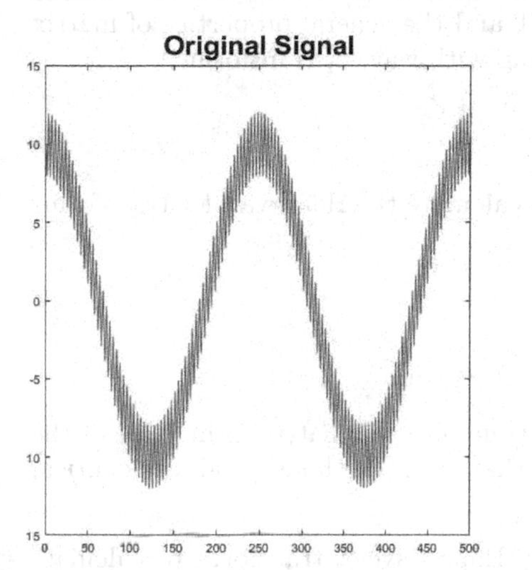

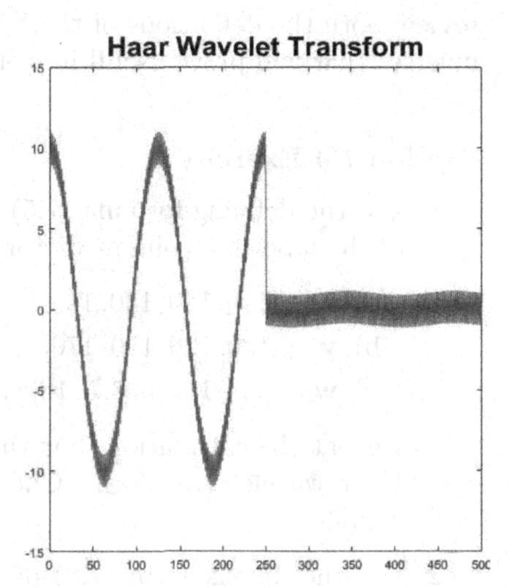

FIGURE 7.1: Haar wavelet transform of a superposition of two discrete cosine waves.

We can clearly see how the "averaging part" of the Haar wavelet transform suppressed the high-frequency component of the signal and how the "differencing part" of the transform suppressed the low-frequency component (just as we expected based on the experience of Chapter 6). In view of Figure 7.1 and using the terminology of Chapter 6, we can describe the Haar wavelet transform as an application of the filter pair consisting of the low-pass *Haar filter*

$$\mathbf{h} = (1/2, 1/2)$$

(the calculation of the averages of successive values) and the high-pass *Haar wavelet filter*

$$\mathbf{g} = (1/2, -1/2)$$

(the calculation of the half-differences of successive values), followed by downsampling (taking every other value of the filtered output).

We can also recognize the matrix H_N in (7.11) as the convolution matrix of the filter $\mathbf{h}$ and the matrix G_N in (7.12) as the convolution matrix of the filter $\mathbf{g}$ with half of their rows removed as part of down-sampling.

This description opens the door to the use of techniques of frequency analysis - a powerful new approach towards developing more sophisticated wavelet transforms. The Daubechies wavelet transforms we will create in subsequent sections using this method will provide performance vastly superior to that of the Haar wavelet transform in such applications as image compression, edge detection, and image denoising.

But meanwhile, the simplicity of the Haar wavelet transform gives us a perfect opportunity to practice those applications unencumbered with oppressive details. In the exercises at the end of this section, the reader will be asked to review both the definitions of the HWT and the general properties of matrix inverses that will prove useful in working with wavelet transforms.

Section 7.4 Exercises.

1. Use the defining formula (7.5) to calculate the Haar wavelet transforms of the following column vectors:

   (a) $\mathbf{u} = [100, 120, 140, 160]^T$,
   (b) $\mathbf{v} = [100, 120, 150, 170]^T$,
   (c) $\mathbf{w} = [3, -11, 15, 7, 5, 13]^T$.

   Repeat the calculation, but this time, use the matrix form (7.9) of the Haar wavelet transform. Check that both methods produce identical results.

2. Use the matrix form (7.8) of the Haar wavelet transform to calculate the HWT of the following row vectors:

   (a) $\mathbf{v} = [100, 120, 150, 170]$,
   (b) $\mathbf{w} = [3, -11, 15, 7, 5, 13]$.

3. Use the inversion formulas (7.6) in order to "reconstruct" the original sequences in Problems 1 and 2.

4. Construct the inverse Haar wavelet transform matrix, W_N^{-1} based on the inversion formulas (7.6) for use in (7.13).

5. In this section, we have discussed transforms defined by matrix multiplication. Inverting such transforms is equivalent to inverting their matrices. This might, therefore, be an appropriate time to review calculation of inverse matrices. Use the formula

$$\begin{bmatrix} a & b \\ c & d \end{bmatrix}^{-1} = \frac{1}{ad - bc} \begin{bmatrix} d & -b \\ -c & a \end{bmatrix}$$

   to find the inverses of the matrices

(a) $\begin{bmatrix} 3 & 2 \\ 8 & 5 \end{bmatrix}$, (c) $\begin{bmatrix} 3 & 2 \\ 8 & -12 \end{bmatrix}$,

(b) $\begin{bmatrix} 2 & 4 \\ -4 & -6 \end{bmatrix}$, (d) $\begin{bmatrix} 2 & 4 \\ -4 & 8 \end{bmatrix}$.

Did you notice anything special about the answers in parts (c) and (d)? Is there a rule you would like to suggest?

6. In order to calculate the inverse of a larger matrix A, one usually performs the necessary row operations to reduce the matrix $[A|I]$ to $[I|A^{-1}]$. Use this method based on row-reduction to find the inverses of the matrices

(a) $\begin{bmatrix} 2 & -3 & 4 \\ 1 & 0 & -2 \\ -3 & 1 & 4 \end{bmatrix}$, (c) $\begin{bmatrix} 0 & 3 & -3 \\ -1 & 2 & 2 \\ 4 & 1 & 1 \end{bmatrix}$,

(b) $\begin{bmatrix} 1 & -2 & 1 \\ 0 & 1 & 2 \\ -5 & -2 & 1 \end{bmatrix}$, (d) $\begin{bmatrix} 1 & 2 & 3 \\ 4 & 5 & 6 \\ 7 & 8 & 9 \end{bmatrix}$.

Did you notice anything special about the answers in parts (b), and (c)? Does this confirm the guess you made in the previous exercise?

7. Take another look at the inverse Haar wavelet transform matrix you constructed in Problem 4. Are you noticing anything remarkable about the relationship between the matrices W_N and W_N^{-1}?

8. Use the inverse matrix you constructed in Problem 4 to "reconstruct" the original sequences in Problems 1 and 2.

MATLAB Exercises.

1. Write a MATLAB function that will create the Haar wavelet transform matrix for the specified even dimension. Your function must check whether the specified dimension is even and give an error message in case it is not.

2. Use the result of Problem 4 to write a MATLAB function that will create the **inverse** Haar wavelet transform matrix for the specified even dimension. Your function must check whether the specified dimension is even and give an error message in case it is not.

3. Recall that a discrete periodic cosine wave with the amplitude A and the period N/k can be written in the form

$$x(n) = A \cos\left(\frac{2\pi k}{N} n\right). \tag{7.14}$$

Use MATLAB to experiment with various values for the parameters A, N, and k by generating and plotting resulting cosine waves.

4. Write a MATLAB program that will do the following:
   (a) Ask the user to specify the number of frequency components for the discrete periodic signal to be constructed.
   (b) Ask the user to specify the values for the frequency and amplitude for each component.
   (c) Generate and plot the superposition $x(n)$ of the components specified above using the formula (7.14).
   (d) Use the MATLAB function you created in Problem 1 to calculate the Haar wavelet transform $(\mathbf{y}|\mathbf{z})$ of $\mathbf{x}$.
   (e) Plot $\mathbf{x}$ and $(\mathbf{y}|\mathbf{z})$ side by side and comment on the filtering properties of the Haar wavelet transform.
   (f) Reconstruct the original discrete signal $x(n)$ faithfully using the MATLAB function you created in Problem 2 and plot the reconstructed signal.
5. Compile the results of the previous MATLAB exercises into a PowerPoint presentation.

7.5 Haar Wavelet Transforms of Digital Images

Having established the basic terminology and the fundamental formulas in Section 7.4, we are now going to return to the main practical objective of this chapter: learning to fit more images into limited storage space by means of image compression. What is the best way to apply the Haar wavelet transform to digital images? Intuition suggests that we should first use the defining formulas (7.5) to transform the columns of the image matrix and then proceed to transform the rows in a similar manner. The cumulative result of the two successive transformations can be seen in Figure 7.2.

It might be worth taking a few minutes here to examine the structure of the HWT of the image (the right side of Figure 7.2). The upper-left corner $\mathcal{B}$ of the transform represents the blur of the image, and it consists of average values over 2×2 blocks of pixels. The upper-right corner $\mathcal{V}$ represents half-differences between adjacent pixels in the horizontal direction and, therefore, is likely to be representative of prominent vertical features of the image. The lower-left corner $\mathcal{H}$ represents half-differences in the vertical direction and is thus likely to correspond to horizontal features of the image, whereas the

FIGURE 7.2: Haar wavelet transform of a standard image.

lower-right corner $\mathcal{D}$ of the image represents the "diagonal" half-differences and is likely to be representative of the diagonal features.

It is often convenient to visualize the Haar wavelet transform of an image in its matrix form both for theoretical and computational purposes. Suppose that the original image A has M rows and N columns. Then the HWT of the columns is achieved by multiplying the image matrix on the left by the HWT matrix W_M of dimension M, whereas the HWT of the rows is achieved by multiplying the image matrix on the right by the transpose W_N^T of the HWT matrix of dimension N. The end result is

$$A \to W_M \cdot A \cdot W_N^T = \left[\begin{array}{c|c} \mathcal{B} & \mathcal{V} \\ \hline \mathcal{H} & \mathcal{D} \end{array}\right]. \tag{7.15}$$

In practice, in the interest of lowering the computational complexity, one usually uses the defining formulas (7.5) to implement the Haar wavelet transforms of digital images as well as one-dimensional discrete signals. However, for theoretical purposes and for laboratory experimentation, the matrix form (7.15) is tremendously useful.

Next, we observe the effect the Haar wavelet transform has on the amount of space required to store the image. The entropy of the original image is 7, which means that the best achievable average binary code length is 7 bits per pixel. The entropy of the transformed image is 1, which indicates a 7-to-1 compression ratio *without any loss of information*!

We can achieve even better results if we apply the same method to the upper-left corner of the transformed image (that is, to the low-pass portion $\mathcal{B}$ of the transform). This can be done if both the number of rows and the number of columns of the image are divisible by 4. In principle, the process can be continued k times as long as the image dimensions are divisible by 2^k

(and even if they are not, the image size can be adjusted accordingly by means of zero-padding). Figure 7.3 shows the results of two and three iterations of the Haar wavelet transform. As we can see, the more iterations are performed, the lower the entropy of the transform, which implies a better compression ratio.

2 Iterations, Entropy is 0.96

3 Iterations, Entropy is 0.94

FIGURE 7.3: Multiple iterations of the Haar wavelet transform.

So far we have only talked about lossless compression, where the original image could be reconstructed faithfully without any loss of information. What if we have to achieve a better compression ratio than what lossless compression methods can provide? We can do that by sacrificing some features of the image that are deemed unnecessary or unimportant. For example, tiny variations in the brightness (and color) of the sky, the fabrics, and of other smooth surfaces do not usually make a critically important contribution to our perception of the image. Nor do the fine shades of color of the lawn and of the tree crowns. Those small variations result in a lot of small values in the high-pass portions $\mathcal{H}$, $\mathcal{V}$, and $\mathcal{D}$ of the wavelet transform, which, we recall, contain half-differences between the values of adjacent pixels. If we set those small values to zero, the corresponding variations will be lost in the reconstructed image, but we will hardly notice. What we will notice is a significant increase in compression ratio. We sum up this technique of **thresholding** as follows:

Thresholding a Wavelet Transform

In order to achieve an even better compression ratio, we set to zero any pixels in the high-pass portions $\mathcal{H}$, $\mathcal{V}$, and $\mathcal{D}$ of the transform whose absolute values are below the threshold λ.

What is the best way to choose the threshold λ below which the values in the high-pass portions of the transform will be set to zero? There seems to be no simple answer to this question. We can start by experimenting with several different thresholds. Figure 7.4 compares the original image of the cameraman to the reconstructions from lossy compression for the values of $\lambda = 10$, $\lambda = 20$, and $\lambda = 50$.

FIGURE 7.4: Comparing lossy compression with different thresholds.

We can see that higher thresholds result in lower entropy and better compression ratios but at the cost of progressively poorer image quality due to the loss of information on small-scale details. It might prove tricky to identify an optimal threshold by trial and error. It would also be quite impractical to ask the user to specify the threshold every time image compression is called for; at the same time, it would be unreasonable to set a universal threshold for all the images to be compressed.

A better approach would be to specify what percentage of the image energy (or, alternatively, the percentage of the energy in the high-pass portion of the transform) must be preserved during the thresholding stage of the compression process. To lay the technical foundation for implementing this approach, we recall that the energy of a vector $\mathbf{v} = [v_1, ..., v_n]^T$ is defined by

$$E(\mathbf{v}) = ||\mathbf{v}||^2 = \sum_{k=1}^{n} |v_k|^2, \tag{7.16}$$

and the energy of an $M \times N$ image (or matrix) A is similarly defined by

$$E(A) = ||A||^2 = \sum_{m=1}^{M} \sum_{n=1}^{N} |A(m,n)|^2. \tag{7.17}$$

As an aside, we mention that our definitions of vector and matrix energy make perfect intuitive sense because the energy dissipated by a unit resistance during one unit of time is V^2, where the V is the voltage applied to the resistor.

We define the **kth partial energy** $E_k(\mathbf{v})$ of the vector $\mathbf{v} = [v_1, ..., v_n]^T$ to be the contribution of the **smallest** k components of the vector $\mathbf{v}$ to $E(\mathbf{v})$. We also define the **kth cumulative energy**

$$C_k(\mathbf{v}) = \frac{E_k(\mathbf{v})}{E(\mathbf{v})} \tag{7.18}$$

of $\mathbf{v}$ to be the **fraction** of the total vector energy $E(\mathbf{v})$ contributed by the **smallest** k components of $\mathbf{v}$.

For example, the energy of the vector $\mathbf{v} = [7, 9, 1, 3]^T$ is

$$E(\mathbf{v}) = 7^2 + 9^2 + 1^2 + 3^2 = 140,$$

its components listed in the order of increasing absolute value are $(1, 3, 7, 9)$, and its partial energies and cumulative energies are calculated as follows:

k	Partial Energy, $\mathbf{E}_k(\mathbf{v})$	$\mathbf{C}_k(\mathbf{v})$
1	$1^2 = 1$	1/140 =0.7%
2	$1^2 + 3^2 = 10$	10/140 =7.1%
3	$1^2 + 3^2 + 7^2 = 59$	59/140 =42.1%
4	$1^2 + 3^2 + 7^2 + 9^2 = 140$	140/140=100%

For this example, setting the threshold $\lambda = 3$ allows us to keep more than 99% of the total energy of $\mathbf{v}$, since the only component that falls below the threshold and gets discarded has the value 1 and contributes only 0.7% (less than 1%) of $E(\mathbf{v})$. Setting the threshold $\lambda = 7$ allows us to keep more than 90% of the total energy $E(\mathbf{v})$, since the only components of $\mathbf{v}$ that fall below $\lambda = 7$ are 1 and 3, which together contribute only 7.1% (less than 10%) of $E(\mathbf{v})$. In the exercises at the end of this section, the reader will be asked to calculate thresholds both manually and with the help of MATLAB.

Figure 7.5 compares the reconstructions from lossy compression to the original image when we preserve 99%, 95%, and 99% of the energy in the high-pass portions of the transform. Figure 7.6 provides a closeup look at the cameraman's head and the camera. We can see that the less energy we preserve, the higher the related threshold is, and, consequently, the lower the quality of the reconstructed image is. The image in the lower right corner (which only preserves 90% of the energy) shows significant distortions, particularly in the areas with multiple important small-scale details.

Original Image

99% Energy, λ=5

95% Energy, λ=46

90% Energy, λ=100

FIGURE 7.5: Comparing lossy compression with different energies.

While we can perceive the distortions visually and can evaluate the quality of the image subjectively, it would be nice to have a numerical measure of image quality that would help us judge how faithful the reconstruction is to the original image. To measure the difference between two $M \times N$ images A

FIGURE 7.6: Comparing lossy compression with different energies - detail.

and B, we introduce the so-called *error function*

$$Err(A,B) = \frac{1}{MN}\sum_{m=1}^{M}\sum_{n=1}^{N}|A(m,n) - B(m,n)|^2. \tag{7.19}$$

If A is the original image and B is its transformed, thresholded, and reconstructed version, then the quality of the reconstruction is often expressed by the quantity known as its Peak Signal-to-Noise Ratio, defined by

$$PSNR(A,B) = 10\log_{10}\left(\frac{255^2}{Err(A,B)}\right), \tag{7.20}$$

under the assumption that the range of possible pixel values is 0 to 255. PSNR is measured in decibels; 20Db is considered acceptable for a low-quality transmission, and 50Db is considered to be fairly good. Table 7.3 contains the PSNR values for the lossy reconstructions in Figure 7.5.

TABLE 7.3: Image Characteristics for Figure 7.5

Preserved Energy	Calculated Threshhold	PSNR	Entropy
90%	100	25.5 Db	0.81
95%	46	27.6 Db	0.82
99%	5	40.4 Db	0.89

From this table, we can conclude that we might have to preserve as much as 99 percent of the energy of the high-pass portion of the Haar wavelet transform in order to achieve decent image quality (albeit at a fairly good compression ratio).

So far, we have only talked about compressing grayscale images. The same techniques apply to color images as well. We recall that MATLAB stores color images as three-layered matrices of dimension $M \times N \times 3$, with the layers representing the red, green, and blue channels. We could transform and threshold the three color channels separately, but it is more common to first convert the image into the YCbCr format and then apply the wavelet-based compression to the Y, Cb, and Cr channels. The reason is that the human eye has a much greater number and density of brightness receptors as compared to color receptors. As a consequence, we are much more sensitive to the subtle changes and small details in the brightness than in the color of an image. This difference in perception provides the opportunity to store and process brightness and color information separately and to use much higher compression ratios when compressing the chrominance channels C_b and C_r as compared to the luminance channel Y.

Section 7.5 Exercises.

1. Find the cumulative energy of the vectors

   (a) $\mathbf{v} = [1, 1, 5, 3, 6, 8, 4, 9, 1, 1, 2, 7]^T$,

   (b) $\mathbf{v} = [5, 4, 1, 7, 3, 5, 1, 2, 1, 2, 3, 1, 2, 4, 5, 11, 1, 2, 17, 12, 3, 5]^T$.

2. Find the thresholds that preserve 80%, 95%, and 99% of the energy of the vectors in Problem 1.

3. This section makes several mentions of partitioned matrices in the context of describing the four portions of the Haar wavelet transform of a digital image. We, therefore, have an occasion to review operations on, and properties of partitioned matrices, which the rest of this set of exercises is devoted to.

Suppose that the matrix A is partitioned as

$$A = \left[\begin{array}{cc|cc} a_{11} & a_{12} & a_{13} & a_{14} \\ a_{21} & a_{22} & a_{23} & a_{24} \\ \hline a_{31} & a_{32} & a_{33} & a_{34} \\ a_{41} & a_{42} & a_{43} & a_{44} \end{array}\right] = \left[\begin{array}{c|c} A_{11} & A_{12} \\ \hline A_{21} & A_{22} \end{array}\right].$$

Verify that

$$A^T = \left[\begin{array}{c|c} A_{11}^T & A_{21}^T \\ \hline A_{12}^T & A_{22}^T \end{array}\right].$$

4. Unless you have already done so in Section 4.1, solve the following partitioned-matrix equations for the matrices X, Y, and Z.

   (a) $\begin{bmatrix} X & 0 \\ Y & Z \end{bmatrix} \begin{bmatrix} A & 0 \\ B & C \end{bmatrix} = \begin{bmatrix} I & 0 \\ 0 & I \end{bmatrix}$,

   (b) $\begin{bmatrix} X & Y \\ 0 & Z \end{bmatrix} \begin{bmatrix} A & B \\ 0 & C \end{bmatrix} = \begin{bmatrix} I & 0 \\ 0 & I \end{bmatrix}$,

   (c) $\begin{bmatrix} A & B \\ 0 & I \end{bmatrix} \begin{bmatrix} X & Y & Z \\ 0 & 0 & I \end{bmatrix} = \begin{bmatrix} I & 0 & 0 \\ 0 & 0 & I \end{bmatrix}$.

5. Write a matrix equation similar to (7.15) that would describe two iterations of the Haar wavelet transform.

6. Similarly, write a matrix equation similar to (7.15) that would describe three iterations of the Haar wavelet transform.

MATLAB Exercises.

1. Write a MATLAB function that will compute the threshold given the fraction of the overall energy in the high-pass portions of the image transform that the user wishes to preserve.

2. Write a MATLAB function that will set to zero all the pixels in the high-pass portion of the image transform whose absolute values are below the specified threshold. Make sure your function does not affect the pixels in the low-pass region of the transform.

3. Write a MATLAB program that will do the following:

   (a) Ask the user how many iterations are desired (1, 2, or 3).

   (b) Ask the user which image needs to be compressed.

   (c) Display the original image.

   (d) Perform the Haar wavelet transform using as many iterations as specified in Part 3a and display the result of each iteration.

(e) If you have successfully completed Problem 1, ask the user what fraction of the energy of the high-pass regions of the transform needs to be preserved. Use the function you created in Problem 1 to calculate the corresponding threshold. If you have not yet completed Problem 1, you can ask the user to specify the threshold.

(f) Set to zero all the entries in the high-pass portions of the transform whose absolute values are below the threshold.

(g) Reconstruct the original picture both faithfully and with the loss of quality caused by the thresholding.

(h) Calculate the relevant entropies and comment on the compression ratios.

Test your program using several different pictures. It should work with any rectangular grayscale images. For color images you can either apply the transforms to each color channel (Red, Green, and Blue) or you can first convert the image to the YCbCr format and then apply the transform to the Y, Cb, and Cr channels.

7.6 Discrete-Time Fourier Transform

At this point, we probably have our hands full with the MATLAB exercises for Section 7.5 and are not quite ready for another MATLAB project, which makes this an opportune time for introducing a bit of new theoretical material. The reason is that as we are practicing compressing, thresholding, and reconstructing our images, we are probably beginning to develop a bit of a disappointment regarding the quality of images compressed with the help of the Haar wavelet transform. It is likely that, already at this early stage, we are beginning to think of finding a possible improvements for the HWT.

The following two sections will indeed be devoted to devising such improvements. However, prior to diving into such a major project, we have to get a couple of preliminaries out of the way in order to avoid having to do that work "on the side of the road". In particular, we need to discuss yet another variety of discrete Fourier transforms.

We recall that the discrete Fourier transform $X(n)$ of a finite or infinite N-periodic sequence $\mathbf{x}$ was defined in Section 6.4 by

$$\hat{x}(n) \equiv X(n) = \sum_{k=0}^{N-1} x(k)e^{-2\pi ikn/N}, \tag{7.21}$$

where N is the length of $\mathbf{x}$ (or the length of one period of $\mathbf{x}$). The DFT has served us well, particularly for the purposes of image restoration. There are, however, three areas of possible complaints about the discrete Fourier transform that we could be making, and they are as follows:

- The DFT is not defined for infinite non-periodic sequences.
- The sequence length N is a prominent element of the definition of the DFT, yet, as we have discovered, it is not really a fixed parameter in any way, because sequences are zero-padded whenever their length needs to be extended to suit the needs of the moment. We might, therefore, feel somewhat uneasy about the fact that the frequency content of a sequence is defined in terms of such a "fickle" parameter. For instance, it might be annoying that both the values and the period of $X(n)$ are affected by zero-padding the sequence $\mathbf{x}$.
- The discrete Fourier transform of a discrete sequence is itself discrete, which is certainly a good thing when it comes to computer applications. However, it might turn into a disadvantage if we decide that we need to use techniques of calculus to study the frequency content of a sequence (as we will in the very next section).

Taking all of those considerations into account, we make a substitution

$$\omega = \frac{2\pi n}{N} \tag{7.22}$$

in the definition (7.21) of the DFT and proceed to define the **discrete-time Fourier transform (DTFT)** of a (finite or infinite) sequence $\mathbf{x}$ by

$$X(\omega) = \sum_{k=-\infty}^{\infty} x_k e^{-ik\omega} \tag{7.23}$$

provided that the series converges.

It is evident that the DTFT defined by (7.23) is 2π-periodic for any sequence, finite or infinite, periodic or non-periodic. Furthermore, there are no references to the parameter N in the definition of the discrete-time Fourier transform. And, most importantly for our purposes in this section, $H(\omega)$ is always continuous and differentiable, at least for physically realizable sequences. The symmetry, shift, and other properties of the discrete Fourier transform are very similar to those of DFT, but they are free from all the technical complications caused by finiteness or periodicity.

For example, the discrete-time Fourier transform of the even-symmetric and real-valued sequence

$$\mathbf{x} = (x_{-1}, x_0, x_1) = \begin{pmatrix} 1/4, & 1/2, & 1/4 \end{pmatrix}$$

is

$$X(\omega) \quad = \quad x_{-1}e^{i\omega} + x_0 + x_1 e^{-si\omega}$$

$$= \frac{1}{4}e^{i\omega} + \frac{1}{2} + \frac{1}{4}e^{-i\omega}$$
$$= \frac{1}{2}(1 + \cos\omega),$$

which is also even-symmetric and real-valued. Similarly, the discrete-time Fourier transform of the odd-symmetric and real-valued sequence

$$\mathbf{y} = (y_{-1}, y_0, y_1) = (1/2, \quad 0, \quad -1/2)$$

is

$$\begin{aligned} Y(\omega) &= y_{-1}e^{i\omega} + y_0 + y_1 e^{-si\omega} \\ &= \frac{1}{2}e^{i\omega} - \frac{1}{2}e^{-i\omega} \\ &= i\sin(\omega), \end{aligned}$$

which is odd-symmetric and purely imaginary.

As we can guess based on the experience of Chapter 6, most important for us are the convolution properties of the discrete-time Fourier transform. The convolution $\mathbf{g} = \mathbf{x} * \mathbf{y}$ of two infinite sequences $\mathbf{x}$ and $\mathbf{y}$ is just the familiar linear convolution defined by

$$g_k = \sum_{m=-\infty}^{\infty} x_m y_{k-m},$$

and this definition also applies to finite sequences (which can always be turned into infinite sequences by zero-padding them). The DTFT of the convolution $\mathbf{g} = \mathbf{x} * \mathbf{y}$ can be calculated as

$$\begin{aligned} G(\omega) &= \sum_{k=-\infty}^{\infty} g_k e^{-ik\omega} \\ &= \sum_{k=-\infty}^{\infty} \sum_{m=-\infty}^{\infty} x_m y_{k-m} e^{-ik\omega} \\ &= \sum_{m=-\infty}^{\infty} x_m e^{-im\omega} \sum_{k=-\infty}^{\infty} y_{k-m} e^{-i(k-m)\omega} \\ &= X(\omega) \cdot Y(\omega), \end{aligned}$$

where the interchange of the order of summation in the last step can be justified using methods of advance analysis. We have proved the following rule, to which we will be referring countless times for the rest of this chapter:

The Convolution Theorem for DTFT

The discrete-time Fourier transform of the convolution of two sequences is the pointwise product of their discrete-time Fourier transform. More formally, if

$$\mathbf{g} = \mathbf{x} * \mathbf{y},$$

then

$$G(\omega) = X(\omega) \cdot Y(\omega).$$

We recall that when a sequence h acts on other sequences by convolution, it is called a "filter". We also recall that the defining property of a low-pass filter is that the values of its DFT for low frequencies (around $n = 0$ or $n = N$) are close to 1, whereas the values of its DFT for high frequencies (around $n = N/2$) are close to 0. The defining properties of a high-pass filter were formulated in a similar manner. However, since we made the substitution $\omega = 2\pi n/N$ during the discussion leading up to the definition of the DTFT, we can now do away with any mention of the fickle parameter N in this context as well. As a consequence, we can conveniently redefine the low-pass and high-pass filters as follows:

Definition 7.3 *A finite or infinite sequence* $\mathbf{h}$ *is called a low-pass filter if its discrete-time Fourier transform* $H(\omega)$ *satisfies the property*

$$|H(0)| = 1 \quad \textit{and} \quad H(\pi) = 0.$$

Similarly, a finite or infinite sequence $\mathbf{h}$ *is called a high-pass filter if its discrete-time Fourier transform* $H(\omega)$ *satisfies the property*

$$H(0) = 0 \quad \textit{and} \quad |H(\pi)| = 1.$$

For example, the sequence

$$\mathbf{x} = (x_{-1}, x_0, x_1) = \big(1/4, \quad 1/2, \quad 1/4\big)$$

that we considered earlier is a low-pass filter, because its discrete-time Fourier transform $X(\omega) = \frac{1}{2}\big(1 + \cos\omega\big)$ satisfied the property

$$X(0) = \frac{1}{2}\big(1 + \cos(0)\big) = 1 \quad \text{and} \quad X(\pi) = \frac{1}{2}\big(1 + \cos(\pi)\big) = 0.$$

In the exercises that follow, the reader will be asked to verify the basic properties of the discrete-time Fourier transform and to practice calculating the DTFT. Most of the exercises will probably seem quite familiar thanks to all the work done in Chapter 6. Upon completing this exercise section, the reader will be fully equipped to move on to the next section, where we will recreate some of the work done in the field of digital image processing as recently as during the 1990s.

Section 7.6 Exercises.

1. Calculate discrete-time Fourier transforms of the following finite and infinite sequences:

   (a) The averaging and "half-differencing" sequences

   $$\mathbf{h} = (h_0, h_1) = (1/2, 1/2) \quad \text{and} \quad \mathbf{g} = (g_0, g_1) = (1/2, -1/2)$$

   that are used in the construction of the Haar wavelet transform.

   (b) The unit impulse $\boldsymbol{\delta}$ and shifted unit impulse $\boldsymbol{\delta}_a$ defined by

   $$\delta(k) = \begin{cases} 1, & \text{if } k = 0 \\ 0, & \text{otherwhise} \end{cases} \quad \text{and} \quad \delta_a(k) = \begin{cases} 1, & \text{if } k = a \\ 0, & \text{otherwhise.} \end{cases}$$

   (c) A symmetric finite constant sequence (also known as the rectangular gate sequence) $\mathbf{x}$ and its shift $\mathbf{y}$ defined by

   $$x(k) = \begin{cases} 1, & -4 \leq k \leq 4 \\ 0, & \text{otherwhise} \end{cases} \quad \text{and} \quad y(k) = \begin{cases} 1, & 0 \leq k \leq 8 \\ 0, & \text{otherwhise.} \end{cases}$$

   (d) The odd-symmetric sequence $\mathbf{x}$ and its shift $\mathbf{y}$ defined by

   $$(x_{-2}, x_{-1}, x_0, x_1, x_2) = (y_0, y_1, y_2, y_3, y_4) = (-2, -1, 0, 1, 2).$$

   (e) The decreasing geometric sequence

   $$x(k) = a^k, \text{ where } k \geq 0 \text{ and } |a| < 1$$

   and its reflection across the vertical axis,

   $$y(k) = b^k, \text{ where } k \leq 0 \text{ and } b = 1/a$$

   for several different values of a.

   Do any of those sequences qualify as low-pass filters? As high-pass filters?

   Use MATLAB to plot the sequences alongside their discrete-time Fourier transforms (or the absolute value of the DTFT if it is complex-valued).

2. Prove that the DTFT of any sequence is 2π-periodic.

3. Prove that discrete-time Fourier transform is a linear transformation from the space of real-valued sequences to the space of 2π-periodic complex-valued functions.

4. Show that if a sequence $x(k)$ is real-valued, then its discrete-time Fourier transform $X(\omega)$ is conjugate-symmetric. Construct several examples illustrating this rule.

5. Show that if a sequence $x(k)$ is even (and real-valued), then its discrete-time Fourier transform $X(\omega)$ is real-valued. Construct several examples illustrating this rule.

6. Show that if a sequence $x(k)$ is odd (and real-valued), then its discrete-time Fourier transform $X(\omega)$ is purely imaginary. Construct several examples illustrating this rule.

7. Show that delaying a sequence $x(k)$ by a positions is equivalent to multiplying its discrete-time Fourier transform by a complex exponential and determine that complex exponential.

8. Calculate the convolution $\mathbf{g}$ of the infinite geometric sequences $\mathbf{x}$ and $\mathbf{y}$ defined by

$$x(k) = a^k \text{ and } y(k) = b^k, \text{ where } |a| < 1 \text{ and } |b| < 1.$$

   Check whether the discrete-time Fourier transforms satisfy the equation $G(\omega) = X(\omega) \cdot Y(\omega)$ as promised by the Convolution Theorem.

9. Show that the component-wise product of two sequences is equivalent (up to a constant factor) to the continuous convolution of their discrete-time Fourier transforms.

10. Show that the convolution of two low-pass filters is a low-pass filter and that the convolution of two high-pass filters is a high-pass filter.

7.7 From the Haar Transform to Daubechies Transforms

When we examine the lossy reconstructions in Figure 7.4, Figure 7.5, and, especially, in Figure 7.6, we cannot help but notice the highly-visible blocky artifacts in the reconstructed images. We can almost see the separate 2×2 blocks of pixels that appear quite disjoint from one another. We might make an educated guess that those artifacts are caused by the way the Haar wavelet transform calculates the averages over *disjoint* 2×2 blocks of pixels, which makes all those averages completely unrelated to each other. The result is the annoying lack of continuity and lack of smoothness.

We can guess the mathematical cause of the blocky artifacts by taking another look at the formula (7.11),

$$H = \begin{bmatrix} 1/2 & 1/2 & 0 & 0 & \cdots & 0 & 0 \\ 0 & 0 & 1/2 & 1/2 & \cdots & 0 & 0 \\ \vdots & \vdots & \vdots & \vdots & \ddots & \vdots & \vdots \\ 0 & 0 & 0 & 0 & \cdots & 1/2 & 1/2 \end{bmatrix},$$

which defines the low-pass part of the Haar wavelet transform matrix. It would seem plausible that the root cause of the blockiness of the images compressed with the use of the HWT is that the non-zero portions of the matrix rows in (7.11) *do not overlap.*

How can we construct a better transform that would eliminate the problem of blockiness while preserving most of the desirable features of the HWT? Let us turn our attention to the matrix form (7.10) of the Haar wavelet transform in search of such an improvement. One idea that immediately comes to mind is to use rows with a larger number of non-zero entries and to ensure that those blocks of non-zero entries "overlap". In other words, it would seem reasonable to try to construct the transform matrix of the form

$$W_N = \left[\frac{H_N}{G_N}\right] = \begin{bmatrix} h_3 & h_2 & h_1 & h_0 & 0 & 0 & \cdots & 0 & 0 \\ 0 & 0 & h_3 & h_2 & h_1 & h_0 & \cdots & 0 & 0 \\ \vdots & \vdots & \vdots & \vdots & \vdots & \vdots & \ddots & \vdots & \vdots \\ h_1 & h_0 & 0 & 0 & 0 & 0 & \cdots & h_3 & h_2 \\ \hline g_3 & g_2 & g_1 & g_0 & 0 & 0 & \cdots & 0 & 0 \\ 0 & 0 & g_3 & g_2 & g_1 & g_0 & \cdots & 0 & 0 \\ \vdots & \vdots & \vdots & \vdots & \vdots & \vdots & \ddots & \vdots & \vdots \\ g_1 & g_0 & 0 & 0 & 0 & 0 & \cdots & g_3 & g_2 \end{bmatrix}, \tag{7.24}$$

in the hope that the transform defined by such a matrix would produce better results than the HWT.

What criteria should we be guided by in the task of determining the coefficients h_k and g_k? Our transform has to be invertible – that is the very first and most critical requirement it must satisfy. After all, we would like to be able to reconstruct the images compressed with the use of our transform! For the transform to be invertible, the matrix W_N in (7.24) must be invertible. Ideally, we would like it to be easy to invert W_N; we would also like to avoid the need to do any row-reduction (with all the rounding errors row-reduction causes)!

Is the Haar wavelet transform matrix invertible? Yes, we recall that it is indeed very easy to invert the HWT by means of the inversion formula (7.6). Moreover, in Problem 4 of the previous section the reader was asked to construct the matrix for the inverse Haar wavelet transform. Can we put our finger on the intrinsic property of the HWT that makes it so easy to invert?

Upon carefully examining the Haar wavelet transform matrix

$$HWT_N = \left[\begin{array}{ccccccc} 1/2 & 1/2 & 0 & 0 & \cdots & 0 & 0 \\ 0 & 0 & 1/2 & 1/2 & \cdots & 0 & 0 \\ \vdots & \vdots & \vdots & \vdots & \ddots & \vdots & \vdots \\ 0 & 0 & 0 & 0 & \cdots & 1/2 & 1/2 \\ \hline -1/2 & 1/2 & 0 & 0 & \cdots & 0 & 0 \\ 0 & 0 & -1/2 & 1/2 & \cdots & 0 & 0 \\ \vdots & \vdots & \vdots & \vdots & \ddots & \vdots & \vdots \\ 0 & 0 & 0 & 0 & \cdots & -1/2 & 1/2 \end{array}\right],$$

we observe that it is *orthogonal* (that is, the pairwise dot-products of its rows are all zero)! From elementary linear algebra, we know that the inverse of an orthogonal matrix is just its transpose (up to scalar multiples)!

So, based on the re-examination of the Haar wavelet transform, it appears that the easiest way to guarantee that our new transform is invertible is to impose an orthogonality condition on its matrix W (whose general form is given in (7.24)).

Actually, we will go one step further and insist that the matrix W is *orthonormal*, which will ensure that its inverse precisely equals its transpose, and that there will be no need to keep track of any scalar multiples in the inversion formula.

In order to ensure that the overlapping rows of the low-pass portion H of the transform matrix are orthogonal to each other, we impose the condition

$$h_1 h_3 + h_0 h_2 = 0, \tag{7.25}$$

and in order to ensure that the matrix is not just orthogonal but is also orthonormal, we impose the additional condition

$$h_0^2 + h_1^2 + h_2^2 + h_3^2 = 1. \tag{7.26}$$

How do we ensure that the rows of G are orthogonal to the rows of H? We recall from elementary linear algebra that in order to produce a vector orthogonal to the given vector $\mathbf{v}$ of an even dimension, all we need to do is reverse the order of the components of $\mathbf{v}$ and alternate their signs. Following this recipe, we can set

$$g_0 = h_3, \quad g_1 = -h_2, \quad g_2 = h_1, \text{ and } g_3 = -h_0, \tag{7.27}$$

which, together with (7.25) and (7.26), ensures that the transform matrix W is orthonormal (as the reader can easily check for themselves).

So, let us see where we are. We must determine the values of the four transform coefficients (h_0, h_1, h_2, and h_3), but so far, we only have two equations, (7.25) and (7.26), to work with. We are still two equations short, which is actually a good thing, since it affords us an opportunity to impose two more desirable conditions on the filters **h** and **g**.

What additional properties do we want the filters **h** and **g** to enjoy? To begin with, we recall that the Haar filter $(1/2, 1/2)$ is a low-pass filter. Therefore, we would like $\mathbf{h} = (h_0, h_1, h_2, h_3)$ to also be a low-pass filter, which means that

$$H(\pi) = 0 \quad \text{and} \quad H(0) \neq 0,$$

where H is the discrete-time Fourier transform of the filter **h** . As we recall from Section 7.6,

$$H(\omega) = h_0 + h_1 e^{-i\omega} + h_2 e^{-2i\omega} + h_3 e^{-3i\omega}$$

and, therefore, the low-pass condition $H(\pi) = 0$ becomes

$$h_0 - h_1 + h_2 - h_3 = 0. \tag{7.28}$$

We still have room for one more equation, and we can use this opportunity to impose the requirement that the low-pass filter **h** should be of a better quality than the Haar filter $(1/2, 1/2)$. In order to see what needs to be done, we recall that the discrete-time Fourier transform of the Haar filter is

$$DTFT\big[(1/2, 1/2)\big] = \frac{1}{2}(1 + e^{-i\omega}) = e^{-i\omega/2}\cos(\omega/2)$$

and that the discrete-time Fourier transform of the Haar wavelet filter $(1/2, -1/2)$ is

$$DTFT\big[(1/2, -1/2)\big] = \frac{1}{2}(1 - e^{-i\omega}) = ie^{-i\omega/2}\sin(\omega/2)$$

with the graphs of the **absolute values** of the two DTFTs plotted in Figure 7.7 below.

We can see that the Haar filter, although technically (barely) passing the test for being a low-pass filter by virtue of reaching 0 at π and $-\pi$, is not doing a particularly good job of suppressing the higher frequencies. Likewise, the Haar wavelet filter **g** is not doing a good job suppressing the low frequencies either. In order to improve on the Haar filter, we would like the graph of the DTFT of our filter $\mathbf{h} = (h_0, h_1, h_2, h_3)$ to be *flat* at $\omega = \pi$ and at $\omega = -\pi$. That is, we are going to require that its derivative $H'(\omega)$ is zero at π.

The derivative of $H(\omega)$ is

$$\begin{aligned} H'(\omega) &= (h_0 + h_1 e^{-i\omega} + h_2 e^{-2i\omega} + h_3 e^{-3i\omega})' \\ &= -ih_1 e^{-i\omega} - 2ih_2 e^{-2i\omega} - 3ih_3 e^{-3i\omega}. \end{aligned} \tag{7.29}$$

Substituting $\omega = \pi$ in (7.29) and setting $H'(\pi)$ to zero gives us

$$h_1 - 2h_2 + 3h_3 = 0. \tag{7.30}$$

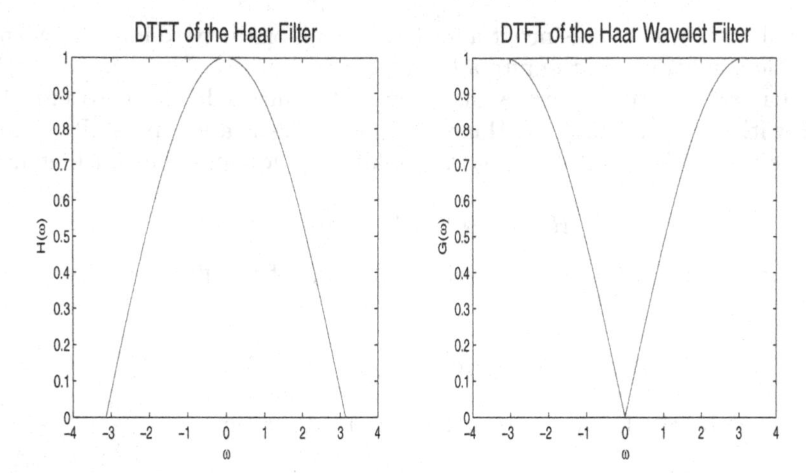

FIGURE 7.7: Discrete-time Fourier transforms of the Haar filter and the Haar wavelet filter.

Putting (7.25), (7.26), (7.28), and (7.30) together, we obtain the following defining system of equations:

System for Finding Daubechies Orthogonal Filter of Length 4

$$\begin{aligned} h_0^2 + h_1^2 + h_2^2 + h_3^2 &= 1 \\ h_1 h_3 + h_0 h_2 &= 0 \\ h_0 - h_1 + h_2 - h_3 &= 0 \\ h_1 - 2h_2 + 3h_3. &= 0 \end{aligned} \tag{7.31}$$

This system is nonlinear and, consequently, it takes quite a bit of effort to solve it. There are several possible approaches we could take, but we will follow the most straightforward (although, admittedly, not the most elegant) one. First, we divide both sides of the second equation by $h_0 h_1$ and obtain

$$\frac{h_3}{h_0} = -\frac{h_2}{h_1} = c,$$

where c is a nonzero parameter, which gives us

$$h_2 = -ch_1 \text{ and } h_3 = ch_0. \tag{7.32}$$

Substituting (7.32) into the first equation of the system (7.31), we obtain

$$h_0^2 + h_1^2 = \frac{1}{1+c^2} \tag{7.33}$$

and substituting (7.32) into the third equation of (7.31) gives us

$$h_1 = \frac{1-c}{1+c} h_0. \tag{7.34}$$

Substituting (7.34) into (7.33), we obtain the equation

$$h_0^2 + \left(\frac{1-c}{1+c}\right)^2 h_0^2 = \frac{1}{1+c^2},$$

which, after simplifications, yields

$$h_0^2 = \frac{(1+c)^2}{2(1+c^2)^2}.$$

For the sake of convenience, we choose the positive value

$$h_0 = \frac{1+c}{\sqrt{2}(1+c^2)}. \tag{7.35}$$

Substituting (7.35) into (7.34) yields

$$h_1 = \frac{1-c}{\sqrt{2}(1+c^2)} \tag{7.36}$$

and substituting (7.35) and (7.36) into (7.32) yields

$$h_2 = -\frac{c(1-c)}{\sqrt{2}(1+c^2)} \text{ and } h_3 = \frac{c(1+c)}{\sqrt{2}(1+c^2)}. \tag{7.37}$$

Finally, substituting (7.36) and (7.37) into the last equation of (7.31) yields (after simplifications) the quadratic equation

$$c^2 + 4c + 1 = 0$$

with the two solutions $c = -2 \pm \sqrt{3}$. Substituting the solution $c = -2 + \sqrt{3}$ into (7.35), (7.36), and (7.37), we obtain the numerical values for the filter coefficients h_0, h_1, h_2, and h_3 and arrive at the following definition:

Definition 7.4 ***The Daubechies orthogonal filter of length 4 (denoted by D4)*** *is the sequence* $\mathbf{h} = (h_0, h_1, h_2, h_3)$*, where the values of* h_0*,* h_1*,* h_2*, and* h_3 *are given by*

$$\boxed{\begin{aligned} h_0 &= \frac{1}{4\sqrt{2}}(1+\sqrt{3}), & h_1 &= \frac{1}{4\sqrt{2}}(3+\sqrt{3}), \\ h_2 &= \frac{1}{4\sqrt{2}}(3-\sqrt{3}), & h_3 &= \frac{1}{4\sqrt{2}}(1-\sqrt{3}). \end{aligned}} \tag{7.38}$$

The related ***Daubechies wavelet filter of length 4*** *is the sequence* $\boldsymbol{g} = (g_0, g_1, g_2, g_3)$ *defined by* $g_0 = h_3$, $g_1 = -h_2$, $g_2 = h_1$, *and* $g_3 = -h_0$. ***The D4-based Daubechies wavelet transform*** *is thus given by the matrix*

$$W_N = \left[\frac{H_N}{G_N}\right] = \left[\begin{array}{ccccccccc} h_3 & h_2 & h_1 & h_0 & 0 & 0 & \cdots & 0 & 0 \\ 0 & 0 & h_3 & h_2 & h_1 & h_0 & \cdots & 0 & 0 \\ \vdots & \vdots & \vdots & \vdots & \vdots & \vdots & \ddots & \vdots & \vdots \\ h_1 & h_0 & 0 & 0 & 0 & 0 & \cdots & h_3 & h_2 \\ \hline -h_0 & h_1 & -h_2 & h_3 & 0 & 0 & \cdots & 0 & 0 \\ 0 & 0 & -h_0 & h_1 & -h_2 & h_3 & \cdots & 0 & 0 \\ \vdots & \vdots & \vdots & \vdots & \vdots & \vdots & \ddots & \vdots & \vdots \\ -h_2 & h_3 & 0 & 0 & 0 & 0 & \cdots & -h_0 & h_1 \end{array}\right], \tag{7.39}$$

with the values of h_0, h_1, h_2, *and* h_3 *specified by* (7.38).

Figure 7.8 shows the discrete-time Fourier transforms of the Haar filter and the Daubechies filter of length 4 side by side. It is evident that D4 is

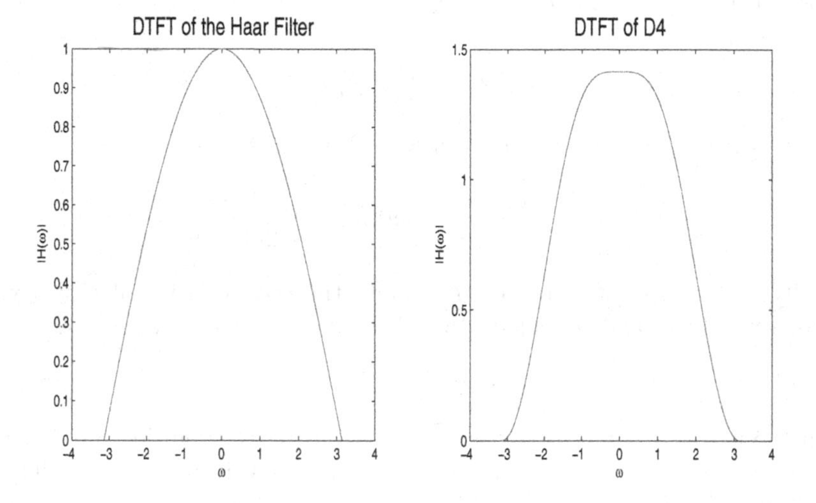

FIGURE 7.8: Discrete-time Fourier transforms of the Haar filter and the Daubechies filter of length 4.

a much better quality low-pass filter than the Haar filter because it does a much better job eliminating the high frequencies and preserving the lower frequencies. That, together with the overlap of the nonzero portions of the rows in the Daubechies wavelet transform matrix, promises that D4 will perform better than the HWT when used for lossy image compression.

Figure 7.9 shows side by side the results of lossy compression using the Haar wavelet transform and the D4-based wavelet transform. The high-pass portions of the transforms were thresholded to preserve 95% of the energy in those portions. It is evident that D4 results in considerably less blockiness and fewer distortions and, consequently, yields a much smoother and better-quality image (even though the peak signal-to-noise ratios of the two images are about the same).

FIGURE 7.9: Comparison of the HWT and the DWT.

What additional steps in the same direction can we take to achieve even better image quality? Since the D4-based transform proved to be an improvement over the Haar wavelet transform, it would be reasonable to expect further improvements from an even longer filter. Let us follow the same strategy that yielded D4 and try to construct an orthogonal filter of length 6. We shall begin by trying to construct the transform matrix of the form

$$W_N = \left[\begin{array}{cccccccccccccc} h_5 & h_4 & h_3 & h_2 & h_1 & h_0 & 0 & 0 & 0 & 0 & \cdots & 0 & 0 & 0 & 0 \\ 0 & 0 & h_5 & h_4 & h_3 & h_2 & h_1 & h_0 & 0 & 0 & \cdots & 0 & 0 & 0 & 0 \\ 0 & 0 & 0 & 0 & h_5 & h_4 & h_3 & h_2 & h_1 & h_0 & \cdots & 0 & 0 & 0 & 0 \\ \vdots & \vdots & \vdots & \vdots & \vdots & \vdots & \vdots & \vdots & \vdots & \vdots & \ddots & \vdots & \vdots & \vdots & \vdots \\ h_1 & h_0 & 0 & 0 & 0 & 0 & 0 & 0 & 0 & 0 & \cdots & h_5 & h_4 & h_3 & h_2 \\ h_3 & h_2 & h_1 & h_0 & 0 & 0 & 0 & 0 & 0 & 0 & \cdots & 0 & 0 & h_5 & h_4 \\ \hline g_5 & g_4 & g_3 & g_2 & g_1 & g_0 & 0 & 0 & 0 & 0 & \cdots & 0 & 0 & 0 & 0 \\ 0 & 0 & g_5 & g_4 & g_3 & g_2 & g_1 & g_0 & 0 & 0 & \cdots & 0 & 0 & 0 & 0 \\ 0 & 0 & 0 & 0 & g_5 & g_4 & g_3 & g_2 & g_1 & g_0 & \cdots & 0 & 0 & 0 & 0 \\ \vdots & \vdots & \vdots & \vdots & \vdots & \vdots & \vdots & \vdots & \vdots & \vdots & \ddots & \vdots & \vdots & \vdots & \vdots \\ g_1 & g_0 & 0 & 0 & 0 & 0 & 0 & 0 & 0 & 0 & \cdots & g_5 & g_4 & g_3 & g_2 \\ g_3 & g_2 & g_1 & g_0 & 0 & 0 & 0 & 0 & 0 & 0 & \cdots & 0 & 0 & g_5 & g_4 \end{array}\right], \tag{7.40}$$

and imposing the requirements that ensure that the matrix W_N is orthonormal and that $\mathbf{h} = (h_0, h_1, h_2, h_3, h_4, h_5)$ is a high-quality low-pass filter. The orthogonality conditions for the upper part of the matrix W_N take the form

$$\begin{aligned} h_0^2 + h_1^2 + h_2^2 + h_3^2 + h_4^2 + h_5^2 &= 1 \qquad (7.41) \\ h_3h_5 + h_1h_3 + h_0h_2 &= 0 \\ h_5h_1 + h_4h_0 &= 0. \end{aligned}$$

In order to ensure that the rows in the lower half of the matrix are orthogonal to the rows in the upper half, we set

$$\begin{aligned} g_0 &= h_5, \quad g_1 = -h_4, \quad g_2 = h_3, \\ g_3 &= -h_2, \quad g_4 = h_1, \quad g_5 = -h_0. \end{aligned} \tag{7.42}$$

We leave it to the reader to verify that the low-pass condition

$$H(\pi) = 0$$

takes the form

$$h_0 - h_1 + h_2 - h_3 + h_4 - h_5 = 0 \tag{7.43}$$

and that the "flatness in the high frequencies" condition

$$H'(\pi) = 0$$

takes the form

$$h_1 - 2h_2 + 3h_3 - 4h_4 + 5h_5 = 0. \tag{7.44}$$

Moreover, thanks to the increased length of $\mathbf{h}$, we now have the luxury of being able to impose additional requirements on its quality as a low-pass filter.

For example, we could insist that $H(\omega)$ is "super-flat" at high frequencies – something that could be achieved by requiring that its **second derivative**

$$H''(\pi) = 0.$$

In terms of the filter coefficients h_k, the last condition takes the form

$$h_1 - 4h_2 + 9h_3 - 16h_4 + 25h_5 = 0, \tag{7.45}$$

which we will ask the reader to verify in the exercises at the end of this section.

As a result, we have a system consisting of the three orthogonality conditions (7.41) together with the three low-pass conditions (7.43) – (7.45). It is very time-consuming to solve by hand, and we recommend using a computer algebra system to obtain the solutions

$$\begin{aligned}
h_0 = & \ -g_5 &= \frac{\sqrt{2}}{32}\left(1+\sqrt{10}+\sqrt{5+2\sqrt{10}}\right), \\
h_1 = & \ g_4 &= \frac{\sqrt{2}}{32}\left(5+\sqrt{10}+3\sqrt{5+2\sqrt{10}}\right), \\
h_2 = & \ -g_3 &= \frac{\sqrt{2}}{16}\left(5-1\sqrt{10}+\sqrt{5+2\sqrt{10}}\right), \\
h_3 = & \ g_2 &= \frac{\sqrt{2}}{16}\left(5-1\sqrt{10}-\sqrt{5+2\sqrt{10}}\right), \\
h_4 = & \ -g_1 &= \frac{\sqrt{2}}{32}\left(5+\sqrt{10}-3\sqrt{5+2\sqrt{10}}\right), \\
h_5 = & \ g_0 &= \frac{\sqrt{2}}{32}\left(1+\sqrt{10}-\sqrt{5+2\sqrt{10}}\right),
\end{aligned} \tag{7.46}$$

which define the Daubechies orthogonal filter of length 6 (denoted by D6) and the corresponding Daubechies wavelet filter of length 6. The D6-based Daubechies wavelet transform is defined by the matrix described in (7.40) with the coefficients h_k and g_k given by (7.46).

In MATLAB exercises at the end of this section, the reader will be asked to implement the wavelet transforms based on the orthogonal filters of length 4 and 6. In a similar way, one can construct the system of equations to obtain the Daubechies filter of any even length L. In the exercise section, the reader will be asked to set up such systems for $L = 8$ and $L = 10$.

Section 7.7 Exercises.

1. Use the technique of reversing the order and alternating the sign to find a vector orthogonal to:

(a) $[1, 2, 3, 4]^T$,
(b) $[1, -2, 3, -4]^T$,
(c) $[3, -1, 2, 5]^T$,
(d) $[-1, 7, 2, -4, 3, 5]^T$,
(e) $[1, 2, 3, 4, 5, 6, 7, 8]^T$,
(f) $[-2, 1, 3, -4, 7, 5, -6, 9]^T$.

2. Which ones of the following matrices are orthogonal?

(a) $\begin{bmatrix} 3 & 2 \\ 4 & -6 \end{bmatrix}$,

(b) $\begin{bmatrix} 4 & 1 \\ 1 & -4 \end{bmatrix}$,

(c) $\begin{bmatrix} 3 & 2 \\ 8 & -12 \end{bmatrix}$,

(d) $\begin{bmatrix} 2 & 4 \\ -4 & 8 \end{bmatrix}$,

(e) $\begin{bmatrix} 2 & 3 & 1 \\ -2 & 3 & 1 \\ 0 & -1 & 6 \end{bmatrix}$,

(f) $\begin{bmatrix} 1 & 2 & 1 \\ -1 & 2 & 1 \\ 0 & -1 & 4 \end{bmatrix}$,

(g) $\begin{bmatrix} 1 & 1 & 2 & 4 \\ -2 & -2 & 1 & -2 \\ 1 & 1 & 0 & -1 \\ 1 & -1 & 0 & 1 \end{bmatrix}$,

(h) $\begin{bmatrix} 2 & -2 & 0 \\ -1 & 2 & 2 \\ 4 & 1 & 1 \end{bmatrix}$.

3. Determine (by trial and error) whether it is possible to construct a Daubechies filter of length 3? How about length 5? How about any odd length?

4. Verify that the conditions (7.25), (7.26), and (7.27), ensure that the transform matrix W is orthonormal.

5. Find the solution of the system (7.31) that corresponds to choosing the negative value in (7.35).

6. Find the solution of the system (7.31) that corresponds to substituting $c = -2 - \sqrt{3}$ into (7.35), (7.36), and (7.37).

7. Verify Equations (7.43), (7.44) and (7.45).

8. Set up systems of equations for calculating the coefficients for the Daubechies orthogonal filters of length 8 and of length 10. Do not attempt to solve the system by hand.

Section 7.7 MATLAB Exercises.

1. Use a computer algebra system to verify the coefficients of the Daubechies orthogonal filter of length 6.

2. Write a MATLAB function that will create the D4-based Daubechies wavelet transform matrix for the specified even dimension. Your function must check whether the specified dimension is even and give an error message in case it is not.

3. Similarly, write a MATLAB function that will create the D6-based Daubechies wavelet transform matrix of the specified dimension.

4. Unless you have already done so in Section 7.5, write a MATLAB function that will compute the threshold given the fraction of the overall energy in the high-pass portions of the image transform that the user wishes to preserve.

5. Write a MATLAB program that will do the following:

   (a) Ask the user which picture must be compressed.

   (b) Display the original picture.

   (c) Ask the user how many iterations of the Daubechies wavelet transform are desired (1, 2, or 3) and whether D4 or D6 should be used.

   (d) Ask the user what fraction of the energy of the high-pass regions of the transform needs to be preserved. Use the function you created in MATLAB Exercise 4 of Section 7.5 to calculate the corresponding threshold.

   (e) Perform the Daubechies wavelet transform using as many iterations as desired and display the result of each iteration. Set to zero all the pixels in the high-pass portions of the transform whose absolute values are below the threshold.

   (f) Reconstruct the original picture both faithfully and with loss of quality caused by the thresholding.

6. Test your program using several different pictures. It should work with any rectangular images (grayscale or color).

7. Compare the results of Daubechies wavelet transform-based compression and reconstruction with those obtained using the Haar wavelet transform.

8. Compile your results into a PowerPoint presentation.

7.8 Biorthogonal Wavelet Transforms

A careful examination of Figure 7.9 and, more importantly, our own experience completing MATLAB exercises for Sections 7.4 and 7.7 have undoubtedly convinced us of the superiority of the Daubechies wavelet transforms over the Haar wavelet transform. However, a mere glance at the formulas that define the Daubechies filter coefficients gives us a pause. There is something deeply unsettling about the long irrational expressions in (7.38) and, especially, in (7.46), which leaves us wondering whether there might be an alternative approach that leads to more pleasing results.

Indeed, there are reasons to be less than completely happy with our filters. To begin with, since irrational coefficients inevitably cause rounding errors, the orthogonal Daubechies transforms appear somewhat less than perfectly suited for lossless compression. Furthermore, on the basis of our informal derivation of the Haar wavelet transform, we expect the low-pass portion of the transform to represent the averages (or at least weighted averages of some sort) of neighboring pixel values. However, there is nothing natural or intuitive about irrational coefficients in expressions used to calculate weighted averages. Moreover, on general grounds, we would expect symmetrically-positioned pixels to make equally-weighted contributions to the weighted average and, consequently, we would expect the filter coefficients to be symmetric. The Daubechies filter coefficients are, however, not symmetric, which runs contrary to our intuition and expectations.

Therefore, due to the concerns mentioned above, this might be a good time to reexamine the approach that guided us in our construction of the Daubechies orthogonal filters. To do so, we compile a list of properties that we would like the "averaging" filter $\mathbf{h}$ and its coefficients to have.

Of course, if possible, we would like the coefficients to be rational. Even more importantly,

1. We would like $\mathbf{h}$ to be a *low-pass* filter;
2. We would like the coefficients of $\mathbf{h}$ to be *symmetric*;
3. We would like the (down-sampled) matrix H defined by the filter $\mathbf{h}$ to be *orthogonal.*

Unfortunately, we are up for a disappointment. A major theorem due to Ingrid Daubechies decrees that the only filter that satisfies all three requirements is the Haar filter $(1/2, 1/2)$. So, if we would like to use a longer filter, we can only have two out of the three properties from our wish list. Which one of the three are we willing to do without?

We have just made an argument (albeit an intuitive and imprecise one) for the importance of symmetry. So, Property 2 is a keeper. We have also made a case for the importance of maintaining the strategy of calculating averages of neighboring pixels, and that is equivalent to low-pass filtering. So, we will have to keep Property 1 as well. It appears increasingly likely that we will have to try to find a suitable replacement for orthogonality. Let us try to recall where the need for orthogonality came from in the first place.

The reason why we need orthogonality for the Haar wavelet transform and the Daubechies wavelet transforms is the imperative necessity for the transform matrix W to be not just invertible, but *easily* invertible. The easiest type of a matrix to invert is an orthonormal matrix, because its inverse equals its transpose.

Since we are looking to replace orthogonality with an alternative property, the following question arises naturally: What would be the "second easiest" type of matrices to invert? And here a breakthrough idea comes to our assistance, namely, the concept of *biorthogonality*:

Biorthogonality

Instead of insisting that the inverse of our transform matrix equal *its own* transpose, we can instead require that its inverse equal the transpose *of some other matrix*! Formally, instead of insisting that

$$W^{-1} = W^T,$$

we are going to require that

$$W^{-1} = \tilde{W}^T \tag{7.47}$$

for a suitable matrix $\tilde{W}$.

This seems easier said than done. As opposed to our task in Sections 7.4 and 7.7 where we only had to construct one filter (and one matrix) for each transform, we now have to construct *two* matrices instead of one. Where would we even begin? Well, let us move one step at a time, guided by our intuition and making reasonable educated guesses along the way.

First, based on our experience with the Haar wavelet transform and the Daubechies wavelet transforms, we are going to assemble W and $\tilde{W}$ out of the (down-sampled) convolution matrices defined by suitable low-pass and high-pass filters. That is, we set

$$W = \left[\frac{H}{G}\right] \text{ and } \tilde{W} = \left[\frac{\tilde{H}}{\tilde{G}}\right]$$

for suitable H and G. It then follows from (7.47) that

$$\begin{aligned} I_N &= W\tilde{W}^T = \left[\frac{H}{G}\right]\left[\ \tilde{H}^T \mid \tilde{G}^T\ \right] \\ &= \left[\begin{array}{c|c} H\tilde{H}^T & H\tilde{G}^T \\ \hline G\tilde{H}^T & G\tilde{G}^T \end{array}\right] = \left[\begin{array}{c|c} I_{N/2} & 0 \\ \hline 0 & I_{N/2} \end{array}\right], \end{aligned}$$

which is equivalent to the following system of four matrix equations:

$$\begin{aligned} H\tilde{H}^T &= I_{N/2}, & H\tilde{G}^T &= 0, \\ G\tilde{H}^T &= 0, & G\tilde{G}^T &= I_{N/2}. \end{aligned} \tag{7.48}$$

Empowered by our success with the Daubechies wavelet transforms, we can feel confident in our ability to come up with a good educated guess for the high-pass filters $\mathbf{g}$ and $\tilde{\mathbf{g}}$, once we have constructed their low-pass counterparts. Therefore, we will begin our effort by trying to determine the appropriate coefficients of the low-pass filters $\mathbf{h}$ and $\tilde{\mathbf{h}}$.

7.8.1 Biorthogonal Spline Filters

We first approach the easier task of constructing $\tilde{H}$. We recall that in order for $\tilde{\mathbf{h}}$ to be a low-pass filter, its discrete-time Fourier transform must satisfy the condition

$$\tilde{H}(\pi) = 0 \quad \text{and} \quad \tilde{H}(0) \neq 0.$$

We note that the function $C(\omega) = \cos(\omega/2)$ satisfies this condition. Let us also recall that the discrete-time Fourier transform

$$DTFT\big[(1/2, 1/2)\big] = \frac{1}{2} + \frac{1}{2}e^{-i\omega} = e^{-i\omega/2}\cos(\omega/2)$$

of the Haar filter is just $C(\omega)$ times the "nuisance factor" $e^{-i\omega/2}$ of absolute value 1. Finally, we note the fact that if we raise the DTFT of a low-pass filter to a positive integer power, the result will be the DTFT of another low-pass filter (and in this connection, the reader might recall Exercise 10 of Section 7.6).

Guided by those three observations, we feel empowered to define

$$\begin{aligned} \tilde{H}(\omega) &= \big[\cos(\omega/2)\big]^2 \qquad (7.49) \\ &= \left[\frac{e^{i\omega/2} + e^{-i\omega/2}}{2}\right]^2 \\ &= \frac{1}{4}\left(e^{-i\omega} + 2 + e^{i\omega}\right). \end{aligned}$$

Recalling the definition of the discrete-time Fourier transform, we see that the low-pass filter $\tilde{\mathbf{h}}$ is

$$\tilde{\mathbf{h}} = (h_{-1}, h_0, h_1) = \left(\frac{1}{4}, \ \frac{1}{2}, \ \frac{1}{4}\right). \qquad (7.50)$$

We call low-pass filters obtained by raising $\cos(\omega/2)$ to integer powers **spline filters**. The graph of the $\tilde{H}(\omega)$ we have just constructed is shown on the right side of Figure 7.10, side by side with the graphs of the DTFTs of the Haar filter and of the Daubechies orthogonal filter of length 4.

We can happily observe that the filter $\tilde{\mathbf{h}}$ meets all our requirements. It satisfies the low-pass conditions by design, and its coefficients are rational and symmetric. It seems to be just as effective as D4 in eliminating the higher frequencies (although not as effective in preserving the low frequencies), and is obviously superior to the Haar filter in that regard.

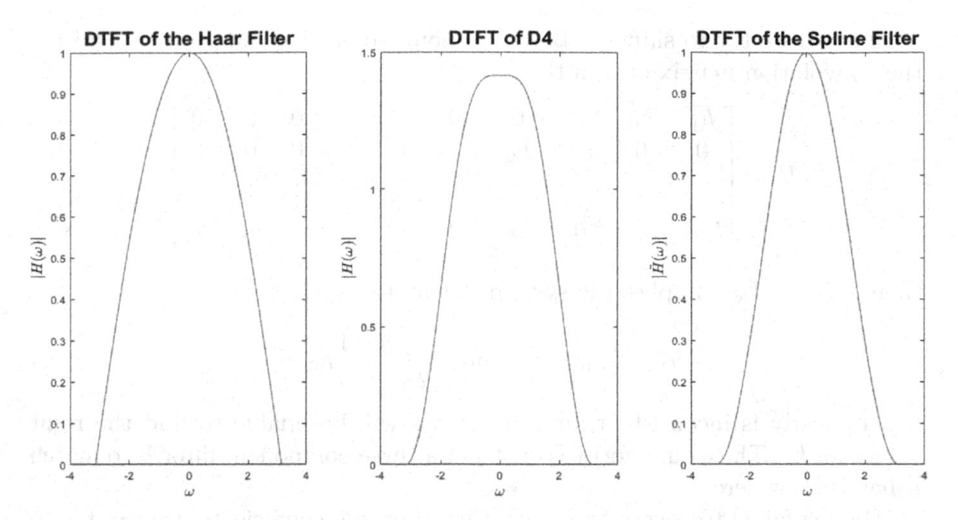

FIGURE 7.10: Discrete-time Fourier transforms of the Haar filter, of the Daubechies orthogonal filter of length 4, and of the spline filter of length 3.

The natural way to set up the (down-sampled) convolution matrix of $\tilde{\mathbf{h}}$ seems to be

$$\tilde{H} = \begin{bmatrix} 1/2 & 1/4 & 0 & 0 & 0 & \cdots & 0 & 0 & 0 & 1/4 \\ 0 & 1/4 & 1/2 & 1/4 & 0 & \cdots & 0 & 0 & 0 & 0 \\ \vdots & \vdots & \vdots & \vdots & \vdots & \ddots & \vdots & \vdots & \vdots & \vdots \\ 0 & 0 & 0 & 0 & 0 & \cdots & 0 & 1/4 & 1/2 & 1/4 \end{bmatrix},$$

and we see that the non-zero portions of the rows do overlap, which promises smoother lossy reconstructions as compared to those obtained with the help of the Haar wavelet transform.

Our next task is to construct the filter $\mathbf{h}$ that would match $\tilde{\mathbf{h}}$ by satisfying the first biorthogonality condition $H\tilde{H}^T = I_{N/2}$ of the system (7.48).

How many nonzero entries do we believe $\mathbf{h}$ ought to have? Clearly, we would like it to have more than two. If we try to endow $\mathbf{h}$ with three components, that is, if we set $\mathbf{h} = (h_{-1}, h_0, h_1)$, then the (down-sampled) convolution matrix of $\mathbf{h}$ will take the form

$$H = \begin{bmatrix} h_0 & h_{-1} & 0 & 0 & 0 & \cdots & 0 & 0 & 0 & h_1 \\ 0 & h_1 & h_0 & h_{-1} & 0 & \cdots & 0 & 0 & 0 & 0 \\ \vdots & \vdots & \vdots & \vdots & \vdots & \ddots & \vdots & \vdots & \vdots & \vdots \\ 0 & 0 & 0 & 0 & 0 & \cdots & 0 & h_1 & h_0 & h_{-1} \end{bmatrix},$$

and the equation $H\tilde{H}^T = I_{N/2}$ will imply that $h_1 = 0$ (and, hence, if we want $\mathbf{h}$ to be symmetric, $h_{-1} = 0$ as well), which is clearly not what we hoped for.

Even if we try to shift the blocks of nonzero entries one position and try the convolution matrix of $\mathbf{h}$ in the form

$$H = \begin{bmatrix} h_1 & h_0 & h_{-1} & 0 & 0 & 0 & \cdots & 0 & 0 & 0 \\ 0 & 0 & h_1 & h_0 & h_{-1} & 0 & \cdots & 0 & 0 & 0 \\ \vdots & \vdots & \vdots & \vdots & \vdots & \vdots & \ddots & \vdots & \vdots & \vdots \\ h_{-1} & 0 & 0 & 0 & 0 & & \cdots & 0 & h_1 & h_0 \end{bmatrix},$$

then $H\tilde{H}^T = I_{N/2}$ implies the system of equations

$$\frac{1}{2}h_1 + \frac{1}{4}h_0 = 1 \quad \text{and} \quad \frac{1}{2}h_1 + \frac{1}{4}h_0 = 0,$$

which clearly is inconsistent and, hence, we will be unable to find the right values for h_k. Thus, our effort to construct a three-component filter $\mathbf{h}$ to match $\tilde{\mathbf{h}}$ has led nowhere.

Having failed to construct $\mathbf{h}$ with three nonzero coefficients, we can try to make a similar attempt with a filter of length four. The reader is encouraged to try and see for themselves that it is not going to work either. We do not lose heart, however, and move on to try a filter of length five, which we denote by $\mathbf{h} = (h_{-2}, h_{-1}, h_0, h_1, h_2)$. Since we would like the filter to be symmetric, we can set $h_{-2} = h_2$ and $h_{-1} = h_1$, which leaves us with just three unknown coefficients. Then the (down-sampled) convolution matrix H is going to have the form

$$H = \begin{bmatrix} h_0 & h_1 & h_2 & 0 & 0 & 0 & \cdots & 0 & 0 & 0 & h_2 & h_1 \\ h_2 & h_1 & h_0 & h_1 & h_2 & 0 & \cdots & 0 & 0 & 0 & 0 & 0 \\ \vdots & \vdots & \vdots & \vdots & \vdots & \vdots & \ddots & \vdots & \vdots & \vdots & \vdots & \vdots \\ h_2 & 0 & 0 & 0 & 0 & 0 & \cdots & 0 & h_2 & h_1 & h_0 & h_1 \end{bmatrix},$$

and, consequently, the biorthogonality condition $H\tilde{H}^T = I_{N/2}$ implies that

$$\frac{1}{2}h_0 + \frac{1}{2}h_1 = 1 \quad \text{and} \quad \frac{1}{2}h_2 + \frac{1}{4}h_1 = 0. \tag{7.51}$$

Since we would also like $\mathbf{h} = (h_2, h_1, h_0, h_1, h_2)$ to be a low-pass filter, we consider its discrete-time Fourier transform

$$\begin{aligned} H(\omega) = \sum_{n=-2}^{2} h_n e^{-in\omega} &= h_2 e^{2i\omega} + h_1 e^{i\omega} + h_0 + h_2 e^{-i\omega} + h_1 e^{-2i\omega} \\ &= h_0 + 2h_1 \cos\omega + 2h_2 \cos(2\omega) \end{aligned}$$

and impose the low-pass condition $H(\pi) = 0$, which takes the form

$$h_0 - 2h_1 + 2h_2 = 0. \tag{7.52}$$

Putting the equations (7.51) and (7.52) together, we obtain the following system of equations:

System for Finding the Biorthogonal Spline (5,3) Filter Pair

$$\begin{aligned} h_0 + h_1 &= 2 \\ h_1 + 2h_2 &= 0 \\ h_0 - 2h_1 + 2h_2 &= 0 \end{aligned} \tag{7.53}$$

Unlike the systems for finding the Daubechies orthogonal filter coefficients, this system is linear and is easy to solve (we leave the details to exercises at the end of this section). Solving the system, we obtain

$$\mathbf{h} = (h_{-2}, h_{-1}, h_0, h_1, h_2) = \left(-\frac{1}{4}, \ \frac{1}{2}, \ \frac{3}{2}, \ \frac{1}{2}, \ -\frac{1}{4}\right). \tag{7.54}$$

As expected, the filter $\mathbf{h}$ is rational and symmetric. The discrete-time Fourier transform of $\mathbf{h}$ is

$$H(\omega) = \frac{3}{2} + \cos\omega - \frac{1}{2}\cos(2\omega),$$

and its graph is shown in Figure 7.11 alongside the graphs of the DTFTs of D4 and of the spline filter $\tilde{h}$ of length 3.

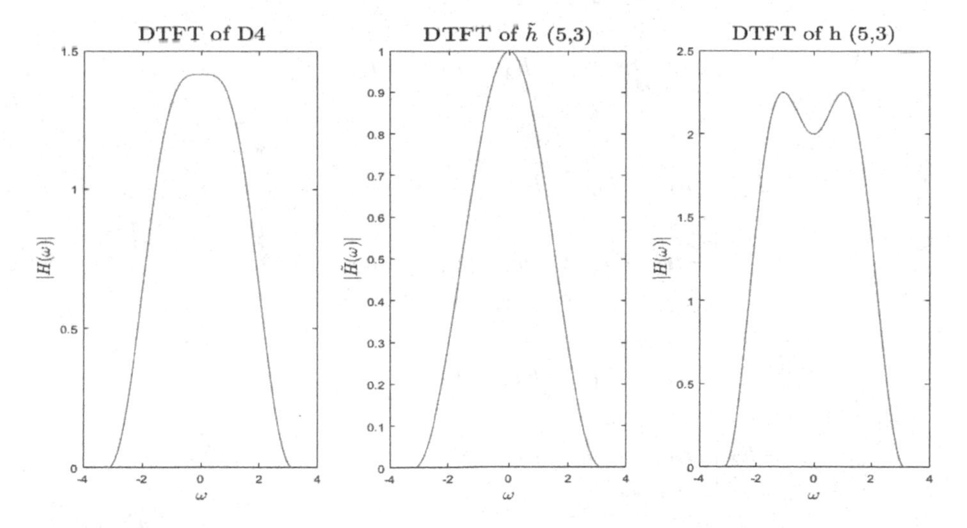

FIGURE 7.11: Discrete-time Fourier transforms of D4 and of the (5,3) spline filter pair.

The first derivative of $H(\omega)$ is

$$H'(\omega) = -\sin(\omega) + \sin(2\omega),$$

and we can expect that because $H'(\pi) = 0$, $\mathbf{h}$ is not going to be inferior to D4 in its ability to filter out higher frequencies. It is also evident from Figure 7.11 that we can expect $\mathbf{h}$ to be superior to D4 in preserving a wide range of lower frequencies.

Even though we have succeeded in determining the coefficients of both $\tilde{\mathbf{h}}$ and $\mathbf{h}$ and have constructed the matrices $\tilde{H}$ and H, we are not completely done yet. We still have to figure out the coefficients of the high-pass filters $\tilde{\mathbf{g}}$ and $\mathbf{g}$ and to assemble the transform matrices W and $\tilde{W}$.

How can we construct $\mathbf{g}$ and $\tilde{\mathbf{g}}$ in a way that ensures that the second equation $H\tilde{G}^T = 0$ and the third equation $G\tilde{H}^T = 0$ of the biorthogonality conditions (7.48) are satisfied? When working with even-length Daubechies filters, all we had to do in order to construct a high-pass filter $\mathbf{g}$ was to reverse the order of the coefficients of the corresponding low-pass filter $\mathbf{h}$ and to alternate the signs. With odd-length vectors, this trick generally does not quite work. However, our filters $\mathbf{h}$ and $\tilde{\mathbf{h}}$ are symmetric, which might help! This is what we can do: use the entries of $\mathbf{h}$ to populate the rows of $\tilde{G}$, shift the non-zero blocks of the rows by one position (to ensure that the dot products all have even numbers of terms), and alternate the signs. Graphically, we would like to see the rows of H to overlap with (transposed) columns of $\tilde{G}$ in the following types of patterns:

	h_2	h_1	h_0	h_1	h_2				
				$-h_2$	h_1	$-h_0$	h_1	$-h_2$	

	h_2	h_1	h_0	h_1	h_2				
		$-h_2$	h_1	$-h_0$	h_1	$-h_2$			

		h_2	h_1	h_0	h_1	h_2			
	$-h_2$	h_1	$-h_0$	h_1	$-h_2$				

				h_2	h_1	h_0	h_1	h_2	
	$-h_2$	h_1	$-h_0$	h_1	$-h_2$				

With this chart as a guide, in order to ensure $H\tilde{G}^T = 0$, we set

$$\tilde{\mathbf{g}} = (\tilde{g}_{-3}, \tilde{g}_{-2}, \tilde{g}_{-1}, \tilde{g}_0, \tilde{g}_1) = (-h_2, h_1, -h_0, h_1, -h_2)$$

and design the matrix $\tilde{G}$ accordingly in the form

$$\tilde{G} = \begin{bmatrix} h_1 & -h_0 & h_1 & -h_2 & 0 & 0 & \cdots & 0 & 0 & 0 & 0 & -h_2 \\ 0 & -h_2 & h_1 & -h_0 & h_1 & -h_2 & \cdots & 0 & 0 & 0 & 0 & 0 \\ \vdots & \vdots & \vdots & \vdots & \vdots & \vdots & \ddots & \vdots & \vdots & \vdots & \vdots & \vdots \\ h_1 & -h_2 & 0 & 0 & 0 & 0 & \cdots & 0 & 0 & -h_2 & h_1 & -h_0 \end{bmatrix}.$$

The reader is asked to verify that the equation $H\tilde{G}^T = 0$ does indeed hold.

In a similar manner, in order to satisfy the third equation $G\tilde{H}^T = 0$ of the biorthogonality conditions (7.48), we can use the entries of $\tilde{\mathbf{h}}$ to populate the rows of G, shift the non-zero blocks of the rows by one position (to ensure that the dot products all have even numbers of terms), and alternate the signs. Graphically, we would like the rows of G to overlap with the (transposed) columns of $\tilde{H}$ in the following pattern:

	$\tilde{h}_1$	$\tilde{h}_0$	$\tilde{h}_1$	
		$\tilde{h}_1$	$-\tilde{h}_0$	$\tilde{h}_1$

	$\tilde{h}_1$	$\tilde{h}_0$	$\tilde{h}_1$	
$\tilde{h}_1$	$-\tilde{h}_0$	$\tilde{h}_1$		

In other words, the filter $\mathbf{g}$ will be defined by

$$\mathbf{g} = (g_{-2}, g_{-1}, g_0) = (\tilde{h}_1, -\tilde{h}_0, \tilde{h}_1),$$

and, correspondingly, the matrix G is going to be of the form

$$G = \begin{bmatrix} 1/4 & -1/2 & 1/4 & 0 & 0 & 0 & \cdots & 0 & 0 & 0 \\ 0 & 0 & 1/4 & -1/2 & 1/4 & 0 & \cdots & 0 & 0 & 0 \\ \vdots & \vdots & \vdots & \vdots & \vdots & \vdots & \ddots & \vdots & \vdots & \vdots \\ 1/4 & 0 & 0 & 0 & 0 & 0 & \cdots & 0 & 1/4 & -1/2 \end{bmatrix}.$$

Once again, we ask the reader to verify that the equations $G\tilde{H}^T = 0$ and $G\tilde{G}^T = I_{N/2}$ do indeed hold. Putting everything together, we obtain the following definition:

Definition 7.5 ***The biorthogonal spline (5,3) filter pair*** *consists of the filters*

$$h = (h_{-2}, h_{-1}, h_0, h_1, h_2) = \left(-\frac{1}{4}, \frac{1}{2}, \frac{3}{2}, \frac{1}{2}, -\frac{1}{4}\right) \tag{7.55}$$

and

$$\tilde{\boldsymbol{h}} = (h_{-1}, h_0, h_1) = \left(\frac{1}{4}, \ \frac{1}{2}, \ \frac{1}{4}\right).$$

The ***wavelet transform matrices for the biorthogonal spline (5,3) filter pair*** *are*

$$W = \left[\begin{array}{cccccccccccc}
h_0 & h_1 & h_2 & 0 & 0 & 0 & \cdots & 0 & 0 & 0 & h_2 & h_1 \\
h_2 & h_1 & h_0 & h_1 & h_2 & 0 & \cdots & 0 & 0 & 0 & 0 & 0 \\
\vdots & \vdots & \vdots & \vdots & \vdots & \vdots & \ddots & \cdots & \vdots & \vdots & \vdots & \vdots \\
h_2 & 0 & 0 & 0 & 0 & 0 & \cdots & 0 & h_2 & h_1 & h_0 & h_1 \\
\hline
\tilde{h}_1 & -\tilde{h}_0 & \tilde{h}_1 & 0 & 0 & 0 & \cdots & 0 & 0 & 0 & 0 & 0 \\
0 & 0 & \tilde{h}_1 & -\tilde{h}_0 & \tilde{h}_1 & 0 & \cdots & 0 & 0 & 0 & 0 & 0 \\
\vdots & \vdots & \vdots & \vdots & \vdots & \vdots & \ddots & \vdots & \vdots & \vdots & \vdots & \vdots \\
\tilde{h}_1 & 0 & 0 & 0 & 0 & 0 & \cdots & 0 & 0 & 0 & \tilde{h}_1 & -\tilde{h}_0
\end{array}\right] \tag{7.56}$$

and

$$\tilde{W} = \left[\begin{array}{cccccccccccc}
\tilde{h}_0 & \tilde{h}_1 & 0 & 0 & 0 & 0 & \cdots & 0 & 0 & 0 & 0 & \tilde{h}_1 \\
0 & \tilde{h}_1 & \tilde{h}_0 & \tilde{h}_1 & 0 & 0 & \cdots & 0 & 0 & 0 & 0 & 0 \\
\vdots & \vdots & \vdots & \vdots & \vdots & \vdots & \ddots & \vdots & \vdots & \vdots & \vdots & \vdots \\
0 & 0 & 0 & 0 & 0 & 0 & \cdots & 0 & 0 & \tilde{h}_1 & \tilde{h}_0 & \tilde{h}_1 \\
\hline
h_1 & -h_0 & h_1 & -h_2 & 0 & 0 & \cdots & 0 & 0 & 0 & 0 & -h_2 \\
0 & -h_2 & h_1 & -h_0 & h_1 & -h_2 & \cdots & 0 & 0 & 0 & 0 & 0 \\
\vdots & \vdots & \vdots & \vdots & \vdots & \vdots & \ddots & \cdots & \vdots & \vdots & \vdots & \vdots \\
-h_1 & -h_2 & 0 & 0 & 0 & 0 & \cdots & 0 & 0 & -h_2 & h_1 & -h_0
\end{array}\right]$$

with the coefficients h and $\tilde{h}$ given by (7.55).

Note: In the theoretical literature, the filters $\mathbf{h}$ and $\tilde{\mathbf{h}}$ in (7.55) are usually given in their normalized form by

$$\mathbf{h} = (h_{-2}, h_{-1}, h_0, h_1, h_2) = \frac{\sqrt{2}}{2}\left(-\frac{1}{4}, \ \frac{1}{2}, \ \frac{3}{2}, \ \frac{1}{2}, \ -\frac{1}{4}\right)$$

and

$$\tilde{\mathbf{h}} = (h_{-1}, h_0, h_1) = \sqrt{2}\left(\frac{1}{4}, \ \frac{1}{2}, \ \frac{1}{4}\right).$$

In practical applications, the irrationality introduced by the $\sqrt{2}$ is clearly undesirable. For some applications, particularly for the JPEG2000 Compression Standard, the filter $\mathbf{h}$ in (7.55) is divided by the factor of 2, and the filter $\tilde{\mathbf{h}}$ is multiplied by the same factor of 2, after which their roles are switched, with the result known as the **LeGall filter pair**

$$\mathbf{h} = (h_{-2}, h_{-1}, h_0, h_1, h_2) = \left(-\frac{1}{8}, \ \frac{1}{4}, \ \frac{3}{4}, \ \frac{1}{4}, \ -\frac{1}{8}\right) \tag{7.57}$$

and

$$\tilde{\mathbf{h}} = (h_{-1}, h_0, h_1) = \left(\frac{1}{2}, \quad 1, \quad \frac{1}{2}\right).$$

Having constructed the matrices W and $\tilde{W}$, we can finally try to obtain the proof of the pudding by compressing an image using our new transform. There are several different orders in which we can perform the compression. For instance, we could compress the columns of our image using $\tilde{W}$ followed by compressing the rows by means of W^T. Denoting our $M \times N$ image by A and its transform by B, this sequence of operations can be formally described by

$$B = \tilde{W}_M \cdot A \cdot W_N^T.$$

Taking advantage of the biorthogonal property (7.47) of the W-$\tilde{W}$ pair, we can afterwards reconstruct the original image A by means of

$$W_M^T \cdot B \cdot \tilde{W}_N = \left(W_M^T \tilde{W}_M\right) \cdot A \cdot \left(W_N^T \tilde{W}_N\right) = A.$$

Alternatively, we could compress the columns of our image using W and the rows using $\tilde{W}$. We could even use W for compressing both the rows and the columns of the image with $\tilde{W}$ used exclusively for reconstruction or, vice versa, we could use solely $\tilde{W}$ for compression and solely W for reconstruction. In the exercises at the end of this section, the reader will be asked to formulate the matrix equations for the compression and reconstruction of the image that correspond to all four approaches, and in MATLAB exercises, the reader will be asked to compare the results of image compression and reconstruction using the four approaches.

Figure 7.12 shows side by side the results of lossy compression using the transforms based on D4 and on the biorthogonal spline (5,3) filter pair. The high-pass portions of the transforms were thresholded to preserve 95% of the energy in those portions. In terms of the quality of the image, (5,3) certainly does not seem inferior to D4, and the entropies of the compressed images appear to be virtually identical. We note that the peak signal-to-noise ratio is slightly higher for the (5,3) pair. This is not surprising because, due to the irrational coefficients, the D4-based transform suffers some loss of information as a result of the "quantization noise" caused by the introduction of rounding errors. This last observation further underscores the benefits of the rational coefficients of the biorthogonal spline filter pairs.

We encourage the reader to review all the details of the derivation of the (5,3) biorthogonal spline filter pair. In the exercises at the end of this session, the reader will also be asked to construct and implement the (7,5) and (9,7) biorthogonal spline filter pairs, to apply them to digital images of their choice, and to compare the quality of the resulting image compression.

FIGURE 7.12: Comparison of the compression performed using the D4 and the biorthogonal spline (5,3) filters.

Subsection 7.8.1 Exercises.

1. Construct several noncollinear vectors with at least four nonzero components orthogonal to the given vector.

   (a) $\mathbf{v} = [0, 1, 2, 3, 4, 5]^T$,

   (b) $\mathbf{v} = [4, 3, -1, 2, 5, 0]^T$.

2. Construct several noncollinear vectors with at least six nonzero components orthogonal to the given vector.
   (a) $[-1, 7, 2, -4, 3, 5, -2, 0, 0]^T$,
   (b) $[0, 1, 2, 3, 4, 5, 6, 7]^T$,
   (c) $[0, -2, 1, 3, -4, 7, 5, -6, 0]^T$.
3. Solve System (7.53).
4. Verify that the conditions $H\tilde{G}$, $G\tilde{H}^T = 0$ and $G\tilde{G}^T = I_{N/2}$ do indeed hold in the biorthogonal spline (5,3) filter pair constructed in this subsection.
5. Explain why it is not possible to design a biorthogonal spline filter pair of lengths 3 and 4.
6. Find several other pairs of integers that could work as possible lengths of biorthogonal spline filter pairs.
7. Set up and solve linear systems for designing biorthogonal spline filter pairs of lengths (7,5) and (9,7) using the method presented in this subsection.
8. Set up and solve linear systems for designing biorthogonal spline filter pairs of lengths determined in Exercise 6 using the method presented in this subsection.
9. Construct wavelet transform matrices W and $\tilde{W}$ for the biorthogonal spline filter pairs designed in the previous exercise using the method presented in this subsection.

7.8.2 Daubechies Theorem for Biorthogonal Spline Filters

In the previous subsection, we used an algebraic approach to find the coefficients for the biorthogonal spline filter pairs. It turns out, however, that it is possible to obtain the discrete-time Fourier transforms of $\mathbf{h}$ and $\tilde{\mathbf{h}}$ without having to solve any systems of algebraic equations by applying the following fundamental theorem (due to Ingrid Daubechies).

Theorem 7.1 (Daubechies Formula for $\mathbf{H}(\omega)$.) *Suppose that for an even integer power $\tilde{N}$, the spline filter $\tilde{h}$ is defined by*

$$\tilde{H}(\omega) = \sqrt{2}\cos^{\tilde{N}}(\omega/2).$$

Denote $\tilde{N} = 2\tilde{l}$ and $N = 2l$ and define

$$H(\omega) = \sqrt{2}\cos^{N}(\omega/2)\sum_{k=0}^{l+\tilde{l}-1}\binom{l+\tilde{l}-1+k}{k}\sin^{2k}(\omega/2). \tag{7.58}$$

Then $\mathbf{h}$ *and* $\tilde{\mathbf{h}}$ *form a biorthogonal spline filter pair in the sense that* $H^T \tilde{H} = I$.

Remark: In Ingrid Daubechies' original paper and in other sources, Theorem 7.1 is usually formulated for both odd and even values of N and $\tilde{N}$. However, in this text, we will only consider the even case, because the odd case does not add to our conceptual understanding of the material and involves unnecessary complications in the form of complex-valued "nuisance factors" like $e^{i\omega/2}$.

The proof of Theorem 7.1 is quite technical and, although not entirely beyond the scope of this text, will not be given here. We will, however, demonstrate the power of Theorem 7.1 by using it as an alternative way to construct the biorthogonal spline (5,3) pair. To do so, we set $\tilde{l} = 1$ and $l = 1$. Then the Daubechies Formula (7.58) implies that

$$\begin{aligned}
H(\omega) &= \sqrt{2}\cos^2(\omega/2)\sum_{k=0}^{1}\binom{1+k}{k}\sin^{2k}(\omega/2) \\
&= \sqrt{2}\cos^2(\omega/2)\left[1 + 2\sin^2(\omega/2)\right] \\
&= \sqrt{2}\left[\cos^2(\omega/2) + 2\cos^2(\omega/2)\sin^2(\omega/2)\right] \\
&= \sqrt{2}\left(\cos^2(\omega/2) + \frac{1}{2}\sin^2\omega\right) \\
&= \sqrt{2}\left(\frac{1+\cos\omega}{2} + \frac{1-\cos(2\omega)}{4}\right) \\
&= \frac{\sqrt{2}}{2}\left(1 + \cos\omega + \frac{1}{2} - \frac{1}{2}\cos(2\omega)\right) \\
&= \frac{\sqrt{2}}{2}\left(\frac{1}{4}e^{-2i\omega} + \frac{1}{2}e^{-i\omega} + \frac{3}{2} + \frac{1}{2}e^{i\omega} - \frac{1}{4}e^{2i\omega}\right),
\end{aligned}$$

and we can read the coefficients of the filter $\mathbf{h}$ off the last line.

In the exercises following this subsection, the reader will have an opportunity to use Theorem 7.1 to derive other biorthogonal spline filter pairs.

Subsection 7.8.2 Exercises.

1. When using Daubechies formula for constructing biorthogonal spline filters (Theorem 7.1) covered in this subsection, we have to rely on a number of trigonometric identities. Unless you have already done so in Section 6.2, this seems to be as good a point as any to review some of the most commonly used ones and to recall how helpful the complex expressions for the sine and the cosine functions are in proving those

identities. In this exercise, we ask the reader to use the formulas

$$\cos t = \frac{e^{it} + e^{-it}}{2} \quad \text{and} \quad \sin t = \frac{e^{it} - e^{-it}}{2i}$$

to establish the following trigonometric identities:

(a) $\cos^2 t = \frac{1}{2}\left[1 + \cos(2t)\right]$,

(b) $\sin^2 t = \frac{1}{2}\left[1 - \cos(2t)\right]$,

(c) $\sin(2t) = 2\sin t \cos t$,

(d) $\cos(2t) = \cos^2 t - \sin^2 t$,

(e) $\cos(3t) = 4\cos^3 t - 3\cos t$,

(f) $\sin(3t) = 3\sin t - 4\sin^3 t$.

2. Derive power-reduction trigonometric identities for reducing the power in $\cos^4 t$, $\cos^6 t$, $\sin^4 t$, and $\sin^6 t$ by using

   (a) The power-reduction identities from the previous exercise.

   (b) The complex expressions for the cosine and the sine functions.

3. Use Theorem 7.1 to find the discrete-time Fourier transforms for the orthogonal spline filter pairs of lengths (5,7) and (7,9).

4. Experiment with using Theorem 7.1 to find the DTFTs for the orthogonal spline filter pairs using different values of l and $\tilde{l}$.

7.8.3 The CDF97 Transform

In order to understand the need to construct yet another filter pair and to develop yet another type of a biorthogonal wavelet transform, it is important to complete the MATLAB exercises at the end of this section. Picking up where we have left off in the previous subsection, we set $\tilde{l} = 3$ and $l = 1$ in Theorem 7.1. Then $\tilde{N} = 6$, $N = 2$,

$$\tilde{H}(\omega) = \sqrt{2}\cos^6(\omega/2), \tag{7.59}$$

and

$$\begin{aligned} H(\omega) &= \sqrt{2}\cos^2(\omega/2)\sum_{k=0}^{3}\binom{3+k}{k}\sin^{2k}(\omega/2) \\ &= \sqrt{2}\cos^2(\omega/2)\left[1 + 4\sin^2(\omega/2) + 10\sin^4(\omega/2) + 20\sin^6(\omega/2)\right], \end{aligned} \tag{7.60}$$

which the reader is encouraged to verify. The graphs of both $\tilde{H}(\omega)$ and $H(\omega)$ are shown below in Figure 7.13.

Even a casual glance at the graphs makes us feel uneasy to say the least.

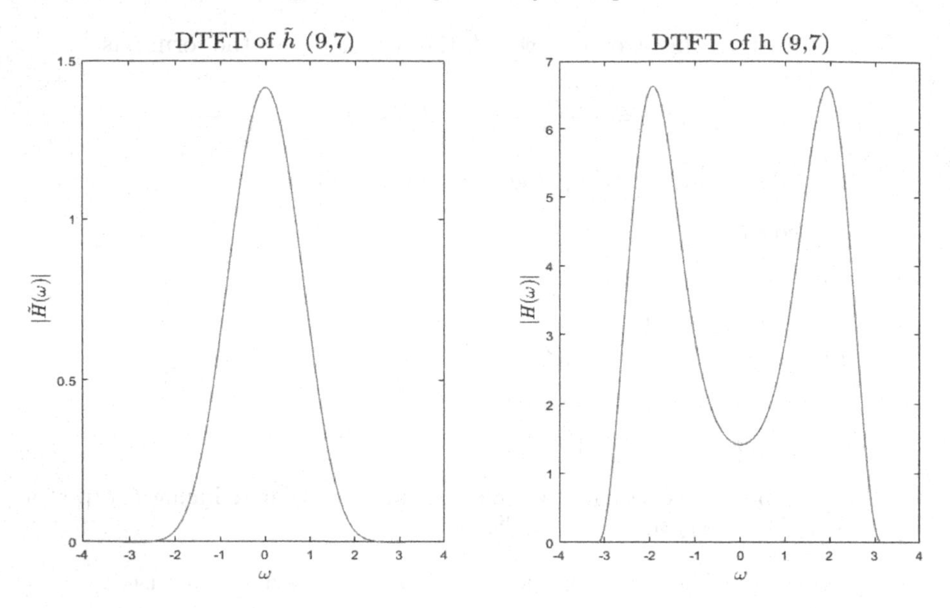

FIGURE 7.13: Discrete-time Fourier transforms of the biorthogonal (9,7) spline filter pair.

Practice shows (and our own experience with MATLAB exercises at the end of this section confirms it) that wavelet transforms work best when the two filters, $\mathbf{h}$ and $\tilde{\mathbf{h}}$ are similar in their characteristics. Unfortunately, that is not the case with the (9,7) biorthogonal spline pair. The two filters seem vastly different. In addition to the purely visual difference, we can point out that thanks to the sixth power of the cosine function, the first five(!) derivatives of $\tilde{H}(\omega)$ vanish at π. On the contrary, only the first derivative of $H(\omega)$ vanishes at π, which explains the relatively poor performance of $\mathbf{h}$ in eliminating the higher frequencies. It is also evident from the graph of $H(\omega)$ that $\mathbf{h}$ can be expected to wreak havoc on the mid-range frequencies.

As we reexamine the formulas (7.59) and (7.60), we may wonder whether it might be possible to rebalance the filter pair by asking $\tilde{H}(\omega)$ to "lend" a couple of powers of the cosine function to $H(\omega)$ and for $H(\omega)$ to "repay" with a factor that involves some powers of the sine function. More formally, we might try to rewrite the part of $H(\omega)$ that involves the sine functions as a factored polynomial in $t = \sin^2(\omega/2)$. As a result, $H(\omega)$ takes the form

$$\begin{aligned} H(\omega) &= \sqrt{2}\cos^2(\omega/2)\left[1 + 4\sin^2(\omega/2) + 10\sin^4(\omega/2) + 20\sin^6(\omega/2)\right] \\ &= \sqrt{2}\cos^2(\omega/2)\left[1 + 4t + 10t^2 + 20t^3\right] \\ &= 20\sqrt{2}\cos^2(\omega/2)(t - t_1)(t - t_2)(t - t_3), \end{aligned} \tag{7.61}$$

where t_1, t_2, and t_3 are the roots of the polynomial $p(t) = 1 + 4t + 10t^2 + 20t^3$.

It now becomes apparent how the "exchange of factors" between $H(\omega)$ and $\tilde{H}(\omega)$ might proceed. There are several practical decisions to be made, mainly whether we would like to "transfer" $\cos^2(\omega/2)$ or $\cos^4(\omega/2)$ from $\tilde{H}(\omega)$ to $H(\omega)$ and which factor(s) we would like to transfer from $H(\omega)$ to $\tilde{H}(\omega)$.

As the details are quite tedious and involve the use of a computer algebra system, we will omit them since they do not add to our understanding of the underlying concept. It is along these and similar lines, guided by the goal of securing the best possible balance between the filters $\mathbf{h}$ and $\tilde{\mathbf{h}}$, that A. Cohen, I. Daubechies, and J.-C. Feauveau developed the CDF97 filter pair and wavelet transform in 1992[3].

For the reader's convenience, we provide the values of the filter coefficients in the box below, even though, as mentioned above, we omit the tedious technical details of carrying through the development of the CDF97.

$$
\begin{aligned}
h_0 &= \; -\tilde{g}_1 &&= 0.852699 \\
h_1 &= \; h_{-1} &&= \tilde{g}_0 = \tilde{g}_2 = 0.377403 \\
h_2 &= \; h_{-2} &&= -\tilde{g}_{-1} = -\tilde{g}_3 = -0.110624 \\
h_3 &= \; h_{-3} &&= \tilde{g}_{-2} = \tilde{g}_4 = -0.023850 \\
h_4 &= \; h_{-4} &&= -\tilde{g}_{-3} = -\tilde{g}_5 = 0.037829 \\
\tilde{h}_0 &= \; -g_1 &&= 0.788486 \\
\tilde{h}_1 &= \; \tilde{h}_{-1} &&= g_0 = g_2 = 0.418092 \\
\tilde{h}_2 &= \; \tilde{h}_{-2} &&= -g_{-1} = -g_3 = 0.040689 \\
\tilde{h}_3 &= \; \tilde{h}_{-3} &&= g_{-2} = g_4 = -0.064539
\end{aligned}
\tag{7.62}
$$

In MATLAB exercises at the end of this section, the reader will be asked to use the coefficients (7.62) to implement their own version of CDF97-based wavelet transform and to use it to compress images of their choice.

We hope that while following our brief overview of the CDF97 construction, the reader was bothered by the following question: by what right can $H(\omega)$ and $\tilde{H}(\omega)$ exchange factors? After all, $\mathbf{h}$ and $\tilde{\mathbf{h}}$ were carefully constructed specifically to meet the biorthogonality condition expressed by the matrix equation $\tilde{H}H^T = I$. Aren't they going to lose that vital property as a result of such an exchange of factors?

Fortunately, the answer is no, as the following theorem (also due to Ingrid Daubechies) assures us:

Theorem 7.2 *Suppose that the discrete-time Fourier transforms $H(\omega)$ and $\tilde{H}(\omega)$ of the filters $\mathbf{h}$ and $\tilde{\mathbf{h}}$ respectively satisfy the condition*

$$\tilde{H}(\omega)\overline{H(\omega)} + \tilde{H}(\omega+\pi)\overline{H(\omega+\pi)} = 2. \tag{7.63}$$

Then $\boldsymbol{h}$ *and* $\tilde{\boldsymbol{h}}$ *form a biorthogonal filter pair in the sense that their (downsampled) convolution matrices satisfy the biorthogonality condition* $\tilde{H}H^T = I$*.*

Thus, according to Theorem 7.2, it is *not the separate expressions* for $H(\omega)$ and $\tilde{H}(\omega)$ but their *product* that is important in ensuring that the all-important biorthogonality relation $\tilde{H}H^T = I$ holds. That is why the biorthogonality relations are safe under any exchange of factors between $H(\omega)$ and $\tilde{H}(\omega)$, since such an exchange does not affect the products $\tilde{H}(\omega)\overline{H(\omega)}$ and $\tilde{H}(\omega+\pi)\overline{H(\omega+\pi)}$.

For more detailed information of the construction of CDF97, we refer the reader to the original paper [3] as well as to [9] and to [43].

Subsection 7.8.3 Exercises.

Since the key step in the construction of the CDF97 is splitting a certain polynomial into linear factors, this seems to be as good a point as any to review polynomial factorization.

1. Find all the roots of the following polynomials and factor them into linear factors.

   (a) $p(t) = t^3 - 5t^2 + 2t - 9$,

   (b) $p(t) = t^4 + t^3 - 7t^2 - 5t + 10$,

   (c) $p(t) = t^3 - 7t^2 + 14t - 6$,

   (d) $p(t) = 5t^3 - t^2 - 35t + 7$,

   (e) $p(t) = 7t^3 - t^2 - 21t + 3$.

2. Factor the following polynomials into linear and quadratic factors.

   (a) $p(t) = t^3 - 4t^2 - 2t + 20$,

   (b) $p(t) = t^3 - 8t^2 + 29t - 52$,

   (c) $p(t) = 5t^3 - 54t^2 + 170t - 104$,

   (d) $p(t) = t^4 - t^2 - 90$,

   (e) $p(t) = t^3 - 8$.

3. Find the numerical values of the roots t_1, t_2, and t_3 of the polynomial

$$p(t) = 20t^3 + 10t^2 + 4t + 1$$

   used in the CDF97 construction (see (7.61)).

Section 7.8 MATLAB Exercises.

1. For each filter pair you designed in Exercise 8 of Subsection 7.8.1 and in Exercise 3 of Subsection 7.8.2, plot the graphs of $\tilde{H}(w)$ and $H(w)$ side by side. Make predictions about the performance of the corresponding wavelet transform based on the (dis)similarity of the graphs.

2. Write MATLAB functions that will create transform matrices W and $\tilde{W}$ of specified dimensions for the following filter pairs:

   (a) the (5,3) biorthogonal spline filter pair,
   (b) the (7,5) biorthogonal spline filter pair,
   (c) the (9,7) biorthogonal spline filter pair,
   (d) the CDF97 biorthogonal filter pair.

   As a reminder, the reader was asked to create the transform matrices for the (7,5) and the (9,7) filter pairs in Exercise 9 of Subsection 7.8.1.

3. Compare the performance of the above transforms, as well as that of the Haar, D4, and D6 transforms as follows:

   (a) Prepare a selection of images to experiment with. The best way to do that is to save a picture in raw format and to then cut out a fragment in order to meet MATLAB memory restrictions.
   (b) Ask the user which image needs to be compressed.
   (c) Perform at least 3 iterations of each transform you have implemented.
   (d) Estimate the lossless compression ratio by comparing the entropies of the original image and of its transform.
   (e) Display the original image, its transform, and the lossless reconstruction. The image of the transform should come with relevant numerical characteristics (entropy, compression ratio, etc.).
   (f) For each image and each transform, calculate the threshold that would preserve 95% of the energy in the high-pass portions of the transform. Set to zero all the entries in the high-pass portions of the transform whose absolute values are below the threshold. Create the lossy reconstruction of the image.
   (g) Display the original image, the transform, and the lossy reconstruction.

4. Compile the results of the previous MATLAB exercises into a PowerPoint presentation.

7.9 An Overview of JPEG2000

In this section, we provide an extremely brief overview of JPEG 2000, which is a wavelet transform-based image compression standard and coding system originally created by the Joint Photographic Experts Group committee in 2000 with the intention of eventually replacing the JPEG standard (in use since 1992).

The old JPEG is based on one of the variants of the Fourier transform called discrete cosine transform (which was developed by Nasir Ahmed, T. Natarajan, and K.R. Rao back in 1974). While JPEG remains the most common format for storing and transmitting still images on the Internet, it does have a few important weaknesses, some of which are as follows:

- As part of the JPEG algorithm, the discrete cosine transform is applied to 8×8 blocks of pixels, which leads to significant blocking artifacts (especially at high compression ratios).
- JPEG is not well-suited for line drawings and iconic graphics with sharp contrasts between adjacent pixels, and, likewise, is not appropriate for storing digitized books or fingerprint databases.
- JPEG is inherently lossy due to the rounding errors, which are unavoidable because the values of the Discrete Cosine Transform are naturally irrational. Thus, JPEG is not suitable for use in medical imaging and in certain scientific applications where image fidelity is of the highest importance.
- JPEG is also not suitable for images that are likely to undergo extensive and/or multiple edits as the image quality is reduced every time the image is cropped, shifted, rotated, compressed, and reconstructed.

JPEG also has technical issues related to image sizes, to the difficulty of separating the image into non-overlapping parts that require different levels of compression quality, and to some difficulties with progressive image transmission. JPEG2000 successfully resolves all of these issues.

Here, we provide a very brief description of the way JPEG2000 works. For simplicity, we will only discuss the compression of a gray-scale $M \times N$ image A. Actually, JPEG2000 can support a wide variety of color models in addition to the RGB and YCbCr models that we have studied in this text. For a much more detailed treatment, we refer the reader to [39].

1. First, the pixel values are shifted down by 128, so that they now occupy the interval from -128 to 127.
2. The image may be separated into regions of interest, which makes it possible to store different parts of the same image at different quality levels.

3. For lossless compression, the algorithm determines the largest value r, so that both M and N are divisible by 2^r. Then, it performs r iterations of the wavelet transform based on the LeGall Filter Pair (7.57). Since the LeGall Filter Pair is based on the biorthogonal spline (5-3) filter pair, it is nicely balanced, produces only rational values in the transform (which can then be easily mapped into integers), and can be effectively inverted producing a completely faithful reconstruction of the original image.

4. For lossy compression, the process is considerably more complex. The wavelet transform of choice is based on the CDF97 filter pair discussed in the preceding section. The number of iterations is determined by a variety of factors, and the image boundaries may need to be adjusted to ensure that the image dimensions are divisible by the appropriate powers of 2. Since the coefficients of CDF97 are irrational, rounding (called quantization) is unavoidable, but it is done separately for different portions of the image transform. Different levels of precision (called the quantization steps) are calculated for different portions of the transform.

5. Instead of Huffman encoding, JPEG2000 uses a version of arithmetic coding called *embedded block coding with optimized truncation* (EBCOT). Unfortunately, a discussion of EBCOT is beyond the scope of this text.

Overall, JPEG 2000 provides a modest but noticeable 20 percent increase in compression ratio as compared to JPEG. At least half of that increase can be attributed to the advantages of EBCOT over Huffman encoding, with the rest coming from the advantages of the wavelet transforms over the discrete cosine transform. However, the JPEG2000 main selling points are not limited to its superior compression performance but also include the availability of a genuinely lossless compression mode, the flexibility of encoding, and the improved quality of reconstructed images with respect to visible compression artifacts.

Unlike JPEG, which produces both ringing and blocking artifacts, JPEG only produces some degree of blurring and slight ringing near the edges in the image, which are much less visible and at high bit-rates become virtually imperceptible.

Another advantage enjoyed by JPEG 2000 is in the area of progressive decoding and signal-to-noise ratio scalability. Thanks to the structure of the wavelet transforms which contain the low-frequency portion (the blur) and the high-frequency portions, it is possible to download the small low-frequency portion prior to downloading the other parts of the transform. The advantage of this method is that the viewer can quickly get to see a blur of the picture with its quality progressively improving as more data is being downloaded.

In spite of all its advantages, as of 2020, JPEG 2000 is not widely supported in web browsers and, hence, is not widely used on the Internet. There seem to be two reasons for that. First, the advantages of JPEG2000 over the old JPEG are least dramatic when it comes to compressing smooth photographic images without sharp edges – the kind that make up the bulk of the pictures taken and transmitted over the web. Even more important, however, are the legal considerations related to patent law. The JPEG Committee has stated that although "... agreement has been reached with over 20 large organizations holding many patents in this area to allow use of their intellectual property in connection with the standard without payment of license fees or royalties", nevertheless, "it is, of course, still possible that other organizations or individuals may claim intellectual property rights that affect implementation of the standard, and any implementers are urged to carry out their own searches and investigations in this area"[39]. As the patents related to various components of JPEG2000 expire, one might expect that the use of JPEG2000 is likely to expand.

Section 7.9 Exercises.

Put yourself into the shoes of a member of the JPEG committee. Based on your experience in this chapter, write a short essay suggesting the use of specific wavelet transforms for lossless and lossy image compression.

7.10 Other Applications of Wavelet Transforms

The chapter devoted to wavelet methods would not be complete without at least a brief mention of two other areas of application for discrete wavelet transforms – edge detection and noise reduction.

7.10.1 Wavelet-Based Edge Detection

We recall that the wavelet transform of a digital image has the structure

$$\left[\begin{array}{c|c} \mathcal{B} & \mathcal{V} \\ \hline \mathcal{H} & \mathcal{D} \end{array}\right],$$

where the upper-right corner $\mathcal{V}$ contains the differences between adjacent pixels in the horizontal direction and, therefore, is likely to be representative of prominent vertical features, the lower-left corner $\mathcal{H}$ contains the differences in the vertical direction and is thus likely to be representative of horizontal features, whereas the lower-right corner $\mathcal{D}$ is comprised of the "diagonal" differences and is likely to correspond to diagonal features of the image.

Let us take a look at the Haar wavelet transform of the familiar photo of the Russian State Library (Figure 7.14(b)). We cannot help but think that the high-pass regions of the transform already contain all the edges of the image, combining what would be the output of the gradient and the Roberts cross edge detectors. All we have to do is put three high-pass regions together. What is the most efficient way to do that? How about just setting the low-pass region $\mathcal{B}$ of the transform to zero and reconstructing the image solely from the high-pass regions?

(a) The Original Image

(b) The Haar Wavelet Transform

(c) The Edges from HWT

(d) The Edges from D4

FIGURE 7.14: Edge detection with the use of the HWT and D4.

The result of this type of "high-pass" reconstruction from the Haar wavelet transform of the image is shown in Figure 7.14(c), and it looks quite encouraging. Can we do even better than that? Would a longer wavelet filter produce more prominent edges by "anticipating" them in advance and "remembering" them for an extra step? Having compiled an entire library of wavelet transforms, we can feel free to have fun and try edge detection with any of them.

Figure 7.14(d) shows the edges produced with the help of Daubechies orthogonal filter of length 4. Unfortunately, they do not seem to be much of an improvement over the ones produced by the Haar wavelet transform. How about trying the biorthogonal filter pairs? After all, they are symmetric, and symmetry might help in edge detection. And what about the CDF97 filter pair? Another possible way to improve wavelet-based edge detection is to use two or more iterations of the chosen wavelet transform and to set only the highest-level blur to zero.

All of those transforms and approaches ought to be tried, and the reader will be asked to do exactly that in the following MATLAB exercises.

MATLAB Exercises.

1. Prepare a selection of images that you would like to turn into drawings, just as you did for the exercises in Section 5.3. An architectural ensemble or a decorative pattern could be two possible themes. If necessary, you can use the MATLAB *rgb2gray* function to convert a color image to grayscale.

2. Write a MATLAB program that would do the following:

   (a) Display the original image.

   (b) Ask the user which edge wavelet transform is to be used for edge detection.

   (c) Ask the user how many iterations of the selected wavelet transform are to be performed.

   (d) Perform the specified number of iterations of the specified wavelet transform and set the low-pass portion of the transformed image to zero. Reconstruct the image edges from the high-pass portions of the transform.

   (e) Display the "drawing" obtained by using this method of edge detection. For comparison, display it side by side with the ones obtained with the use of Roberts cross, Prewitt, Sobel, and Laplacian edge detectors.

   Test your program using several different images and several different wavelet transforms. Compare the results.

3. Compile the results of the previous MATLAB exercises into a PowerPoint presentation.

7.10.2 Wavelet-Based Image Denoising

We recall that high-frequency components of an image are those that correspond to rapid changes on a small scale. It would appear evident that random noise is one of them and, consequently, most of the noise is going to end up in the high-pass regions of any wavelet transform of a noisy image.

We illustrate this situation in Figure 7.15(a), where a noisy image of the Russian State Library (corrupted by additive white Gaussian noise) is shown side by side with its D4-based wavelet transform. The noise level was estimated to be $\sigma = 30$. It is evident that the low-pass regions of the transform is relatively free from the noise, most of which got caught in the high-pass regions. What is the best way to get rid of the noise now that we have it "cornered"?

We could, of course, just set the entire high-frequency area of the transform to zero, but that would be throwing the baby out with the bathwater. The high-frequency components of the image are also important; without them the image is going to lose sharpness and will risk becoming too blurry.

What we can do is identify the most prominent high-frequency components? Some of the edges in the high-pass regions are still visible through the noise – keep those. More precisely, preserve the parts of the high-frequency components that stand out above the noise. Mathematically, this is achieved by means of the *soft threshold rule.*

To apply the soft threshold rule to the wavelet transform of an image, we first choose a threshold $\lambda > 0$. For example, we could set λ at the noise level (measured by its standard deviation). If the absolute value of a pixel in the high-pass portion of the transform is below the threshold, then we set that pixel to zero. If it is above the noise level, then we calculate by how much it exceeds the threshold. Formally, the soft threshold rule is given by the piecewise-linear function

$$f_\lambda(t) = \begin{cases} t + \lambda, & t < -\lambda \\ 0, & -\lambda \le t \le \lambda \\ t - \lambda, & \lambda < t \end{cases} \tag{7.64}$$

Unfortunately, a comprehensive discussion of methods to determine the optimal threshold is beyond the scope of this text. For now, we will make the unrealistic assumption that the noise level σ is known to us and will use trial and error in setting the threshold to integer multiples of σ.

According to the so-called "68 - 95 - 99.7 Rule" from general statistics, setting $\lambda = \sigma$ is going to eliminate 68% of the noise, whereas setting $\lambda = 2\sigma$ should get rid of as much as 95% of the noise (as long as the wavelet filter is properly normalized). Figure 7.15(c) and Figure 7.15(d) show the result of soft threshold function applied to the high-pass regions of the wavelet transform with $\lambda = \sigma$ and $\lambda = 2\sigma$ respectively.

(a) A Noisy Image

(b) D4 of the Noisy Image

(c) Soft Threshold, $\lambda=\sigma$

(d) Soft Threshold, $\lambda=2\sigma$

FIGURE 7.15: Wavelet-based image denoising.

As compared to the method of noise suppression discussed in Section 5.1, which consisted of merely applying a low-pass filter to the image, the wavelet-based strategy appears more sophisticated and less wasteful. In addition to blurring the image in the hope that the noise gets "averaged out", this method also involves an attempt to extract as much useful information on high-frequency components as possible, particularly on those that are not completely obliterated by the noise.

To conclude this section, we outline a more systematic method of selecting the optimal value of the threshold λ proposed by D. Donoho and I. Johnstone in 1994 [8]. It is known as SureShrink and is based on the well-known Stein's unbiased risk estimator (SURE) [34].

In its simplified form, C. Stein's famous theorem can be stated as follows:

Theorem 7.3 *Suppose that the signal* $\boldsymbol{w} = (w_1, w_2, \dots, w_n)$ *can be expressed by*

$$\boldsymbol{w} = \boldsymbol{x} + \boldsymbol{\epsilon},$$

where $\boldsymbol{x} = (x_1, x_2, \dots, x_n)$ *represents the useful information and* $\boldsymbol{\epsilon} = (\epsilon_1, \epsilon_2, \dots, \epsilon_n)$ *is white additive Gaussian noise with the standard deviation* $\sigma = 1$. *Suppose that we are trying to estimate the useful part of the signal* $\boldsymbol{x}$ *by means of*

$$\hat{x}_k(w_k) = w_k + g_k(w_k), \tag{7.65}$$

where the functions g_k *are piecewise-differentiable. Then the estimation error*

$$E(||\hat{\boldsymbol{x}} - \boldsymbol{x}||^2) = E\left(N + \sum_{k=1}^{N} g_k(w_k)^2 + 2\sum_{k=1}^{N} g_k'(w_k)\right). \tag{7.66}$$

It is our task to construct the functions g_k that would minimize the estimation error expressed in the form (7.66). In the context of wavelet-based denoising, **w** represents the pixel values in the high-pass regions of the wavelet transform, and we estimate the "useful" part x_k of each pixel by means of the soft thresholding function

$$\hat{x}_k(w_k) = f_\lambda(w_k) = \begin{cases} w_k + \lambda, & w_k < -\lambda \\ 0, & -\lambda \le w_k \le \lambda \\ w_k - \lambda, & \lambda < w_k \end{cases}.$$

In view of (7.65),

$$g_k(w_k) = \begin{cases} \lambda, & w_k < -\lambda \\ -w_k, & -\lambda \le w_k \le \lambda \\ -\lambda, & \lambda < w_k \end{cases},$$

which implies that

$$g_k(w_k)^2 = \begin{cases} |w_k|^2, & |w_k| \le \lambda \\ \lambda^2, & |w_k| > \lambda \end{cases} = \min\{\lambda^2, w_k^2\} \tag{7.67}$$

and, consequently,

$$\sum_{k=1}^{N} g_k(w_k)^2 \quad = \quad \sum_{k=1}^{N} \min\{\lambda^2, w_k^2\}$$

$$= \sum_{|w_k|\leq\lambda} w_k^2 + \big|\{k : |w_k| > \lambda\}\big|\lambda^2, \tag{7.68}$$

where the vertical bars denote the number of elements in the set. Furthermore,

$$g_k'(w_k) = \begin{cases} -1, & |w_k| \leq \lambda \\ 0, & |w_k| > \lambda \end{cases} \tag{7.69}$$

and, consequently,

$$\sum_{k=1}^{N} g_k'(w_k) = -\big|\{k : |w_k| \leq \lambda\}\big|. \tag{7.70}$$

Substituting (7.68) and (7.70) into (7.66) yields

$$\begin{aligned} Err(\lambda) &= E\left(||\hat{\mathbf{x}} - \mathbf{x}||^2\right) \qquad (7.71) \\ &= N - 2\,|\{k : |w_k| \leq \lambda\}| + \sum_{|w_k|\leq\lambda} w_k^2 + \big|\{k : |w_k| > \lambda\}\big|\lambda^2. \end{aligned}$$

A careful examination of this expression reveals that $Err(\lambda)$ is piecewise-quadratic and, consequently, it reaches its absolute minimum value at one of the points w_k. Therefore, in order to find the optimal value for the soft threshold λ, all we have to do is evaluate $Err(\lambda)$ at all the pixel values in the high-pass region of the image wavelet transform and determine the λ that corresponds to the smallest value.

Example 7.1 *Suppose that the (reshaped) high-pass regions of the wavelet transform form the (admittedly unrealistic) sequence* $\boldsymbol{w} = (0, 0, 0, 0, 0, 1, 1, 1, 1, 2, 3, 4)$. *Then, according to* (7.71),

$$\begin{aligned} Err(0) &= 12 - 2 \cdot 5 = 2 \\ Err(1) &= 12 - 2 \cdot 9 + 4 + 3 \cdot 1^2 = 1 \\ Err(2) &= 12 - 2 \cdot 10 + 8 + 2 \cdot 2^2 = 8 \\ Err(3) &= 12 - 2 \cdot 11 + 17 + 1 \cdot 2^2 = 11 \\ Err(4) &= 12 - 2 \cdot 12 + 33 = 21, \end{aligned}$$

which yields the optimal soft threshold $\lambda = 1$.

For a much more detailed description of SureShrink (complete with plots, examples, and background information), we refer the reader to [9] as well as to the original paper [8].

In the MATLAB exercises at the end of this section, the reader will be asked to apply the wavelet transform designed in this chapter to the task of noise reduction and to compare the results of wavelet-based noise suppression to those obtained with the use of naive averaging filters and the Gaussian filters described in Section 5.1.

Subsection 7.10.2 Exercises.

1. Verify the formulas (7.67) through (7.71).

2. Suppose that the pixels w_k in the high-pass region of the wavelet transform have been arranged to form a non-decreasing sequence in the sense that
$$|w_1| \leq |w_2| \leq \cdots \leq |w_N|.$$
Derive simplified versions of the formula (7.71) for the approximation error $Err(\lambda)$ in the following three cases:

   (a) $\lambda < |w_1|$,

   (b) $\lambda \geq |w_N|$,

   (c) $w_k \leq \lambda < w_{k+1}$.

   You will likely find those simplified expressions useful in the next exercise.

3. Use the SureShrink method discussed in this subsection to calculate (by hand) the optimal value of the soft threshold λ for the following sequences:

   (a) $\mathbf{w} = (0, 1, 2, 3)$,

   (b) $\mathbf{w} = (0.1, -1, 1.2, 3)$,

   (c) $\mathbf{w} = (0, 0, 1, 1, 1, 2, 2, 2, 2, 3, 4, 5)$,

   (d) $\mathbf{w} = (0, 0, 1, -1, 1, -1, 2, 2, -2, 3, -4, 5)$,

   (e) $\mathbf{w} = (0, 0, 1, 1, 1.1, 1.1, 1.2, 1.2, 1.3, 2, 2, 3)$.

   Based on your experience with parts (a), (c) and (d), is it possible to construct a sequence using the values 0, 1, 2, and 3 so that the optimal threshold λ equals 2? How about $\lambda = 3$?

4. While following Example 7.1 and while working on the previous exercise, the reader may have detected a strong possibility that a recursion formula might exist for the sequence $Err(k)$. Try to discover and prove this recursion formula. If successful, use it to check your answers in the previous exercise.

5. Stein's theorem specifically assumes that the noise standard deviation $\sigma = 1$. Is that something we need to be concerned about? Do we need to adjust all our formulas and procedures if (as is usually the case in practice) the noise level is different from $\sigma = 1$?

6. In this subsection, we mentioned the "68-95-99.7" Rule from general statistics. This might be as good a point as any to review this topic. It is known that the heights of adult males are normally distributed with the mean of 175 centimeters and the standard deviation of 7 centimeters. Use the "68-95-99.7" Rule to provide **approximate** answers to the following questions:

   (a) Determine a range of heights that contains 68 percent of the adult male population.
   (b) Determine a range of heights that contains 95 percent of the adult male population.
   (c) Determine a range of heights that contains 99.7 percent of the adult male population.
   (d) What percentage of adult males are taller than 189 cm?
   (e) What percentage of adult males are shorter than 168 cm?
   (f) What percentage of adult males are taller than 196 cm?
   (g) What percentage of adult males are taller than 168 cm. but shorter than 189 cm?

MATLAB Exercises.

1. Write a MATLAB function that would apply the soft threshold rule with the specified threshold to the high-pass regions of the wavelet transform of the specified grayscale image.

2. Next, create an improved version of the function you created in MATLAB Exercise 1 with the additional capability to apply the soft threshold rule with **three different** specified thresholds to the three high-pass regions of the wavelet transform of the specified grayscale image.

3. Write a MATLAB function that would use the SureShrink method discussed in this subsection to calculate the optimal value of the soft threshold λ for the specified sequence.

4. Write a MATLAB program that would do the following:

   (a) Ask the user which image is to be imported into the MATLAB programming environment.
   (b) Display the original image.
   (c) Ask the user to specify the noise level σ.

(d) Create a noisy version of the selected image by adding to it white Gaussian noise of the specified level.

(e) Ask the user which wavelet transform is to be used for noise suppression.

(f) Call the function you created in MATLAB Exercise 1 to apply the soft threshold rule to the high-pass region of the specified wavelet transform of the selected image. Try both the threshold value $\lambda = \sigma$ and $\lambda = 2\sigma$.

(g) Next, use the function you created in MATLAB Exercise 3 to calculate optimal soft thresholds separately for each high-pass region of the wavelet transform. Call the function you created in MATLAB Exercise 2 to apply the soft threshold rule separately to each high-pass region.

(h) Display all three versions of the denoised image alongside its noisy version. Comment on the difference in the image quality when using naive methods versus using SureShrink to determine the value of the soft threshold λ.

Test your program using several different images and all the available wavelet transforms. Comment on the results.

5. Next, perform image noise reduction by means of naive averaging filters and Gaussian filters of your choice. Compare the effectiveness of Gaussian filters and wavelet transforms in reducing the noise levels in digital images.

6. Compile the results of the previous MATLAB exercises into a PowerPoint presentation.

Bibliography

[1] S. Attaway. *MATLAB: A Practical Introduction to Programming and Problem Solving.* Wiley, 5 edition, 2018.

[2] Robert G. Bartle and Donald R. Sherbert. *Introduction to Real Analysis.* Wiley, 4 edition, 2011.

[3] A. Cohen, I. Daubechies, and J.-C. Feauveau. Biorthogonal bases of compactly supported wavelets. *Comm. Pure Appl. Math.,*, 45:485–560, 1992.

[4] I. Daubechies. Orthogonal bases of compactly supported wavelets. *Comm. Pure Appl. Math.*, 41:909–996, 1988.

[5] I. Daubechies. *Ten Lectures on Wavelets.* SIAM, 1992.

[6] I. Daubechies. Orthogonal bases of compactly supported wavelets: Ii. variations on a theme. *SIAM J. Math. Anal.*, 24(2):499–519, 1993.

[7] Harold Davis. *The Photographer's Black and White Handbook: Making and Processing Stunning Digital Black and White Photos.* Monacelli Studio, 2011.

[8] D. Donoho and I. Johnstone. Adapting to unknown smoothness via wavelet shrinkage. *J. Amer. Stat. Assoc.*, 90(432):1200–1224, 1995.

[9] Patrick J. Van Fleet. *Discrete Wavelet Transformations: An Elementary Approach with Applications.* Wiley, 2008.

[10] Michael W. Frazier. *An Introduction to Wavelets Through Linear Algebra.* Springer, 4 edition, 2011.

[11] Rafael C. Gonzalez and Richard E. Woods. *Digital Image Processing.* Pearson, 4 edition, 2017.

[12] Rafael C. Gonzalez, Richard E. Woods, and Steven L. Eddins. *Digital Image Processing Using MATLAB.* Pearson, 2004.

[13] Joel R. Hass, Christopher E. Heil, and Maurice D. Weir. *Thomas' Calculus: Early Transcendentals.* Pearson, 14 edition, 2017.

[14] S. Haykin. *Communication Systems.* Wiley, 3 edition, 1994.

[15] Robert V. Hogg, Elliot Tanis, and Dale Zimmerman. *Probability and Statistical Inference.* Pearson, 9 edition, 2014.

[16] David A. Huffman. A method for the construction of minimum-redundancy codes. *Proc. Inst. Radio Eng.*, 40:1098–1101, 1952.

[17] Deborah Hughes-Hallett. *Calculus: Single and Multivariable.* Wiley, 7 edition, 2017.

[18] David W. Kammler. *A First Course in Fourier Analysis.* Prentice Hall, 2000.

[19] Linda Almgren Kime and Judy Clark. *Explorations in College Algebra,.* Wiley, 5 edition, 2011.

[20] B.P. Lathi. *Signal Processing and Linear Systems.* Oxford University Press, 2017.

[21] B.P. Lathi and Roger Green. *Linear Systems and Signals.* Oxford University Press, 3 edition, 2017.

[22] David C. Lay. *Linear Algebra and its Applications.* Pearson, 4 edition, 2011.

[23] A. McAndrew. *A Computational Introduction to Digital Image Processing.* CRC Press, 2 edition, 2016.

[24] Julie Miller and Donna Gerken. *Precalculus.* McGraw-Hill Education, 2016.

[25] J. R. Parker. *Algorithms for Image Processing and Computer Vision.* Wiley, 2 edition, 2010.

[26] William B. Pennebaker and Joan L. Mitchell. *JPEG: Still Image Data Compression Standard.* Springer, 1993.

[27] Maria M. P. Petrou and Costas Petrou. *Image Processing: The Fundamentals.* Wiley, 5 edition, 2018.

[28] William K. Pratt. *Introduction to Digital Image Processing.* CRC Press, 2013.

[29] John G. Proakis and Dimitris G Manolakis. *Digital Signal Processing.* Pearson, 4 edition, 2007.

[30] Gary K. Rockswold. *Essentials of College Algebra with Modeling and Visualization.* Pearson, 4 edition, 2011.

[31] E. B. Saff and A.D. Snider. *Fundamentals of Complex Analysis: with Applications to Engineering and Science.* Pearson, 3 edition, 2014.

[32] David Salomon. *A Concise Introduction to Data Compression.* Springer, 2008.

[33] Chris Solomon and Toby Breckon. *Fundamentals of Digital Image Processing: A Practical Approach with Examples in Matlab.* Wiley-Blackwell, 2011.

[34] C.M. Stein. Estimation of the mean of a multivariate normal distribution. *Ann. Stat.*, 9(6):1135–1151, 1981.

[35] James Stewart. *Calculus: Early Transcendentals.* Brooks/Cole, 8 edition, 2015.

[36] Gilbert Strang. *Introduction to Linear Algebra.* Wellesley - Cambridge Press, 5 edition, 2016.

[37] Dirk J. Struik. *Lectures on Analytic and Projective Geometry.* Dover, 2011.

[38] Michael Sullivan. *Statistics: Informed Decisions Using Data.* Pearson, 5 edition, 2016.

[39] David Taubman and Michael Marcellin. *JPEG2000: Image Compression Fundamentals, Standards and Practice.* Springer, 2001.

[40] Scott E Umbaugh. *Digital Image Processing and Analysis: Applications with MATLAB and CVIPtools.* CRC Press, 3 edition, 2017.

[41] Richard D. De Veaux, Paul F. Velleman, and David E. Bock. *Intro Stats.* Pearson, 5 edition, 2017.

[42] Dennis Wackerly, William Mendenhall, and Richard L. Scheaffer. *Mathematical Statistics with Applications.* Pearson, 7 edition, 2008.

[43] David F. Walnut. *An Introduction to Wavelet Analysis.* Birkhauser, 2004.

[44] Leonid P. Yaroslavsky. *Theoretical Foundations of Digital Imaging Using MATLAB.* CRC Press, 2012.

Index

Printed in the United States
By Bookmasters